STRENGTH AND STRUCTURE OF SOLID MATERIALS

Published Proceedings of a Joint Japan-U.S.A. Seminar
on Strength and Structure of Solid Materials
held at Minnowbrook Conference Center of Syracuse University,
Blue Mountain Lake, New York, U.S.A., October 7-11, 1974.

Sponsored by The National Science Foundation of the United States of
America, Washington, DC, U.S.A., and The Japanese Society for the
Promotion of Sciences, Tokyo, Japan.

STRENGTH AND STRUCTURE OF SOLID MATERIALS

Edited by

Prof. Dr. H. Miyamoto
Dept. of Precision Engineering, Univ. of Tokyo, Tokyo, Japan

Prof. Dr. T. Kunio
Dept. of Mech. Engineering, Keio Univ., Yokohama, Japan

Prof. Dr. H. Okamura
Univ. of Tokyo, Tokyo, Japan

Prof. V. Weiss
Dept. of Materials Science, Syracuse University, Syracuse, NY, USA

Prof. M. Williams
School of Engineering, Univ. of Pittsburgh, Pittsburgh, PA, USA

Prof. H. Liu
Syracuse University, Syracuse, NY, USA

NOORDHOFF INTERNATIONAL PUBLISHING   LEYDEN

ISBN 90 286 0286 0

Printed in The Netherlands by Drukkerij Vonk Zeist B.V.

PREFACE

This volume contains the proceedings of the U.S.A.-Japan Seminar on
"The Strength and Structure of Solid Materials", held at the Minnowbrook
Conference Center of Syracuse University, at Blue Mountain Lake, New York,
October 7 to 11, 1974. The Seminar was sponsored jointly by the U.S. National
Science Foundation and the Japanese Society for the Promotion of Sciences
under the auspices of the United States-Japan Scientific Exchange Program.
Professors Miyamoto, Kunio and Okamura of Japan and Professors Weiss, Williams
and Liu of the U.S.A. coordinated and planned the overall program. Thirty-six
researchers, eighteen from the United States and eighteen from Japan, partici-
pated in the conference. Six topical sessions dealt with continuum mechanics,
fracture of metals, effects of elevated temperature, fatigue, non-metals and
composites. A general discussion session and a summary session concluded the
program. There were thirty presentations, seven in continuum mechanics, seven
in fracture of metals, four on effects of elevated temperature, eight on
fatigue and four on non-metals and composites. The present proceedings contain
twenty-eight of these which were submitted by the authors. The abstracts of
all the presentations have been published in the August 1975 issue of the
International Journal of Fracture.[*] The organization and the schedule of the
conference stimulated considerable interaction between all participants both
during the sessions and on an informal basis.

In addition to the presentation of new information, the program included
discussions of critical issues in research on the mechanical behavior of solids,
the assessment of the state of research in the field and suggestions for future
research. Among the needs for future research efforts the participants pointed
to studies in the areas of fatigue and creep interaction, non-linear fracture

---

[*] A Journal of Noordhoff International Publishing, Leyden, The Netherlands.

FROM LEFT TO RIGHT: DRS. OKAMURA, TIEN, MIYAMOTO, KUNIO, WEISS AND LIU

AT THE RECEPTION

AT THE RECEPTION

MEAL TIME

VIII

mechanics, modeling of material inhomogeneities, and viscoelastic and composite materials. The need to apply the knowledge obtained from research programs towards practical uses in design was also emphasized.

The Organizing Committee gratefully acknowledges the farsighted support of the National Science Foundation and the Japanese Society for the Promotion of Sciences which made this Seminar possible. Special thanks are also due to Professor M.L. Williams for his effort in connection with the publication of the abstracts of these proceedings.

The Editors

CONTENTS

X

SESSION V   NON-METALS AND COMPOSITES

SESSION I

Continuum Mechanics

Metallurgical Mechanics and Its
Applications to Composite Materials

by   Hiroshi Miyamoto

1.Introduction

In the research of the fracture process from the viewpoint of the continuum mechanics, the approach which treats the body being isotropic and homogeneous is limited.

When we pursue such phenomena as the yielding or failure of the solid body, which initiates at the local point and propagates all over the structure, we expect that we could solve the yielding or the fracture in overall by knowing those of their constituents. These constituents are grains and grain boundaries in the polycrystalline metals, and grains, bond materials and pores in the grinding wheel materials.

In the research of the strength of materials we treat the yielding and fracture of the whole structure. But the present approach is used in the research of the metallurgy and the method of the research is reverse. In this field Taylor, Sachs, and Hill have already tried and obtained some successful results as is well known.

This approach, which will be stated here, used the finite element method and we can obtain in a straightforward way the mechanical behavior of the structure using those of its constituents. We shall explain this more in detail. As is well known, in order to connect the mechanical behavior of single crystals with those of polycrystals, the following should be considered.

(1) The measurements of characteristics of a single crystal
(2) The condition of a grain boundary
    (i) The characteristics of a grain boundary
    (ii) The difference of the characteristics between a single
        crystal and those of a crystal grain
    (iii) The interaction of grains

Dr., Prof., Department of Precision Machinery Engineering,
Faculty of Engineering, University of Tokyo

(3) The information on the aggregate structure

(4) The averaging process based on (1) - (3)

(5) The measurements of characteristics of a polycrystal

Up to now, the mathematical treatments of this problem belong to the averaging process (4) in this table. But here, instead of the averaging process, we should like to follow the behaviors of crystal grains minutely using the finite element method. Then it is important how to make a model considering the condition (2).

There are 2 difficulties in modelling the grain boundary:

(a) The mechanical behavior of the crystal grain boundary

(b) The shape of grains

As to the item (a), we should like to assume; "The grain boundary is the boundary of grains, which have different orientations." As to the item (b), we already discussed in another paper [1]. About the condition (2)(ii), we used the characteristics of a single crystal to obtain those of crystal grains. Therefore, the finite element calculations incorporate the condition (iii) on the assumption of (i) and (ii). The condition of continuity of nodal forces and displacements at the grain boundary is satisfied. In the above discussion polycrystalline metals were taken up but this approach can be applied of course to composite materials.

Using this approach, I showed in other papers the results on

(1) Initiation of the slip lines of coarse grained aluminum [1,2].

(2) Elastic constants of polycrystalline metals[2], grinding wheel materials [1], cast irons, and composite materials [3].

In this paper, I should like to present some results about the mechanical behavior of the tricrystal of 99.99% Al using the calculation and experimentation, and also some calculated and experimental results about elastic fractures of composite materials.

2.Mechanical Behavior of an Aluminum Tricrystal [4]

2.1 Simulation Technique Using Finite Element Method

If the global coordinates and the local coordinates with respect to the slip direction of a crystal are taken as shown in Fig.1, stresses $\{\sigma\}$ and elastic strains $\{\varepsilon_e\}$ with respect to specimen coordinates system can be expressed

$$\{\sigma\} = [\Phi]^T [C] [\Phi] \{\varepsilon_e\} = [D_e] \{\varepsilon_e\} \tag{1}$$

where suffix T denotes transpose.

$$[\Phi] = \begin{bmatrix} l_1^2 & m_1^2 & n_1^2 & l_1m_1 & m_1n_1 & n_1l_1 \\ l_2^2 & m_2^2 & n_2^2 & l_2m_2 & m_2n_2 & n_2l_2 \\ l_3^2 & m_3^2 & n_3^2 & l_3m_3 & m_3n_3 & n_3l_3 \\ 2l_1l_2 & 2m_1m_2 & 2n_1n_2 & l_1m_2+l_2m_1 & m_1n_2+m_2n_1 & n_1l_2+n_2l_1 \\ 2l_2l_3 & 2m_2m_3 & 2n_2n_3 & l_2m_3+l_3m_2 & m_2n_3+m_3n_2 & n_2l_3+n_3l_2 \\ 2l_3l_1 & 2m_3m_1 & 2n_3n_1 & l_3m_1+l_1m_3 & m_3n_1+m_1n_3 & n_3l_1+n_1l_3 \end{bmatrix}$$

$[C]$ denotes elastic coefficients of a crystal grain, and for such FCC crystals as Al, $[C]$ is expressed as

$$[C] = \begin{bmatrix} c_{11} & c_{12} & c_{12} & 0 & 0 & 0 \\ c_{12} & c_{11} & c_{12} & 0 & 0 & 0 \\ c_{12} & c_{12} & c_{11} & 0 & 0 & 0 \\ 0 & 0 & 0 & c_{44} & 0 & 0 \\ 0 & 0 & 0 & 0 & c_{44} & 0 \\ 0 & 0 & 0 & 0 & 0 & c_{44} \end{bmatrix}$$

2.2 Elastic Analysis of Aluminum Tricrystal

stresses and strains of the cross section (X-Y plane in Fig.2) of a tricrystal was calculated. The model specimen studied was a tricrystal composed of two kinds of crystals; crystal M and crystal S. Crystal M, orientations of which differ from crystal S is sandwiched between two crystal S's, with their grain boundaries being in the direction parallel to the tensile axis. The directions of these crystals are as follows: (Fig.3 and Fig.4)

(a) Series I   : The direction of the crystal M is changed by 4 degrees stage by stage for 5 stages with its rotating axis being normal to the surface plane of the crystal

(b) Series II   : The direction of the crystal M is changed by 4 degrees stage by stage for 3 stages, with its rotating axis being normal to the slip plane of the crystal

(c) Series III : The direction of the crystal M is changed by 4 degrees with its rotating axis being the intersection between the specimen surface plane and the slip plane of the crystal S.

The calculated results of elastic stresses and strains of these crystals are shown in Fig.5 through Fig.7. Fig.5 shows the stresses, strains, and Young's moduli averaged across the whole cross section at the point of yielding against the misorientation in the abscissa. Fig.6 shows the stress-strain distribution in the case of θ =4 degrees for Series I. The abscissa represents the length perpendicular to the grain boundary. Fig.7 shows the stress and strain distribution in the tensile direction (z) for each Series. The abscissae of these figures are the same as that of Fig.6. The distribution of the averaged stresses, strains and Young's moduli of Fig.5 across the section can be accounted for by the distribution of 6_z in the tensile direction of Fig.7. That is, considering that the yielding depends on the Schmid factor between the tensile axis and the dominant slip system, and the value of the Schmid factor for each Series being such as shown in Table 1, the crystal M controls the strains at the point of yielding and the stresses show such a distribution as shown in Fig.7 in the case of Series I. For Series II, III the crystal S controls the yielding so that the stress distribution of the crystal S do not differ appreciably in Fig.7. Accordingly, the averaged stress-strain distribution in Fig.5 becomes minimum at θ =12, 16 degrees in the case of Series I and it does not change in Series II, III. It is observed from Fig.7 that the effect of grain boundary reaches to the extent of the order of the plate thickness (approximately 1 mm), and that the differences of stresses across the grain boundaries have a tendency to become greater with the increase of the rotating angle.

3.Elastic Fracture of Composite Materials

3.1 Simulation of Fracture Process of the Grinding Wheel Materials [5]

3.1.1 Simulation of Fracture Process by Photoelasticity

Simulations of fracture by photoelasticity were carried out in order to clarify the correlation between the crack propagation direction and the microstructure.

Materials of the test specimen used in the simulation are epoxy resins of Araldite, type B and type CY 230, which are equivalent to the grain and the bond, respectively. The mechanical properties of each constituent element are listed in Table 2. In the table, the ratios of Young's modulus of the two resins are 4. They are chosen as much as possible to correspond with the ratios of the effective

grinding wheel constituent elements. Fig.8 shows the geometry of the
test specimen. After the preparation of the test specimens was
completed, uniaxial tensile stress was applied to the specimens in
order to obtain the principal stress direction and the stress
distribution from the isoclinic line and isochromatic fringe pattern,
respectively. The analysis is defined in the range of two dimensional
photoelasticity. Situation after penetration of a crack in one element
were also observed. From the knowledge of the first breakdown of an
element, an artificial crack of a certain direction was prepared in
order to examine the correlation between the principal stress
direction and the crack propagation direction. Fig.9 illustrates the
isoclinic line in a part of the modeled structure. In Fig.10, the
first crack initiation appeared at the point of maximum tensile
stress of the minimum cross section. The fact that the crack extends
through 'A' is important in order to know the mechanism of the
composite characteristics of the material. In the case of a crack 'A',
the crack extends along the tensile principal stress direction and
be arrested at a weaker stress region. From a series of the
experimental results, it can be concluded a) that a crack passes
through the composite boundary, provided that the strengths between
two materials are in comparable order with each other, and b) that
the direction of the crack propagation corresponds to the principal
stress direction. It should be noted that the principal stress
direction is important to the crack propagation of the brittle
composite material. The fracture processes of the effective grinding
wheel material might be clarified by the simulation using the
photoelastic material.

3.1.2 Simulation of Fracture Processes by Numerical Stress Analysis

The simulation of the fracture processes of the grinding wheel
will be analyzed when the information of each of the configuration
of the grinding wheel structure, strength of each constituent
element, and of the composite boundary parts are given. In this
analysis, the finite element analysis are applied to the same
configuration model used in the photoelasticity. The analysis of the
successive fracture process of the modeled structure is compared to the
experiment described before.

(1) Numerical Procedure

Griffith's theory is applied to the analysis, since both the

8

grain and the bond (including those of the epoxy resins) show the
brittle behavior. It is assumed that Griffith's criterion holds
completely for each element. The fracture criterion in terms of
Griffith's theory is expressed in the form: Introducing
$\sigma_m = \frac{1}{2}(\sigma_1 + \sigma_2)$, $\tau_m = \frac{1}{2}(\sigma_1 - \sigma_2)$, $\tau_m \sim \sigma_m$ curves for the fracture are;
$\tau_m^2 = -4T_0 \sigma_m$ when $2\sigma_m + \tau_m \gtrless 0$, $\tau_m = \sigma_m - T_0$ when $2\sigma_m + \tau_m < 0$
in which σ_1 and σ_2 ($\sigma_1 > \sigma_2$) are principal stresses, T_0 is the
tensile strength under uniaxial tension. The equation of Mohr's
envelope to Griffith's criterion is: $\tau^2 = -4T_0(\sigma - T_0)$
If the maximum stress of the element traverses the parabola in $\tau \sim \sigma$
curve, it can be assumed that the element has fractured. Then the
stiffness of the element is reduced to one hundredth. Calculations
are repeated until the rupture of the structure occurs. The numbers
of elements and nodes are 376 and 231, respectively. Loading conditions
were defined according to the uniaxial tension and the uniaxial
compression.
(2) Results of the Analysis
a) Fracture process of the photoelastic material, using its material
constants
　　　The situation of fracture proceeding under applied tensile
stress is illustrated in Fig.11. These fractures are initiated
simulatneously in the element of maximum tensile stress, because of the
symmetry of the figure. After the equilibrium breaks, the fracture
initiated from 'B' proceeds to 'A' and finally leads to 'B'. An
increase of applied external load did not occur during the fracture
proceeding from 'B' to 'B'. We can conclude by reference to Fig.9
that the fracture proceeds roughly along the principal stress
direction. Fig.11 suggests that fracture initiated from other parts
of the element propagates simultaneously to a certain length. Stresses
in the neighborhood of the element are released by the rupture of one
element, so that new fracture extensions in other elements are
accelerated. At this moment, an increase of a new external load is
necessary to cause the next rupture. The pore, therefore, serves
as the crack arrester. Fracture proceeding until the final breakdown
of this structure, is shown in Fig.12.
b) Fracture proceeding processes using material constants of the
grinding wheel
　　　The mechanical properties of the grinding wheel used are

illustrated in Table 3. In the table, the values of Young's modulus
and Poisson's ratio of the grain were quoted from the results for
aluminum obtained by other researchers. The values of the material
constants of the bond were measured by the bending test of the
preshaped block. The tensile strengths of the grain were measured by
the method of crashing proposed by Yoshikawa. The equation of the
tensile strength is $\delta = \frac{1}{0.32} \frac{P}{A}$
in which P is the applied load, and A is the maximum cross section,
respectively. The values in Table 3 are the mean values for 50 samples
with grain size #16. In order to eliminate the influence of the size
effect between the grain and the bond, the strengths of the bonds are
also measured in the same way and at same size mentioned above. The
analysis was carried out for the same configuration as was used in the
photoelasticity technique using the material constants listed in the
table.

b-1) Uniaxial tension

Since the strength ratios between the grain and the bond
obtained by the experiment are about 3, the fracture proceeds through
the bond area without extending to the grain. Since the fracture
does not occur in the grain, the result is different from the
experiment of the photoelasticity material. The reason might be
explained as follows: Although the fracture described in this analysis
is defined as a considerable decrease of stiffness of the element
with a finite width, the continuance of the real fracture is caused
by the crack propagation. The crack initiated at the bond might
penetrate the grain, provided that the ratio of strength between
the grain and the bond is comparatively small. By combining the
finite element method with probability, which takes the distribution
of the defects of the elements, the fracture processes of the
grinding wheel might be far more clarified using the simulation
technique.

b-2) Uniaxial compression

The results for the case of uniaxial compression are the same
as the results of uniaxial tension. There is no fracture that goes
through the grain. The cause is obvious from the preceding discussion
of the uniaxial tension.

3.2 <u>Simulation of Fracture Process of the Cast Iron</u> [6]

For the purpose of the discussion, we will regard spherical

graphite cast iron as a porous body since the mechanical strength of graphite is quite weak compared to that of matrix, and we will attempt to develop the simulation technique of its fracture process.

3.2.1 Experiments

The test specimen is made of polyester and has shape shown in Fig.13. Its parallel part (50 x 50 mm^2) is divided into 100 squares (5 x 5 mm^2). Of these we chose randomly 50 squares and drilled a circular hole of 3 mm diameter into each of them. In this simulation the pore rate is 13.4%. There are two types of models; (a) RN-3145 (b) RN-1973. The experiments are carried out on 8 (a) models and 5 (b) models. Fig.14 shows the stress-strain curve of the matrix. From this curve it is known that the matrix shows linear mechanical behavior.

3.2.2 Analysis by Substructure

In order to make a precise analysis of the stress distribution of such a complicated structure as this model (Fig.15), it is better to use the substructure technique of the finite element method. The model (a) (14,400 elements, 8560 nodes) is constructed using 2 kinds of substructure, 1 and 2. The practical calculation is carried out in 2 steps. In the first step we treat 100 elements and 561 nodes and in the second step 124 elements and 69 nodes.

3.2.3 Comparison of Calculations and Experiments

The results of the calculations are shown in Table 4. The $\bar{\sigma}/\sigma_N$ at the sides of the model (Fig.16) are shown in this table, where σ_N is the nominal stress and $\bar{\sigma}$ is the equivalent stress obtained by nodal force. In the (a) model $\bar{\sigma}/\sigma_N$ is high in the sections a $\sim$ a, b $\sim$ b, c $\sim$ c, d $\sim$ d. In the experiment fracture occur twice in the section b $\sim$ b and 6 times in d $\sim$ d. In the (b) model $\bar{\sigma}/\sigma_N$ is high in the section e $\sim$ e, and fracture occurs in the section e $\sim$ e during the experiment.

4.Summary

I call such an approach which connects the metallurgy and mechanics from the mechanics side "metallurgical mechanics", while "mechanical metallurgy" is the corresponding approach from the metallurgy side, and works carried out in our laboratory in this field have been introduced.

The second chapter describes applications to tricrystals and the technique mentioned herein can be directly extended to polycrystalline metals. If an appropriate fracture criterion is given, it can be extended to crack propagation problems.

The third chapter describes the elastic fracture of composite materials. We are intending to extend these works to elasto-plastic and fracture problems.

I should like to connect micro and macro mechanics of fracture, based on these crystal grain order mechanics.

In closing, I should like to thank Dr.Shuichi Fukuda, Mr.Kyoji Homma, Mr.Shinji Sakata (University of Tokyo), Prof.Isohachi Oda (Kanazawa University) and Mr.Saburo Matsuoka (National Research Institute for Metals) for their assistance.

5.References

1) Miyamoto, H. et al., "The Application of the Finite Element Method to Fracture Mechanics" Japan - U.S. Seminar 1972

2) Miyamoto, H., "Application of Finite Element Method to Fracture Mechanics" 1st SMiRT, Berlin, 1971

3) Miyamoto, H. et al., "Recent Developments in the Application of Finite Element Method to Fracture Mechanics in Japan" Meeting on Finite Element Technology in Fracture Mechanics, I.S.D., Stuttgart, 1973

4) Miyamoto, H., and Matsuoka, S., "Elasto-plastic Analysis Based on Slip Theory", Preprint of 21st SMSJ, pp.217-218, 1972 (in Japanese)

5) Miyamoto, H., and Homma, K., "Fracture Process of the Grinding Wheel", Proc. of I.C.P.E., Tokyo, 1974

6) Miyamoto, H. et al., "Study of Fracture Behavior of a Porous Body" Preprint of JSME, No.740-11, pp.271 - 274, 1974 (in Japanese)

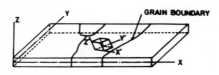

Fig.1 Global and
      local coordinates

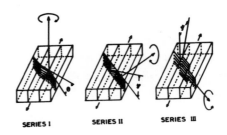

SERIES I   SERIES II   SERIES III

Fig.4 Three models of tricrystal

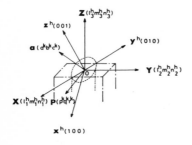

(XYZ)   SPECIMEN COORDINATES
$(x_i^h y_i^h z_i^h)$ h CRYSTAL COORDINATES
$(a^k b^k c^k)$ SLIP PLANE NORMAL VECTOR OF K SLIP SYSTEM
                OF h CRYSTAL
$(p^k q^k r^k)$ SLIP DIRECTION VECTOR

Fig.2 Relation among specimen
      and slip plane and
      direction of crystal

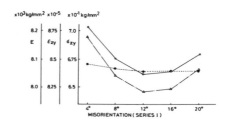

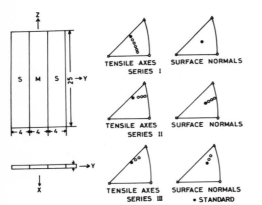

TENSILE AXES   SURFACE NORMALS
  SERIES I

TENSILE AXES   SURFACE NORMALS
  SERIES II

TENSILE AXES   SURFACE NORMALS
  SERIES III   • STANDARD

                    ○ MISORIENTATION

Fig.3 Orientation of tricrystal

Fig.5 Yield stress, yield strain
      and Young's modulus versus
      misorientation

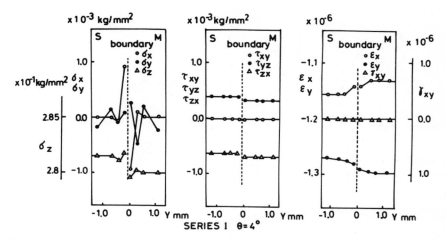

Fig.6 Stress and strain distribution near the grain boundary

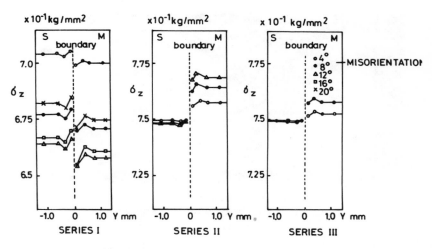

Fig.7 Stress distribution in the tensile direction
near the grain boundary

SERIES	MISORIENTATION				
	4°	8°	12°	16°	20°
S	0.4361				
I	0.4646	0.4829	0.4911	0.4899	0.4799
II	0.4310	0.4299	0.4122		
III	0.4329	0.4289			

Table 1 Schmid factor

Epoxy Resin	Young's Modulus (kg/mm²)	Poisson's Ratio	Tensile Strength (kg/mm²)	Photoelastic Sensitivity (mm/kg)
A	350	0.38	8	0.8
B	90	0.42	9	1.5

Table 2 Mechanical properties
of the epoxy resin

14

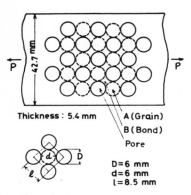

Thickness : 5.4 mm   A (Grain)
                      B (Bond)
                      Pore

D = 6 mm
d = 6 mm
ℓ = 8.5 mm

Dimension of the grain and the bond
and distances between centers of them

Fig.8 Specimen used in
      photoelastic
      experiment

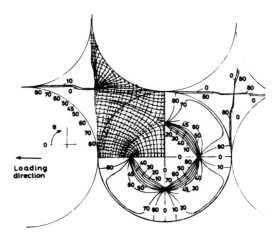

Fig.9 Isoclinic line in a part
      of the modeled structure

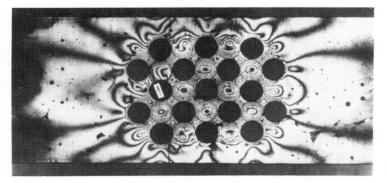

Fig.10 Isochromatic fringe pattern
       after first breakdown of an element

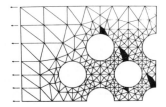

Fig.11 Fracture before
       breakdown, using
       photoelastic
       material constant

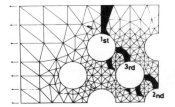

Fig.12 Fracture at the
       breakdown of the
       structure

Material	Young's Modulus (kg/mm²)	Poisson's Ratio	Tensile Strength (kg/mm²)
Grain	35000	0.2	22.4
Bond	6000	0.2	7.4

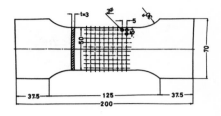

Table 3 Mechanical properties of
the grinding wheel material

Fig.13 Test specimen

No.	RN=3145 Model $\overline{\sigma}/\sigma_h$	Order	RN=1973 Model $\overline{\sigma}/\sigma_h$	Order
1	3.256		3.512	
2	2.981		3.554	
3	2.838		3.156	
4	2.971		2.928	
5	3.514		2.767	
6	3.623		3.566	
7	3.832	3	3.582	8
8	3.881	1	2.710	
9	3.047		2.983	
10	3.091		3.721	6
11	3.519		3.793	4
12	3.623		3.603	9
13	3.742	7	3.683	7
14	3.761	5	3.970	2
15	3.722	9	3.991	1
16	3.665		2.683	
17	3.606		2.881	
18	3.440		3.174	
19	3.045		2.954	
20	3.052		3.501	
21	3.753	6	3.523	
22	3.729	8	3.012	
23	3.663		2.714	
24	3.601		3.808	3
25	3.704	10	3.752	5
26	3.642		3.324	
27	3.505		3.254	
28	3.532		3.141	
29	3.354		3.248	
30	3.402		3.567	10
31	3.802	4	3.372	
32	3.861	2		

Table 4 Calculated results

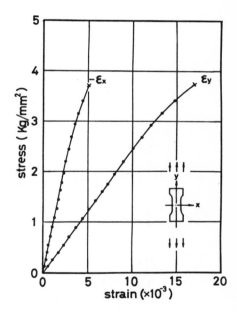

Fig.14 Stress-strain curve
of the matrix

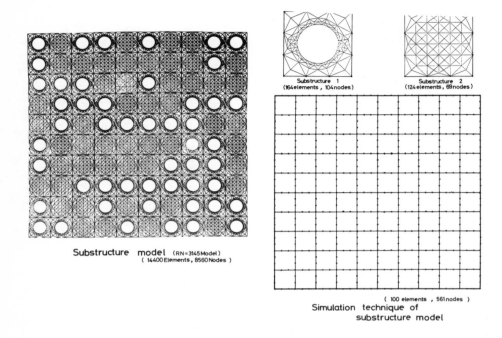

Substructure model (RN=3145 Model)
(14400 Elements , 8560 Nodes)

Substructure 1
(164 elements , 104 nodes)

Substructure 2
(124 elements , 69 nodes)

(100 elements , 561 nodes)
Simulation technique of substructure model

Fig.15 Substructure model

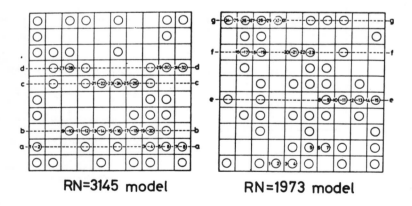

RN=3145 model

RN=1973 model

Fig.16 Calculated models

Mechanics of Metal Fracture

G. C. Sih[*]

1. INTRODUCTION

Perhaps, one of the least understood problems in fracture mechanics is that of crack propagation in a material that deforms
beyond its elastic limit. The assumption made in many of the
existing two-dimensional models is that the crack propagates
through a plastic zone as implied in the models of circular en-
clave, plastic fan, narrow strip, etc. The application of these
models to explain metal fracture of plate specimens raises some
serious questions since fracture of metal plates is basically a
three-dimensional phenomenon and cannot be adequately explained
by any two two-dimensional analyses which ignore the thickness
effect. (1). Too often, agreement between theory and experiment is
purely coincidental or contrived. A classical example is the
plasticity strip model for predicting the yielded zone size in
thin metal sheets or plates. The so-called observed "plasticity"
ahead of the crack is due to necking of the material in the thickness direction while the analysis considers no such effect and is
strictly a planar one. A refined two-dimensional elastic-plastic
stress analysis will reveal that the material along the line of
expected crack extension will not distort sufficiently to cause
yielding.

In metal alloys, the amount of energy consumed in developing
the commonly observed shear lips is caused by distortion or necking of the materials near the plate surfaces. On a macroscopic

[*]Institute of Fracture and Solid Mechanics, Lehigh University,
Bethlehem, Pennsylvania 18015.

level, fracture can only occur in those portions of the material
with sufficient stiffness and hence dilatation tends to be the
dominant mode of energy dissipation. Experimentally, it is often
difficult to determine the precise locations where brittle fracture terminated and ductile fracture started or vice versa. However, the events of initiation of stable crack growth[*] (local instability) and onset of rapid crack propagation (global instability) can be identified on the non-linear stress-strain curve. In
this case, the classical linear fracture mechanics approach is by
definition not valid. The conventional failure theories are also
unable to explain the problem of initiation and propagation, either qualitatively or quantitatively.

This paper emphasizes some of the essential features of metal
fracture. The strain energy density theory which effectively
weighs both dilatation and distortion seems to provide a satisfactory explanation for the previously unexplained phenomenon of
slow crack growth in metals.

2. STRAIN ENERGY DENSITY THEORY

Successful prediction of metal fracture depends on the effec-
tiveness of stress analysis, soundness of failure criterion, and
an understanding of the actual fracture process. Repeated careful experiment and observation have revealed that crack propagation is not a continuous process but occurs discretely by breaking elements of material in regions ahead of the crack. Hence,
any theories that assume continuous advancement of the crack such

---

[*] Stable crack growth can also occur under a constant sustained
load (creep) or cyclic loadings (fatigue). The present discussion is limited to metal specimens under monotonically rising
load prior to unstable crack extension.

as the energy release rate concept will have limited application.
Such an assumption does not lend itself to cracks that do not
propagate in a self-similar manner and can lead to analyses involving unwarranted and unprofitable labor. Too much emphasis
cannot be placed on predicting the direction[*] of crack propagation in three dimensions which is a prerequisite for analyzing
metal fracture.

One of the basic concepts of the S-criterion is that fracture
initiates from an *interior element* located at a finite distance
r_o from the crack front (3,4). Sih has referred to r_o as the
radius of the core region within which the material can be highly
distorted and its mechanical properties can be very different
from those in the bulk (4). Moreover, investigation of three-
dimensional fracture behavioral prediction should be carefully
distinguished from two-dimensional crack behavior. In two dimensions, fracture initiates from a point element ahead of the crack
where the strain energy density factor S reaches a critical value,
S_{cr}, and if the fracture is considered as three-dimensional, then
there is a whole line of critical elements parallel to the
straight crack edge. In truly three-dimensional problems such a
line cannot be expected to be parallel with the crack border and
the position (r_o, θ_o, ϕ_o) of each isolated point element must be
determined separately, Figure 1. The basic assumptions of the S-
criterion may be summarized as follows (5):

---

[*] Until recently, the only widely known criterion for predicting
the direction of crack propagation was based on the maximum normal stress. It is common knowledge that this theory can be applied with confidence only if one of the normal stresses dominate while the others are absolutely smaller. Refer to a discussion given by Sih (2).

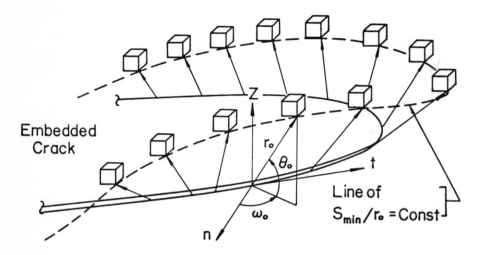

Figure 1 - Variable radius core region

(1) *Crack initiation takes place in a direction of minimum strain energy density factor, S, i.e.,* $\partial S/\partial \theta = 0$ ($\partial^2 S/\partial \theta^2 > 0$) *at* $\theta = \theta_o$.

(2) *Crack extension occurs when the minimum strain energy density factor,* S_{min}*, reaches a critical value,* S_{cr}.

(3) *The radius of core region,* r_o*, locating the points of initial fracture is assumed to be proportional to* S_{min} *such that* S_{min}/r_o *remains constant.*

Assumptions (1) and (2) are sufficient for determining where and when crack propagation occurs in a two-dimensional problem in which all the elements are assumed to fail at the same distance r_o from the straight crack front. In three dimensions, the crack front is generally curved and the distance r_o may vary from point to point along the crack border as illustrated in Figure 1. The direction of maximum strain energy density factor corresponds to the direction of maximum yielding. Further insights into the S-criterion may be gained by resolving S into two component parts, one associated with volume change and the other with shape change as discussed in (5,6).

With this brief introduction on the S-criterion, it is now more pertinent to discuss the physical aspects of the problem of a through crack growing in a ductile material.

3. BIFURCATION NEAR FREE SURFACE

Nonlinear response observed on the stress-strain curve for the problem of a crack passing transversely through a metal plate is commonly referred to as plasticity. Because of the complexity of the problem, the interaction of crack growth with material deformation local to the crack front and the free surfaces of the plate is not understood. A purely elastic analysis will shed light on the problem since it is necessary to distinguish kinematic nonlinearity due to change in crack shapes from the physical or material nonlinearity. Referring to Figure 2, as the thumbnail crack grows the global stress-strain response could be nonlinear even in the absence of plastic deformation.

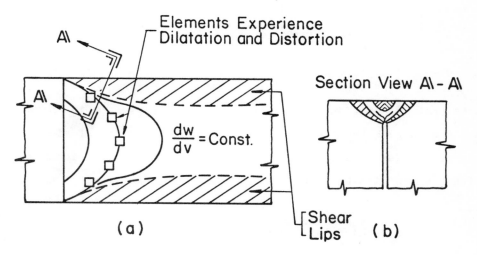

Figure 2 - Cross sectional view of crack growth behavior

Theoretical treatment of elastic-plastic crack problems has been limited to analyses which are based on the two-dimensional models of plane strain (or plane stress) and which incorporate a continuum theory of plasticity. These analyses, though not satisfactory for treating the problem of ductile fracture, can offer some information of a descriptive nature. For example, a plane strain analysis of the problem of a central crack whose tips are close to the specimen boundary indicates a plastic region[*] in front of the crack near the free surface as shown in Figure 3(a).

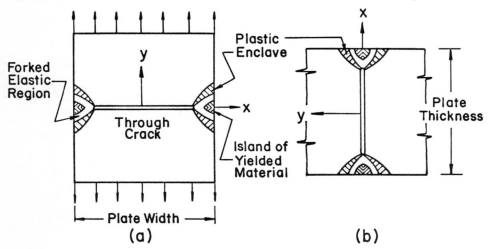

Figure 3 - Plastic zones for crack edges near specimen boundary.

Note that in addition to the usual plastic enclaves, which have now merged with the specimen edge, there is an island of yielded material, leaving a forked region of elastic material. The same situation applies to a thumbnail crack whose edge is close to the

[*] The sizes of the plastic zones in Figure 3 are exaggerated since the specimen boundary will normally be highly distorted or curved. This means that less energy will be available to yield the material.

top and bottom surfaces of a finite thickness plate. A cross
sectional view of the thumbnail crack accompanied by plastic deformation can be obtained simply by rotating Figure 3(a) ninety
degrees and replacing the plate width in the plane strain problem
by plate thickness in the three-dimensional problem as indicated
in Figure 3(b). The strain energy density theory will actually
predict branching of the main crack with a chip of material fall-
ing off the edge, Figure 4(a). This is consistent with the experimentally observed departure of crack direction in the last ligament of growth. In reality, the crack never forks symmetrically.
A slanted fracture surface on the separated pieces, Figure 4(b),
are always observed because of nonalignment of load and specimen.
It is believed that the formation of shear lips which occur as a
thumbnail crack breaks through to the plate surfaces has the same

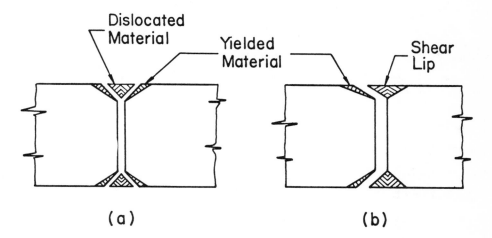

Figure 4 - Slanted fracture near a free surface

physical cause as the cup and cone fracture of a round bar. Figure 4(b) can also be viewed as a broken tensile bar. The central

portion of the fracture is normal to the direction of tension while the outer portions of it form a cone inclined at an angle of approximately 45 degrees with respect to the axis of the specimen, "cup and cone" fracture. In fact, the shear lips formed in through thickness cracking are also inclined at approximately 45 degrees to the direction of loading.

There is evidence to the effect that the main crack forks in the elastic portion of the material. The continuum crack always appear to avoid running into the yielded material and hence releases only elastic energy. The plastic or yielded portion of the material is relatively soft and can only sustain mechanical damage at the microscopic level. This argument is based on a detailed elastic-plastic stress analysis coupled with the strain energy density criterion.

4. GROWTH CHARACTERISTICS OF THUMBNAIL CRACKS

In order to reproduce the fracture growth patterns discussed earlier, the strain energy density criterion can be applied to forecast thumbnail profiles having a constant strain energy density, $dW/dV = S/r$. The material elements along these profiles experience more volume change than shape change. A series of possible new crack front shapes as illustrated in Figure 5(a) can be determined for various values of the parameters involved, say σ (applied stress), σ_{ys} (yield strength), n (strain hardening exponent), a (half crack length), h (plate thickness), ρ (crack tip radius of curvature), etc. For a particular set of parameters, the new crack front shape as predicted by the strain energy density criterion will be the basis of a second set of stress calculations. If the analysis for the extended crack indicates that

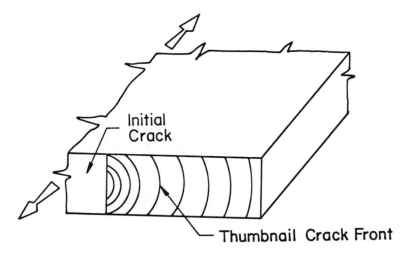

Figure 5 - Development of thumbnail cracks

an increased load is required to obtain further crack extension
using the same growth criterion, then the crack extension will be
considered to be stable and the analysis will be repeated to study
further growth. The compliance values for the various stages of
crack growth can then be determined as the crack front tunnels
through the mid-section of the plate and possibly branches near
the plate surface, Figure 5(b). The results may be summarized in
mathematical form as

$$S_{cr} = S(\frac{\sigma}{\sigma_{ys}}, \; n, \; \frac{\rho}{a}, \; \frac{h}{a})$$

where the critical value of some strain energy density factor,
S_{cr}, can be used as a material constant. Initially, the motion
of the thumbnail crack, Figure 5(a), will be constrained by the
shear lips and the crack grows slowly as the load increases. As
soon as the size of the thumbnail overcomes the thickness constraint, unstable rapid crack propagation begins. This can be
identified on the global stress-strain curve.

Finally, it should be said that additional work is required before fracture in more ductile materials can be understood. The continuum plasticity theory is fundamentally inadequate for analyzing elastic-plastic fracture since it was only intended to model the gross material response. The theory assumes continuity of material excluding any microprocesses that occur within the yielded portion of the material such as the plastic enclaves ahead of a macrocrack. One of the major inconsistencies in the case of continuum plasticity is a consequence of the procedure which must be employed to determine the values for the parameters such as hardening exponent, yield stress, etc., which characterize the material model. In particular, these parameters are obtained from an experimental load-deflection curve which is obtained from a specimen whose size can be measured in centimeters, size scale of $0(1)$. This record is then used to develop a stress-strain curve which is assumed to describe material behavior on a different scale, that of a differential element $0(10^{-3})$. Clearly, the amount of energy dissipated in these two cases differing by three orders of magnitude can be drastically different. For this reason, the analysis of a cracked member based on continuum plasticity cannot be expected to give an adequate description of the state of the material in the vicinity of the crack front. Discontinuity representing physical damage in the yielded material must be included for determining the so-called "energy dissipated in plastic flow", which is not available to drive the continuum, or macro, crack. Such an approach will be left for future discussion.

5. REFERENCES

(1) G. C. Sih, "Elastic-Plastic Fracture Mechanics", Proceedings on the Prospects of Fracture Mechanics, edited
by G. C. Sih, H. C. van Elst and D. Broek, Noordhoff
International Publishing, Leyden, 1974, **pp.** 613-620.

(2) G. C. Sih, Discussion on "Some Observations on Sih's
Strain Energy Density Approach for Fracture Prediction",
International Journal of Fracture, Vol. 10, 1974, pp.
279-283.

(3) G. C. Sih, "A Special Theory of Crack Propagation",
Methods of Analysis and Solutions of Crack Problems,
edited by G. C. Sih, Noordhoff International Publishing,
Leyden, 1973, pp. 21-45.

(4) G. C. Sih, "Surface Layer Energy and Strain Energy Den-
sity for a Blunted Crack or Notch", Proceedings on the
Prospects of Fracture Mechanics, edited by G. C. Sih,
H. C. van Elst and D. Broek, Noordhoff International
Publishing, Leyden, 1974, pp. 85-102.

(5) G. C. Sih, "A Three-Dimensional Strain Energy Density
Factor Theory of Crack Propagation", Three-Dimensional
Crack Problems, edited by G. C. Sih, Noordhoff Inter-
national Publishing, Leyden, 1975, pp. 15-53.

(6) G. C. Sih and B. Macdonald, "Fracture Mechanics Applied
to Engineering Problems - Strain Energy Density Fracture
Criterion", Journal of Engineering Fracture Mechanics,
Vol. 6, 1974, pp. 361-386.

The Deformation of the Multi-cracked
Material under Multiple Loads

by Hiroyuki OKAMURA [*]

1. INTRODUCTION

This paper presents a preliminary approach to the deformation analysis of

the material which contains many distributed cracks or crack-like flaws and

undergoes multiple loads.

It might be very important to know an effective modulus of elasticity or

plasticity in the strength analysis, since the crack tip of our interest is often

located in a crazed zone or in a kind of multi-cracked material such as a gray

cast iron with flaky graphites. The fatigue crack propagation in cast iron,

for example, is greatly influenced by the distribution and the morphology of

graphite in the strain-controlled condition as in the case of cracking thermally

induced in ingot mould[1].

The analysis of the energy release by crack extention and the Maxwell's

reciprocal theorem applied to the compliances of an extended meaning enable

us to get the combined effect of multiple loads on the deformation of the cracked

material and to evaluate the effective moduli of the deformation.

2. DEFORMATION OF CRACKED ELASTIC BODY

To begin with, it seems better to cite here some results in a previous

paper[2] on the deformation of the multi-cracked body shown in Fig.1 , in an

alternative form. The displacement component u_i in the direction of the load

P_i at the loading point is expressed by the sum of the contribution of all loads

$P_1, P_2, \cdots P_i \cdots$ in the form

$$u_i = \sum_j \lambda_{ij} P_j \tag{1}$$

using so-called influence coefficient λ_{ij} which may be called the compliance

---

[*] Dr. of Engineering, Professor of Mechanical Engineering,

Univ. of Tokyo, 7-3-1 Hongo, Bunkyo-ku, Tokyo 113, JAPAN

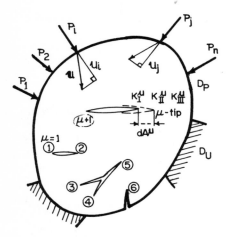

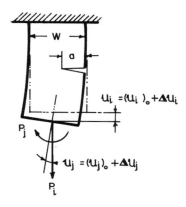

Fig.1 Multi-cracked body under multiple loads

Fig.2 Deformation of the edge-cracked bar

of an extended meaning. And the Maxwell's reciprocal theorem certifies the reciprocal relation on the compliances

$$\lambda_{ij} = \lambda_{ji} \tag{2}$$

which is conveniently used in several analyses as shown later. The stress intensity factor K^{μ} at the μ-th crack tip is the sum of the contribution of the stress intensity factor K_i^{μ} of load P_i ;

$$K_I^{\mu} = \sum_i K_{Ii}^{\mu} \quad , \quad K_{II} = \sum_i K_{IIi}^{\mu} \quad , \quad K_{III} = \sum_i K_{IIIi}^{\mu} \tag{3}$$

By considering the strain energy release due to the infinitesimal increment of the crack area dA^{μ} at the μ-th crack tip, the following relation is obtained.

$$\frac{P_i P_j}{2} \frac{d\lambda_{ij}^{\mu}}{dA^{\mu}} = \frac{1}{E'} \left(K_{Ii}^{\mu} K_{Ij}^{\mu} + K_{IIi}^{\mu} K_{IIj}^{\mu} \right) + \frac{1+\nu}{E} \left(K_{IIIi}^{\mu} K_{IIIj}^{\mu} \right) \tag{4}$$

$$E' = \begin{cases} E & \text{(plane stress)} \\ E/(1-\nu^2) & \text{(plane strain)} \end{cases}$$

where E and ν are Young's modulus and Poisson's ratio respectively. The above relation is reduced to a well-known one in the case where $i = j$.

Let the increments of displacement and of the compliance due to the presence of cracks denote Δu_i and $\Delta \lambda_{ij}$ respectively, and then $\Delta \lambda_{ij}$ is given by integrating the above equation and summing up the contribution of the all crack tips as follows;

$$\Delta u_i = \sum_j \Delta \lambda_{ij} P_j \qquad \text{(superposition of the deformation)} \qquad (5)$$

$$\Delta \lambda_{ij} = \Delta \lambda_{ji} \qquad \text{(reciprocal theorem)} \qquad (6)$$

$$\Delta \lambda_{ij} = 2 \iint_A \sum_\mu \left\{ \frac{1}{E'} \left(\frac{K_{\mathrm{I}i}^\mu}{P_i} \frac{K_{\mathrm{I}i}^\mu}{P_j} + \frac{K_{\mathrm{I\!I}i}^\mu}{P_i} \frac{K_{\mathrm{I\!I}i}^\mu}{P_j} \right) + \frac{1+\nu}{E} \frac{K_{\mathrm{I\!I\!I}i}^\mu}{P_i} \frac{K_{\mathrm{I\!I\!I}i}^\mu}{P_j} \right\} \frac{dA^\mu(t)}{dt} \right] dt \qquad (7)$$

where the path of integration $A^\mu = A^\mu(t)$ for the parameter t can be chosen arbitrarily. The simplest illustrative example of $\Delta \lambda_{ij}$ is the edge-cracked bar shown in Fig.2, which undergoes the axial force P_i and the bending moment P_j . The additional elongation Δu_i and the additional angular deformation Δu_j due to the presence of the crack are given as follows;

$$\left. \begin{array}{l} \Delta u_i = \Delta \lambda_{ii} P_i + \Delta \lambda_{ij} P_j \\ \Delta u_j = \Delta \lambda_{ij} P_i + \Delta \lambda_{jj} P_j \end{array} \right\} \qquad (8)$$

where $\Delta \lambda_{ii}$ and $\Delta \lambda_{jj}$ are the ordinary compliances of the crack. We may call $\Delta \lambda_{ii}$ the self-compliance and $\Delta \lambda_{ij}$ the mutual-compliance.

3. ELASTIC DEFORMATION OF TWO-DIMENSIONAL CRACKED MATERIAL

If we consider the deformation of the square of unit area of two-dimensional cracked material, the stress and strain can be regarded as the force and the displacement respectively of the treatment in the preceding section, and the compliance corresponds to the inverse of the modulus of elasticity in this case. For the stress state shown in Fig.3(a), the effective moduli, $E^*, G^*, \nu^*/E^*$, are expressed as

$$\left. \begin{array}{ll} \dfrac{1}{E^*} = \dfrac{1}{E} + \Delta\left(\dfrac{1}{E}\right) , & \Delta\left(\dfrac{1}{E}\right) = \Delta \lambda_{\sigma\sigma} \\[2mm] \dfrac{\nu^*}{E^*} = \dfrac{\nu}{E} + \Delta\left(\dfrac{\nu}{E}\right) , & \Delta\left(\dfrac{\nu}{E}\right) = -\Delta \lambda_{\sigma\sigma'} \\[2mm] \dfrac{1}{G^*} = \dfrac{1}{G} + \Delta\left(\dfrac{1}{G}\right) , & \Delta\left(\dfrac{1}{G}\right) = \Delta \lambda_{\tau\tau} \end{array} \right\} \qquad (9)$$

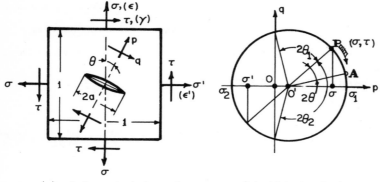

(a)   Orientation of crack   (b)   Mohr's circle

Fig.3   Orientation of crack and stress state

where G is the shear modulus.

Suppose that the number of the crack of length $2a$ is β_2 and that the interaction among cracks can be neglected, and the stress intensity factor of the
crack tip is given as

$$K_I = p\sqrt{\pi a} \quad , \quad K_{II} = q\sqrt{\pi a} \tag{10}$$

where p and q are the normal stress component and the shearing stress component to the crack plane:

$$\left.\begin{array}{l} p = \sigma\cos^2\theta + \sigma'\sin^2\theta + \tau\sin 2\theta \\ q = -\sigma\sin\theta\cos\theta + \sigma'\sin\theta\cos\theta + \tau\cos 2\theta \end{array}\right\} \tag{11}$$

From the equation (7), the compliance increase due to cracks is given by the
relation

$$\Delta\lambda_{ij} = \frac{2\beta_2}{E'}\int_0^a\left(\frac{\partial K_I}{\partial\sigma_i}\frac{\partial K_I}{\partial\sigma_j} + \frac{\partial K_{II}}{\partial\sigma_i}\frac{\partial K_{II}}{\partial\sigma_j}\right)2da \ , \ (\sigma_i,\sigma_j=\sigma,\sigma',\tau) \tag{12}$$

which yields the following results

$$\frac{1}{\dfrac{2\pi a^2\beta_2}{E'}}\begin{bmatrix}\Delta\lambda_{\sigma\sigma}\\[4pt]\Delta\lambda_{\sigma\sigma'}\\[4pt]\Delta\lambda_{\tau\tau}\\[4pt]\Delta\lambda_{\sigma\tau}\\[4pt]\Delta\lambda_{\sigma'\sigma'}\end{bmatrix} = f_I(p)\begin{bmatrix}\cos^4\theta\\[4pt]\sin^2\theta\cos^2\theta\\[4pt]\sin^2 2\theta\\[4pt]\cos^2\theta\sin 2\theta\\[4pt]\sin^4\theta\end{bmatrix} + \begin{bmatrix}\sin^2\theta\cos^2\theta\\[4pt]-\sin^2\theta\cos^2\theta\\[4pt]\cos^2 2\theta\\[4pt]-\sin\theta\cos\theta\cos 2\theta\\[4pt]\sin^2\theta\cos^2\theta\end{bmatrix} \tag{13}$$

The factor $f_1(p)$ is put 1 or 0 depending the stress singularity of mode I as follows;

$$f_1(p)= \begin{cases} 0 & (\text{ crack for } p < 0, \text{ dislocation pile-up etc. }) \\ 1 & (\text{ crack for } p \gtrless 0, \text{ crack-like void \quad etc. }) \end{cases} \qquad (14)$$

When the crack-like flaw corresponds to the dislocation array freely generated from its source at the center, there is no singularity caused by the normal component p. It is also the case with the crack under the compressive p.

For the material which contains many cracks or crack-like flaws with various orientation, the effective compliance variation $\Delta\lambda_{ij}^*$ is given by the weighted mean

$$\Delta\lambda_{ij}^* = \frac{1}{\pi} \int_{-\pi/2}^{\pi/2} g(\theta)\cdot\Delta\lambda_{ij} \, d\theta \qquad (15)$$

where $g(\theta)$ is the probability density of the distribution of crack orientation θ.

In the case of the randomly-oriented cracks, $g(\theta)=1/\pi$. In addition, $f_1(p)$ is always 1 or 0 for any value of θ, when both of the principal stresses σ_1 and σ_2 are positive or negative. Then the effective compliance variations are given as follows;

$$\begin{bmatrix} \Delta\lambda_{\sigma\sigma}^* \\ \Delta\lambda_{\sigma\sigma'}^* \\ \Delta\lambda_{\tau\tau}^* \end{bmatrix} = \begin{bmatrix} \Delta(1/E) \\ -\Delta(\nu/E) \\ \Delta(1/G) \end{bmatrix} = \frac{2\pi a^2 \beta_2}{E'} \begin{bmatrix} \overset{\begin{subarray}{c}\text{all-round} \\ \text{tension} \\ (\sigma_1,\sigma_2 \gtrless 0)\end{subarray}}{1/2} & \sim & \overset{\begin{subarray}{c}\text{all-round} \\ \text{compression} \\ (\sigma_1,\sigma_2 < 0)\end{subarray}}{1/8} \\ 0 & \sim & -1/8 \\ 1 & \sim & -1/2 \end{bmatrix} \qquad (16)$$

When $\sigma_2 < 0 < \sigma_1$, $\Delta\lambda_{ij}^*$ varies with the stress state between the two extremes shown in the above equation. $f_1(p)$ is non-zero and equal to 1 only within the range $-\theta_1 \leqq \theta \leqq \theta_2$ on the Mohr's circle shown in Fig.3(b), and the integration fo the first term of $\Delta\lambda_{ij}$ in the equation (13) should be executed in this range. Referring to the Mohr's circle, we have $\Delta\lambda_{\sigma\sigma}^*$, for example, when $\tau=0$ and $\sigma' < 0 < \sigma$;

$$\frac{\Delta\lambda_{\sigma\sigma}^*}{2\pi a^2\beta_2/E'} = \frac{1}{\pi}\int_{-\theta_1}^{\theta_2} \cos^4\theta \, d\theta \; + \; \frac{1}{\pi}\int_{-\pi/2}^{\pi/2} \sin^2\theta\cdot\cos^2\theta \, d\theta$$

$$= \frac{1}{4\pi} \left\{ 3\theta_0 + 2\sin 2\theta_0 \left(1 + \frac{\cos 2\theta_0}{4} \right) \right\} + \frac{1}{8} \qquad (17)$$

where $\qquad \theta_0 \equiv \theta_1 = \theta_2 = \frac{1}{2}\arccos\left(\frac{\sigma+\sigma'}{\sigma-\sigma'}\right)$

In addition, another factor on the compliance variation should be mentioned

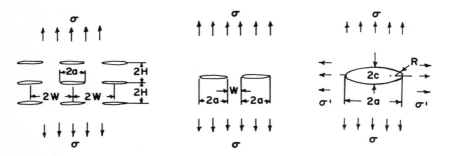

(a) Doubly-periodic cracks   (b) Two adjacent cracks   (c) Elliptical crack

Fig. 4 The effects of the interaction and the flaw configuration

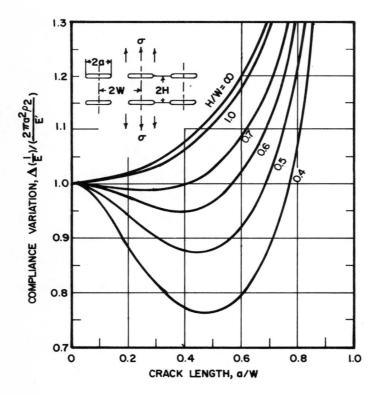

Fig. 5 Compliance variation of doubly-periodic cracks

briefly, the configuration factor of flaws and the interaction factor among flaws, which were neglected in the preceding section. The compliance variation for the case of doubly-periodic cracks in Fig.4(a) is computed from the stress intensity factor given by Isida[3], and is shown in Fig.5, which indicates the interaction effect among cracks or crack-like flaws. The result for $H/W \gg 1$ is approximately expressed by the exact solution for $H/W = \infty$, namely

$$\Delta \lambda_{\sigma\sigma}^* = \Delta\left(\frac{1}{E}\right) = \frac{2\pi a^2 \beta_2}{E'}\left(\frac{2}{\pi}\frac{W}{a}\right)^2 \log\left\{\sec^2\left(\frac{\pi}{2}\frac{a}{W}\right)\right\}, \quad \left(\frac{W}{H}\ll 1\right) \qquad (18)$$

The interaction factor of two adjacent collinear cracks in Fig.4(b) is only 2 when the distance W approaches to zero: $\lim\limits_{W/a \to 0} \Delta \lambda_{\sigma\sigma}^* = \frac{2\pi a^2 \beta_2}{E'}\cdot 2$, $\quad (W/a \ll 1) \quad (19)$

In the case of the elliptical void with radii a and c (the radius of tip $R = c^2/a$) in Fig.4(c), the compliance variations are

$$\begin{bmatrix} \Delta \lambda_{\sigma\sigma}^* \\ \Delta \lambda_{\sigma\sigma'}^* \\ \Delta \lambda_{\tau\tau}^* \end{bmatrix} = \begin{bmatrix} \Delta(1/E) \\ -\Delta(\nu/E) \\ \Delta(1/G) \end{bmatrix} = \frac{2\pi a^2 \beta_2}{E'}\begin{bmatrix} 1 + \frac{1}{2}\frac{c}{a} \\ -\frac{1}{2}\frac{c}{a} - \left(\frac{c}{a}\right)^2 \\ (1+\frac{c}{a})^2 \end{bmatrix} = \frac{2\pi a^2 \beta_2}{E'}\begin{bmatrix} 1+\frac{1}{2}\sqrt{\frac{R}{a}} \\ 0-\frac{1}{2}\sqrt{\frac{R}{a}}\cdot\frac{R}{a} \\ 1+2\sqrt{\frac{R}{a}}+\frac{R}{a} \end{bmatrix} \qquad (20)$$

This relation reveals that the correction is small and of the order of $\sqrt{R/a}$. When $c/a = 1/20$, say, for graphite flake, the correction factor of the configuration for $\Delta(1/E)$ is only 1.025.

4. ELASTIC DEFORMATION DUE TO ELLIPTICAL CRACKS

The three-dimensional problems can be also analyzed similarly, when the stress intensity factor at any point on the leading edge of the crack. As an example of the three-dimensional flaws, elliptical crack will be treated here.

Suppose that the stress components σ, σ' and τ in the (x_1, x_2, x_3) coordinate system are applied to the cube of the unit volume which contains β_3 elliptical cracks with the radii a and b ($b/a \leqq 1$) as shown in Fig.6, and that the interaction among cracks can be neglected. The stress intensity factors at the point A denoted by the angle ξ in the figure are given as follows[4],[5] in terms of the stress components σ_{ij}' , in the (x_1', x_2', x_3') coordinate system as for the principal axes of the cracks;

36

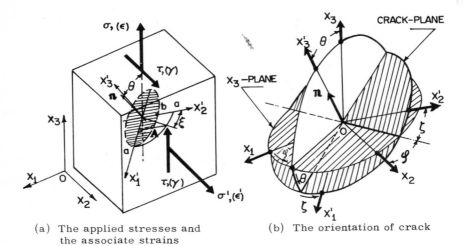

(a) The applied stresses and
the associate strains

(b) The orientation of crack

Fig.6 The elliptical crack contained in a body

$$K_I = f_I(p) \cdot \frac{\sigma'_{33}\sqrt{\pi b}}{E(k)} (1-k^2 \sin^2\xi)^{1/4}, \quad K_{II} = \frac{\sqrt{\pi b}\, k^2}{(1-k^2\sin^2\xi)^{1/4}} \left(\frac{\sigma'_{13}\, k' \sin\xi}{B} + \frac{\sigma'_{23}\cos\xi}{C} \right),$$

$$K_{III} = \frac{\sqrt{\pi b}(1-\nu)\, k^2}{(1-k^2\sin^2\xi)^{1/4}} \left(\frac{\sigma'_{13}\cos\xi}{B} - \frac{\sigma'_{23}\, k' \sin\xi}{C} \right) \tag{21}$$

where $K(k)=\int_0^{\pi/2}(1-k^2\sin^2\xi)^{-1/2}d\xi$ and $E(k)=\int_0^{\pi/2}(1-k^2\sin^2\xi)^{1/2}d\xi$ are the first and the second kind of

perfect elliptical integrals respectively, $k=\sqrt{1-k'^2}$, $k'=b/a$, $B=(k^2-\nu)E(k)+\nu k'^2 K(k)$

and $C=(k^2+\nu k'^2)E(k)-\nu k'^2 K(k)$. Substituting these factors into the equation (7), we have

the compliance variation $\Delta \lambda'_{ij}$ in the (x'_1, x'_2, x'_3) system. For example, we have

$$\Delta \lambda'_{13} \equiv \frac{\Delta \gamma'_{13}}{\tau'_{13}} \equiv \frac{2\Delta \varepsilon'_{13}}{\sigma'_{13}} = \frac{2(1-\nu^2)\beta_3}{E} \int_0^A \left[\left(\frac{\partial K_{II}}{\partial \sigma'_{13}} \right)^2 + \frac{1}{1-\nu} \left(\frac{\partial K_{III}}{\partial \sigma'_{13}} \right)^2 \right] dA$$

$$= \frac{8\pi(1-\nu^2)\, k^2 a b^2 \beta_3}{3EB^2} \int_0^{\pi/2} \frac{(\nu-k^2)\sin^2\varphi + (1-\nu)}{\sqrt{1-k^2\sin^2\varphi}} d\varphi \tag{22}$$

where γ'_{kl} and ε'_{kl} are the engineering shear strain and the tensor component of

shear strain respectively. In this manner, we have the following result;

$$\begin{bmatrix} \Delta \lambda'_{13} \\ \Delta \lambda'_{23} \\ \Delta \lambda'_{33} \end{bmatrix} \equiv \begin{bmatrix} \Delta \gamma'_{13}/\tau'_{13} \\ \Delta \gamma'_{23}/\tau'_{23} \\ \Delta \varepsilon'_{33}/\sigma'_{23} \end{bmatrix} = \frac{8\pi(1-\nu^2)\beta_3 a b^2}{3E} \begin{bmatrix} k^2/B \\ k^2/C \\ f_I(p)/E(k) \end{bmatrix} \tag{23}$$

All the other components of compliance variation due to the presence of cracks are zero because of the symmetry. By the way, the increase in strain energy due to the cracks W_1 per unit volume is given by the relation

$$W_1 = \frac{\tau_{13}'^2}{2}\Delta\lambda_{13}' + \frac{\tau_{23}'^2}{2}\Delta\lambda_{23}' + \frac{\sigma_{33}'^2}{2}\Delta\lambda_{33}' \tag{24}$$

Expressing the orientation of the crack by the angles θ, φ and ζ as shown in Fig.6, and converting the stress and strain components, σ_{ij}' and ε_{kl}', in (x_1', x_2', x_3') system, we have finally the following compliance variation in (x_1, x_2, x_3) system;

$$\begin{bmatrix} \Delta\lambda_{\sigma\sigma} \\ \Delta\lambda_{\sigma\sigma'} \\ \Delta\lambda_{\tau\tau} \end{bmatrix} \equiv \begin{bmatrix} \Delta\varepsilon/\sigma \\ \Delta\varepsilon'/\sigma \\ \Delta\gamma/\tau \end{bmatrix} = \begin{bmatrix} \Delta(1/E) \\ -\Delta(\nu/E) \\ \Delta(1/G) \end{bmatrix} = [A]\begin{bmatrix} \Delta\lambda_{13}' \\ \Delta\lambda_{23}' \\ \Delta\lambda_{33}' \end{bmatrix} \tag{25}$$

where the conversion matrix $[A]$ is given by

$$(26)$$

$$[A] = \begin{bmatrix} (\sin\theta\cos\theta\cos\zeta)^2, & (\sin\theta\cos\theta\sin\zeta)^2, & \cos^4\theta \\ \begin{array}{l}-(\sin\varphi\cos\theta\cos\zeta+\cos\varphi\sin\zeta)\\ \cdot\sin\varphi\sin^2\theta\cos\theta\cos\zeta,\end{array} & \begin{array}{l}-(\sin\varphi\cos\theta\sin\zeta-\cos\varphi\cos\zeta)\\ \cdot\sin\varphi\sin^2\theta\cos\theta\sin\zeta,\end{array} & (\sin\varphi\sin\theta\cos\theta)^2 \\ (\sin\varphi\cos2\theta\cos\zeta+\cos\varphi\cos\theta\sin\zeta)^2, & (\sin\varphi\cos2\theta\sin\zeta-\cos\varphi\cos\theta\cos\zeta)^2, & (\sin\varphi\cos2\theta)^2 \end{bmatrix}$$

The other compliances $\Delta\lambda_{\tau\sigma}$, $\Delta\lambda_{\tau\sigma'}$ ··· can be also obtained similarly, but are not shown here because of less importance. When the probability density of the distribution of crack orientation is given, the variation in the effective compliances $\Delta\lambda_{ij}^*$ (where i, j denote σ, σ' and τ) are obtained from the weighted mean by the density function.

In the case where the orientation is randomly distributed, $\Delta\lambda_{ij}^*$ is given by the mean

$$\Delta\lambda_{ij}^* = \frac{1}{2\pi^2}\int_0^\pi\int_0^\pi\int_0^\pi \Delta\lambda_{ij}\sin\theta\, d\theta\, d\varphi\, d\zeta \tag{27}$$

In the cases of voids and dislocation pile-ups together with the cracks under all-round tension or compression, $f_i(p)$ has the same value for any orientation of flaws, and then the compliance variation is given as follows;

$$\begin{bmatrix} \Delta\lambda_{\sigma\sigma}^* \\ \Delta\lambda_{\sigma\sigma'}^* \\ \Delta\lambda_{\tau\tau}^* \end{bmatrix} = \frac{32(1-\nu^2)\beta_3}{45(2-\nu)E}\left\{\frac{\pi}{2}\frac{ab^2}{E(k)}\right\}\cdot\left\{\frac{2-\nu}{2}f_I(p)\begin{bmatrix} 3 \\ 1 \\ 4 \end{bmatrix} + f_{II}(k,\nu)\begin{bmatrix} 2 \\ -1 \\ 6 \end{bmatrix}\right\},$$

$$\Delta\lambda_{\tau\sigma}^* = \Delta\lambda_{\tau\sigma'}^* = 0 \tag{28}$$

38

where $f_{II}(k,\nu)=(2-\nu)^2k^4E(k)/4BC$, which can be put approximately to be 1.00 under the condition that $\nu \lesssim 1/3$ and $b/a \gtrsim 0.2$. The factors $\pi ab^2/2E(k)$ and $f_{II}(k,\nu)$ are both unity for the penny-shaped flaws (b/a = 1). When one of the sign of the principal stresses is different from the others in the case of cracks, the effective factor of $f_I(p)$ has some value between 0 and 1 , which can be calculated in a similar manner as shown in the process of derivation of the equation (17). For example, the value of $f_I(p)$ for $\Delta(1/G)$ is 1/2 in the case of simple shear ($\sigma=\sigma'=0$). The following is another expression of the above results for cracks in terms of the apparent elastic moduli E^* , ν^* and G^*.

$$
\begin{bmatrix} \frac{E}{E^*}-1 \\ \frac{\nu^*/E^*}{\nu/E}-1 \\ \frac{G}{G^*}-1 \end{bmatrix} = \begin{bmatrix} E\cdot\Delta(\frac{1}{E}) \\ \frac{E}{\nu}\cdot\Delta(\frac{\nu}{E}) \\ G\cdot\Delta(\frac{1}{G}) \end{bmatrix} = \alpha\cdot\begin{bmatrix} \overset{\begin{pmatrix} \text{all-round} \\ \text{tension} \end{pmatrix}}{1} & \overset{\begin{pmatrix} \text{simple} \\ \text{shear} \end{pmatrix}}{\sim} & \overset{\begin{pmatrix} \text{all-round} \\ \text{compression} \end{pmatrix}}{\frac{4}{7}} \\ \frac{1}{9} & \sim & \frac{2}{3} \\ \frac{7}{9} & \sim & \frac{23}{36} & \sim & \frac{1}{2} \end{bmatrix} \quad \nu=\frac{1}{3} \tag{29}
$$

where $\alpha=1.71\beta_3\{\pi ab^2/2E(k)\}$ ($=1.71\beta_3 a^3$ for $b/a=1$). It is interesting to note that the apparent Poisson's ratio under tension decreases due to the presence of cracks but the compliance variation for lateral strain $\Delta(\nu/E)$ is quite small compared with the

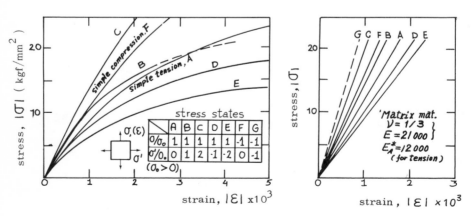

(a) 3% C gray cast iron (after L.F.Coffin) (b) Elastic cracked body

Fig.7 The stress-strain relation in various biaxial stress states

other variations. These facts were also observed in the expriment on the cast iron with flaky graphites[6]. By the way, the lateral contraction does not vary in the two-dimensional crack model as is shown in the equation (16) for tension.

Fig.7(a) shows the experimental stress–strain relation of 3% C gray cast iron with flaky graphites in various biaxial stress states[6], and Fig.7(b) shows the calculated elastic behavior in the same states, postulating $\nu = 1/3, E = 21,000$ kgf/mm^2 for the matrix material and $E_A^* = 12,000$ kgf/mm$^2 = (4/7)E$ for the apparent Young's modulus in tension. A reasonable correspondence with the experiment can be seen, as far as the elastic deformation concerns. In the above analysis, the interaction effect among flaws is assumed negligible. However, in the case where the expected spacing between the nearest neighbours $d_m = (\sqrt{2}/\beta_3)^{1/3}$ is comparable to the flaw size, it is supposed that the factor β_3 in the relation (28) should be multiplied by a certain correction factor for the interaction effect. By the way, d_m corresponds to $1.5a$ in the above example if the penny–shaped graphite is assumed.

5.   ELAST–PLASTIC DEFORMATION DUE TO THE DUGDALE CRACKS

The Dugdale model[7] of crack enables us to evaluate the elasto–plastic strain of cracked materials. In the present analysis, the reciprocal theorem on the mutual compliance is conveniently applied to derive the deformation, as well as the plastic energy dispersion rate during the crack growth.

Consider the elastic deformation of the square of unit volume in Fig.8 containing β_2 cracks of the Dugdale type, and suppose that the stress in the plastic zone $\sigma''(x)$ is uniform yield stress σ_{ys}. When the interaction among cracks can be neglected, the displacement of traction–free surface of the crack of length $2a^*$ by the external stress σ is well–known to be

$$w''(x) = \frac{2\sigma}{E}\sqrt{a^{*2} - x^2} \tag{30}$$

which is directed to the direction of $-\sigma''(x)$. Thus the mutual compliance between σ and $\sigma''(x)$ is given by $-w''(x)/\sigma$, and therefore we have the strain $\Delta\varepsilon$ caused by the application of σ and $\sigma''(x)$ as follows;

40

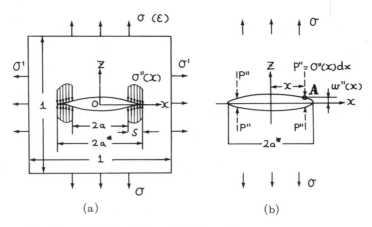

Fig.8   The Dugdale model of crack

$$\Delta \varepsilon = \sigma \cdot (\Delta \lambda_{\sigma\sigma})_{a^*} - \int_a^{a^*} \sigma''(x) \frac{2}{E'} \sqrt{a^{*2} - x^2}\, dx \times 4P_2$$

$$= \frac{2\pi a^{*2} P_2 \sigma}{E'} - \frac{8 P_2 \sigma_{ys}}{E'} \int_a^{a^*} \sqrt{a^{*2} - x^2}\, dx$$

$$= \frac{2\pi a^{*2} P_2 \sigma}{E'} \frac{\sin\left(\frac{\pi \sigma}{2\sigma_{ys}}\right) \cos\left(\frac{\pi \sigma}{2\sigma_{ys}}\right)}{\left(\frac{\pi \sigma}{2\sigma_{ys}}\right)} \tag{31}$$

On the other hand, the nonsingular stress postulate at the crack tip by Dugdale

$$K_\sigma + K_{\sigma''(x)} = \sigma\sqrt{\pi a^*} - \frac{2\sqrt{\pi a^*}}{\pi} \int_a^{a^*} \frac{\sigma''(x)}{\sqrt{a^{*2} - x^2}}\, dx$$

$$= \sigma\sqrt{\pi a^*} - \frac{2\sigma_{ys}\sqrt{\pi a^*}}{\pi} \cos^{-1}\frac{a}{a^*} = 0 \tag{32}$$

yields the relation[7]

$$a = a^* \cos\left(\frac{\pi}{2}\frac{\sigma}{\sigma_{ys}}\right) \tag{33}$$

Thus we have finally the stress–strain relation

$$\Delta \varepsilon = \frac{2\pi a^2 P_2 \sigma_{ys}}{E'} \frac{2}{\pi} \tan\left(\frac{\pi}{2}\frac{\sigma}{\sigma_{ys}}\right) \tag{34}$$

which is shown by the curve (2) in Fig.10.

In addition, we obtain the increase in the stored strain energy W_I due to the presence of the Dugdale cracks as follows:

$$W_I = \Delta \varepsilon \cdot \sigma - \int_0^\sigma \Delta \varepsilon \cdot d\sigma = \frac{8 a^2 P_2 \sigma_{ys}^2}{\pi E'}\left[\frac{\pi}{2}\frac{\sigma}{\sigma_{ys}} \tan\left(\frac{2\sigma}{\pi \sigma_{ys}}\right) - \log\left\{\sec\left(\frac{2}{\pi}\frac{\sigma}{\sigma_{ys}}\right)\right\}\right] \tag{35}$$

Therefore the plastic work rate for the crack growth is

$$\frac{dW_P}{dA} = \frac{dW_I}{dA} = \frac{1}{2a P_2}\frac{dW_I}{da} = \frac{8 a \sigma_{ys}^2}{\pi E'}\left[\frac{\pi}{2}\frac{\sigma}{\sigma_{ys}} \tan\left(\frac{2}{\pi}\frac{\sigma}{\sigma_{ys}}\right) - \log\left\{\sec\left(\frac{2}{\pi}\frac{\sigma}{\sigma_{ys}}\right)\right\}\right] \tag{36}$$

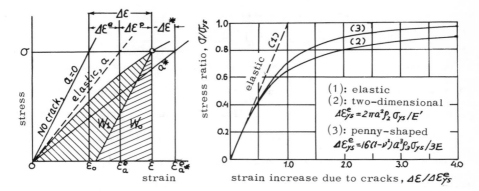

Fig.9   Stress-strain curve

Fig.10   Stress-strain curve of Dugdale model

which, of course, coincides with the result by Goodier and Field[8].

In the case of penny-shaped crack of the Dugdale model, we replace x in Fig.9 by the radius r. From the postulate of nonsingularity, that is,

$$K_\sigma + K_{\sigma''(t)} = \frac{2\sigma\sqrt{\pi a^*}}{\pi} - \frac{2\sigma_{ys}}{\sqrt{\pi a^*}}\sqrt{a^{*2} - a^2} = 0 \tag{37}$$

we have for the outside radius of plastic zone

$$a^* = a \Big/ \sqrt{1 - (\sigma/\sigma_{ys})^2} \tag{38}$$

Since the displacement of the crack surface by the external stress σ is[9]

$$w''(r) = \frac{4(1-\nu^2)\sigma}{\pi E}\sqrt{a^{*2} - r^2} \tag{39}$$

strain increase of the cracked body due to the Dugdale cracks is given by the equation

$$\Delta\varepsilon = \sigma(\Delta\lambda_{\sigma\sigma})_{a^*} - \frac{4(1-\nu^2)}{\pi E}\int_a^{a^*}\sqrt{a^{*2} - r^2}\,\sigma''(r)\,2\pi r dr \times 2\beta_3$$

$$= \frac{16(1-\nu^2)a^{*3}\beta_3\sigma}{3E}\left\{1 - \left(\frac{\sigma}{\sigma_{ys}}\right)^2\right\}$$

$$= \frac{16(1-\nu^2)a^3\beta_3\sigma_{ys}}{3E}\frac{\sigma/\sigma_{ys}}{\sqrt{1 - (\sigma/\sigma_{ys})^2}} \quad , \tag{40}$$

which is shown by the curve (3) in Fig.10.

By the way, the elastic strain energy stored due to the cracks is

$$W_1 = \Delta\varepsilon\cdot\sigma - \int_0^\sigma \Delta\varepsilon\,d\sigma = \frac{16(1-\nu^2)a^3\beta_3\sigma_{ys}^2}{3E}\left\{\frac{1}{\sqrt{1-(\sigma/\sigma_{ys})^2}} - 1\right\} \tag{41}$$

which gives the same plastic work rate as that obtained by Olesiak et al.[10];

$$\frac{dW_P}{dA} = \frac{1}{2\pi a \rho_3} \frac{dW_I}{da} = \frac{8(1-\nu^2)a\sigma_{ys}^2}{\pi E} \left\{ \frac{1}{\sqrt{1-(\sigma/\sigma_{ys})^2}} - 1 \right\} \qquad (42)$$

The above treatment through the mutual—compliance does not need the analysis

of the crack surface displacement under the combined loading of the stresses on the

outer surface and on the crack surface. This approach is thought to be extensively

applicable to many other crack problems of the Barenblatt—Dugdale type.

Combining the treatments in the preceding sections, it is hopeful to obtain the

elasto—plastic behavior in triaxial stress state of the materials which contain

cracks of various orientation.

6. CONCLUDING REMARKS

The compliance of an extended meaning and the related reciprocal theorem are

tentatively applied to analyse the deformation of cracked material. This kind of

approach could be helpfull not only to the analysis of the deformation itself, but also

to the strength analysis of several phases; the strength of cracked or crazed

material, the behavior of the prominent crack within crazed zone, the strength of

the cracked member contained in a statically indeterminate structure and so on.

The author wishes to express his gratitude to Dr. Y. Naito of Komatsu Ltd. for

the preparation of manuscripts and Dr. K. Watanabe of University of Tokyo for

usefull discussions.

REFERENCES

1) T.Namai, J. Japan Foundrymen's Soc. ,43,11(1971) 970
2) H. Okamura et al., ASTM, STP-536 (1973) 423
3) M.Isida, Mechanical Behavior of Materials, 1 (1972) 394, SMC Japan
4) A.E.Green et al., Proc. Cambridge Phil. Soc., 46 (1950) 159
5) M.K.Kassir et al., Trans. ASME, J. Appl. Mech., 33 (1966) 601
6) L.F.Coffin et al., J. Appl. Mech., 17 (1950) 257
7) D.S.Dugdale, J. Mech. Phys. Solids, 8 (1960) 100
8) J.N.Goodier et al., Fracture of Solids, 103 (1963) Wiley, New York
9) I.N.Sneddon, Proc. Royal Soc. London, A, 187 (1946) 229
10) Z.Olesiak et al., Int. J. Fracture Mech., 4 (1968) 383

Stress Intensity Factors For Some
Surface Flaw Problems

By

A. S. Kobayashi[*]

1.   INTRODUCTION

Published examples of actual failed parts (see, for example, Reference 1)
show that failure often originates from surfaces which can be modeled by part-
elliptical or part-circular cracks. Substantial efforts have thus been mounted
to estimate the stress intensity factor of semi-elliptical cracks by semi-em-
pirical solutions after Irwin (2), and by the alternating technique in three-
dimensional elasticity which is the subject of this paper.

2.   THEORETICAL BACKGROUND

The iterative procedure in the alternating technique for solving three-
dimensional problems in fracture mechanics has been well documented in papers
by Smith (3, 4), Shah (5, 6), and Hartranft and Sih (7) and will not be repeated here. Smith and his colleagues and Hartranft and Sih used the mathematically more convenient solution of a totally embedded penny-shaped crack in their
alternating technique. Shah and Kobayashi, on the other hand, used the solution for a totally elliptical crack, originally suggested by Segedin (8), for
their analysis. The latter solution and the half-space solution by Love (9)
have been repeatedly described in the above papers and thus only the numerical
techniques necessary to obtain a reasonable estimate of the stress intensity
factors in surface flaws together with some solutions will be described in
this paper.

[*]Department of Mechanical Engineering, University of Washington, Seattle,
Washington 98195.

One of the major drawbacks of the elliptical crack solution based on the potential function of Segedin is its polynomial representation of the crack pressure distribution which is limited to third order terms due to mathematical complexities in deriving higher order terms. Such crack pressure distribution cannot readily represent the rapidly varying residual crack surface tractions generated during the course of applying the alternating technique to three-dimensional crack analysis. As a result, the surface flaw problem was not analyzed directly in a series of problems solved by Shah and Kobayashi. Nevertheless the stress intensity factor at the deepest penetration of a semi-elliptical crack, was approximated (6) by combining an empirical front surface magnification factor with the calculated back surface magnification factor.

In a recent paper, Kobayashi (10) described a procedure with which the residual surface traction in the iteration process could be reduced significantly and thus extend the usefulness of Shah's alternating technique. In essence, the procedure consists in prescribing a fictitious pressure on the fictitious portion of the elliptical crack surface which protrudes into the free space. This prescribed fictitious pressure is determined from its two-dimensional analog, which yields in a totally embedded two-dimensional crack, the correct single edge-crack stress intensity factor.

Another numerical inconvenience is the calculation of the surface tractions acting on the free bounding planes using Segedin's potential function. Mathematical complexities in deriving the six stress components, i.e.,

$$\sigma_{xx}\big]_{x=0}, \ \tau_{xz}\big]_{x=0}, \ \tau_{xy}\big]_{x=0}, \ \sigma_{yy}\big]_{y=0}, \ \tau_{yx}\big]_{y=0}, \ \text{and} \ \tau_{yz}\big]_{y=0}$$

in terms of the potential function as well as actual computer programming were reduced by numerically differentiating the analytical expressions of,

$$\frac{\partial^2 \phi}{\partial x^2}, \ \frac{\partial^2 \phi}{\partial x \partial y}, \ \frac{\partial^2 \phi}{\partial y^2} \ \text{and} \ \frac{\partial^2 \phi}{\partial z^2}$$

in order to obtain the third partial derivatives of ϕ with respect to x, y and z.

The second step in the alternating technique is to eliminate the residual surface traction on the bounding free surfaces computed by the finite difference procedure described above. A criterion for maximum rectangle size set forth in Reference 10 was used to determine sizes of the rectangles on the free bounding surfaces which reduced the number of rectangles to approximately 60 to 70 from the original 540 used by Smith in 1969 (3).

3. TYPICAL RESULTS

3.1 Semi-Elliptical Surface Flaw

The improved numerical procedure for the alternating technique was then used to determine stress intensity factors in semi-elliptical surface flaws in finite thickness plates and subjected to uniform and linearly varying pressure distribution. Since the two bounding front and back free surfaces interact with each other in this surface flaw problem, running summations of the residual surface tractions on each of the three surfaces, i.e. the semi-elliptical* crack surface and the two free bounding surfaces, due to each removal of residual surface tractions from any other two surfaces, were maintained at all times. These current values of residual surface tractions were used in each erasure process. The rectangle mesh spacing described previously together with 32 to 40 almost evenly spaced points on the quarter-elliptical* crack surface for least square fitting of σ_{zz} were used in each iteration of the alternating technique. Five cycles of such iteration required approximately 1500 seconds of CPU time on the CDC 6400 computer.

Figures 1 and 2, the stress intensity magnification factor, M_K, which is the ratio of the actual stress intensity factor to that of a fully embedded elliptical crack along the crack periphery for two ellipticity of b/a = 0.2

---

*x-axis symmetry in the semi-elliptical crack problem was accounted for in actual computation and thus only quarter of the elliptical crack surface was considered.

and 0.98 at a crack depth of b/t = 0.8. Figure 1 shows the stress intensity magnification factor, M_K, for a semi-elliptical crack pressurized by constant pressure of $\sigma_{zz} = \sigma_o$ and Figure 2 shows that for a semi-elliptical crack pressurized by a linearly varying pressure of $\sigma_{zz} = \sigma_o$ (1-y/b). The residual surface tractions at the end of the fifth iteration on the semi-elliptical crack surface, ranged from a high of 0.17 σ_o at an isolated region, where the crack periphery intersects the free front surface, to low values of $10^{-3}\sigma_o$. The maximum residual surface tractions on the free front and back surfaces were of the order of $10^{-3}\sigma_o$, and $10^{-4}\sigma_o$, respectively.

One time eliminations of the high residual surface tractions on the semi-elliptical crack surface were conducted following the procedure, which used Tada's solution (11), and is outlined in Reference 12. The result is a definite downward trend, which is not as pronounced as that described in Reference 7, in the stress intensity magnification factor near the free front surface.

Also shown in Figures 1 and 2 are stress intensity factors for other semi-elliptical cracks, i.e. for b/a = 0.4, 0.6 and 0.8. These M_K values in Figure 1 were estimated by interpolating the computed M_K values for b/a = 0.2 and 0.98 using the proportionalities of the estimated results in Reference 6 for $\sigma_{zz} = \sigma_o$. In the linearly varying load of $\sigma_{zz} = \sigma_o$ (1-y/b) in Figure 2, the proportionalities of the estimated results in Figure 6 in Reference 10 were used to estimate the M_K values for b/a = 0.4, 0.6, and 0.8.

3.2   Quarter-Elliptical Surface Flaw

Stress intensity factors of quarter-elliptical surface flaws in quarter infinite solids were determined by procedures similar to those described above. Numerically, this computation yielded higher residual surface tractions on the two free bounding surfaces as well as the crack surface due to the close proximity of all three surfaces. Nevertheless, the alternating technique was

terminated at 3 iteration cycles due to restriction in available computer
funds. Three cycles of such iterations required CPU time of 650, 703 and 783
seconds on the CDC 6400 computer for crack aspect ratios of b/a = 0.98, 0.4
and 0.2 respectively. The isolated high local residual surface tractions in
regions where the crack front penetrates the free bounding surfaces, were also
reduced by this procedure described previously. The result of this incomplete
erasure is a more definite trend in the reduction of stress intensity magnification factor as the crack front approaches the two free-bounding surfaces.

The resultant stress intensity magnification factors for three elliptical
cracks with aspect ratios of b/a = 0.98, 0.4 and 0.2 and with a constant prescribed pressure of $\sigma_{zz} = \sigma_o$ are shown in Figure 3. Also shown in Figure 3 is
the finite element results for b/a = 1.0 by Tracey (13). The significant deviations between finite element results for b/a = 1.0 and the results obtained
by the alternating technique for b/a = 0.98 could be attributed to the coarse-
ness of the finite element breakdown.

Figure 4 shows the stress intensity factors of a quarter-elliptical crack
with a linearly decreasing pressure gradient in the direction of the minor axis
of the ellipse, $\sigma_{zz} = \sigma(1 - y/b)$. The average residual surface tractions in
this problem were 1.00, .833 and 0.329 percent of the maximum value of the
linearly varying pressures for crack aspect ratios of b/a = 0.98, 0.4 and 0.2,
respectively. Since the residual tractions in the regions where the crack
front intersects the free bounding surfaces were small, the procedure used to
erase isolated higher values of residual surface tractions did not show signi-
fican drop in the stress intensity factors as the crack front approached the
two free bounding surfaces.

4. CONCLUSIONS AND DISCUSSION

The improved numerical procedure in the alternating technique for three-dimensional fracture mechanics was successfully used to determine the stress intensity magnification factors in deep semi-elliptical surface flaws in a finite-thickness plate and subjected to constant and linearly varying crack pressures. The procedure was also used to determine the stress intensity magnification factors in quarter-elliptic surface flaws in quarter infinite solids and subjected to constant and linearly varying crack pressures.

Linear combination of the two solutions involving constant and linearly varying crack pressures can be used to estimate the stress intensity magnification factors in regions of high stress gradients following the procedure outlined in Reference 14.

5. ACKNOWLEDGEMENT

The work reported here was supported by the U. S. Army Research Office--Durham. Some of the ideas for this work were conceived during the course of the author's consulting engagements with The Boeing Company. The author wishes to express his appreciation to Mr. J. N. Masters, The Boeing Aerospace Company, for his support and particularly to Dr. R. C. Shah, The Boeing Aerospace Company, for his assistance on the many details of the theoretical and numerical analyses.

6. REFERENCES

(1) Fracture Mechanics of Aircraft Structures, (edited by H. Liebowitz), AGARDograph No. 176, (January 1974).

(2) G. R. Irwin, "Crack Extension Force for a Part-Through Crack in a Plate," Journal of Applied Mechanics, Vol. 29, Trans. ASME, Vol. 84, (1962), pp. 651-654.

(3) F. W. Smith, A. F. Emery, and A. S. Kobayashi, "Stress Intensity Factors for Semi-Circular Cracks, Part 2--Semi-Infinite Solid," Journal of Applied Mechanics, Vol. 34, Trans. ASME, Vol. 89, (1967), pp. 953-959.

(4)   R. W. Thresher and F. W. Smith, "Stress Intensity Factors for a Sur-
      face Crack in a Finite Solid," Journal of Applied Mechanics, Vol.
      39, Trans. of ASME, Vol. 95, (March 1972), pp. 195-200.

(5)   R. C. Shah and A. S. Kobayashi, "Stress Intensity Factor for an Elli-
      ptical Crack Under Arbitrary Normal Loading," Journal of Engineering
      Fracture Mechanics, Vol. 3, No. 1, (July 1971), pp. 71-96.

(6)   R. C. Shah and A. S. Kobayashi, "On The Surface Flaw Problem," The
      Surface Crack: Physical Problems and Computational Solutions,
      (edited by J. L. Swedlow), ASME, (1972), pp. 79-124.

(7)   R. J. Hartranft and G. C. Sih, "Alternating Method Applied to Edge
      And Surface Crack Problems," Mechanics of Fracture I--Methods of
      Analysis and Solutions of Crack Problems, (edited by G. C. Sih),
      Noordhoff International, (1973), pp. 179-238.

(8)   C. M. Segedin, "A Note on Geometric Discontinuities in Elasto-
      Static," International Journal of Engineering Sciences, Vol. 6,
      (1968), pp. 601-611.

(9)   A. E. H. Love, A Treatise on the Mathematical Theory of Elasticity,
      Dover Publications, New York, (1944), pp. 241-245.

(10) A. S. Kobayashi, A. N. Enetanya, and R. C. Shah, "Stress Intensity
      Factors of Elliptical Crack," to be published in the Proceedings of
      Conference on the Prospects of Advanced Fracture Mechanics, Delft,
      The Netherlands, (June 20-29, 1974).

(11) H. Tada, P. Paris, and G. Irwin, The Stress Analysis of Cracks,
      Handbook, Del Research Corporation, (1973), pp. 24.2.

(12) A. S. Kobayashi and A. N. Enetanya, "Stress Intensity Factors of a
      Corner Crack," to be published in the Proc. of the 8th National
      Symposium on Fracture Mechanics, ASTM STP, (1975).

(13) D. M. Tracey, "3-D Elastic Singular Element for Evaluation of K
      Along an Arbitrary Crack Front," International Journal of Fracture,
      Vol. 9, (1973), pp. 340-343.

(14) A. S. Kobayashi, "A Simple Procedure for Estimating Stress Intensity
      Factors in Region of High Stress Gradient," Significance of Defects
      in Welded Structures, (edited by T. Kanazawa and A. S. Kobayashi),
      University of Tokyo Press (1974), pp. 127-143.

50

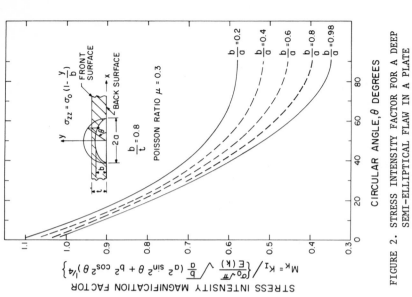

FIGURE 2. STRESS INTENSITY FACTOR FOR A DEEP
SEMI-ELLIPTICAL FLAW IN A PLATE
SUBJECTED TO LINEAR LOAD.

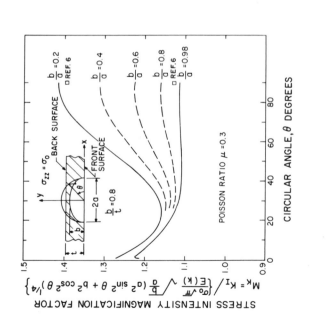

FIGURE 1. STRESS INTENSITY FACTOR FOR A DEEP
SEMI-ELLIPTICAL FLAW IN A PLATE
SUBJECTED TO UNIAXIAL TENSION.

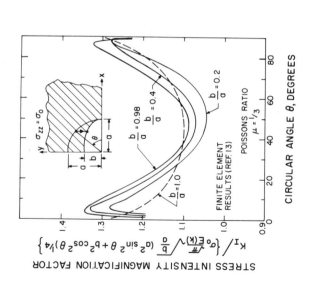

FIGURE 4. STRESS INTENSITY FACTOR FOR A CORNER
FLAW IN A QUARTER INFINITE SOLID SUB-
JECTED TO LINEAR LOAD.

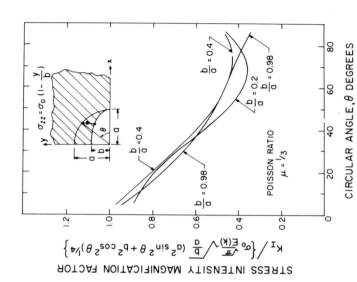

FIGURE 3. STRESS INTENSITY FACTOR FOR A CORNER
FLAW IN A QUARTER INFINITE SOLID SUB-
JECTED TO UNIAXIAL TENSION.

A Method for Calculating Stress Concentrations
and Its Applications

by Hironobu NISITANI*

Abstract

A method for calculating stress concentration factors or stress intensity
factors are presented and its applications are shown.

The method is based on the stress field due to a point force acting in an
infinite plate or an infinite body.

The given boundary conditions are satisfied by applying body force
(continuously embedded point forces) along imaginary boundaries.

The relative errors of the values obtained are smaller than 0.1 % in
two-dimensional cases and 1 % in three-dimensional cases.

1. Introduction

Various methods for calculating stress concentration factors or intensity
factors have been developed for two-dimensional problems. There is, however,
few methods of analysis for three-dimensional problems[1].

In this paper, a method of analysis[2] applicable to two- and three-
dimensional problems is explained and the numerical results obtained by the
method are shown.

2. Principle of the Present Method

The present method for two-dimensional problems is based on the stress
field due to a point force acting in an infinite plate without any hole or
crack. For three-dimensional problems, the method is based on the stress
field due to a point force acting in an infinite body.

---

* Dr., Professor, Faculty of Engineering,
  Kyushu University, Fukuoka

The given boundary conditions are satisfied by applying body force (continuously embedded point forces) along the imaginary boundaries imagined in an infinite plate without any hole or crack (two-dimensional problems) or in an infinite body without any cavity or crack (three-dimensional problems). The density of the body force can be determined from the boundary conditions.

The density of body force can be obtained in a closed form for simple problems, as shown in the following sections. In general problems, however, the density of body force has to be determined by a numerical procedure. Taking as an example the tension of a semi-infinite plate with a crack, the procedure will be explained.

2.1 Tension of an Infinite Plate with an Elliptical Hole

Take an infinite plate with an elliptical hole subjected to a tensile stress σ_∞, as shown in Fig. 1. Then, the x- and y-component of displacement u, v at the boundary of the elliptical hole (ξ , η) are[3]

$$u = \frac{\sigma_\infty}{E}(1+2\frac{b}{a})\,\xi = \frac{\sigma_\infty}{E}(2+\frac{a}{b})\sqrt{b^2-\eta^2}, \qquad v = -\frac{\sigma_\infty}{E}\eta \qquad (1)$$

Now, imagine a plate having the same shape as the elliptical hole. If the elliptical plate is subjected to the same uniform deformation as given by Eq. (1), the strain components ε_x , ε_y at an arbitrary point of the elliptical plate become constant respectively and are expressed as follows.

$$\varepsilon_z = \frac{\sigma_\infty}{E}(1-2\frac{b}{a}), \qquad \varepsilon_y = -\frac{\sigma_\infty}{E} \qquad (2)$$

The stress components σ_x , σ_y corresponding to ε_x , ε_y become

$$\sigma_x = \left\{(1+2\frac{b}{a})-\nu\right\}\frac{\sigma_\infty}{1-\nu^2}, \qquad \sigma_y = -\left\{1-\nu(1+2\frac{b}{a})\right\}\frac{\sigma_\infty}{1-\nu^2} \qquad (3)$$

Then, the elliptical plate subjected to the surface stresses given by Eq. (3) is just to fill up the infinite plate having an elliptical hole subjected to the stress σ_∞ at infinity. Therefore, the solution of the infinite plate with an elliptical hole can be obtained by superposing the stress fields due to a point force acting in an infinite plate without any hole (Fig. 2).

That is, the problem of a plate with a hole is reduced to the problem of an infinite plate without any hole.

The densities of body force ρ_x , ρ_y in this case are obtained from Eq. (3) and become

$$\rho_x = \left\{ (1+2\frac{b}{a}) - \nu \right\} \frac{\sigma_\infty}{1-\nu^2} , \qquad \rho_y = -\left\{ 1 - \nu(1+2\frac{b}{a}) \right\} \frac{\sigma_\infty}{1-\nu^2} \qquad (4)$$

The definition of ρ_x , ρ_y is given by the following equation

$$\rho_x = \frac{dF_x}{d\eta} , \qquad \rho_y = -\frac{dF_y}{d\xi} \qquad (5)$$

where dF_x , dF_y are the x- and y-component of the embeded point forces acting on the infinitesimal element $dS = \sqrt{(d\xi)^2 + (d\eta)^2}$ of an imaginary ellipse which must become free from stress.

The values of stresses at an arbitrary point can be calculated by integrating the stress fields due to point forces $\rho_x \, d\eta$, $\rho_y \, d\xi$.

2.2 Tension of an Infinite Plate with a Crack

A crack can be considered as an extremely slender ellipse (b/a $\rightarrow \infty$). According as b/a $\rightarrow \infty$, the points where point forces should be applied become close to each other. Therefore, the derivative of the stress field due to a point force is fundamental for the crack problems.

Since a $\rightarrow$ 0 in crack problem, ρ_x in Eq. (4) is expressed as follows (in the present problem, we can put ν = 0).

$$\rho_x = (1+2\frac{b}{a})\sigma_\infty \cong \frac{2b}{a}\sigma_\infty = \frac{2\sigma_\infty\sqrt{b^2-\eta^2}}{\xi} \qquad (6)$$

Then, the value of ρ_y is negligiblely small.

When a point force P is applied at a point (ξ , η) of an infinite plate in the x-direction, the stress σ_x^P at (x, y) is[3]

$$\sigma_x^P = -\frac{(x-\xi)\{3(x-\xi)^2+(y-\eta)^2\}}{4\pi\{(x-\xi)^2+(y-\eta)^2\}^2} P \qquad (7)$$

The stress $d\sigma_x$ which appears at (0, y) by a pair of body forces acting on the infinitesimal element $d\eta$ (see Fig. 3) can be calculated from Eq. (7).

$$d\sigma_x = \lim_{x,\xi \to 0} \left[(P_x d\eta) \cdot \frac{\partial \sigma_x^P}{\partial \xi}\Big|_{P=1} \cdot 2\xi \right] = \frac{\partial \sigma_x^P}{\partial \xi}\Big|_{\substack{P=1 \\ \xi=\iota=0}} \cdot (P_x 2\xi)\Big|_{\xi \to 0} \cdot d\eta \tag{8}$$

As seen from Eq. (8), $(P_x 2\xi)\big|_{\xi \to 0}$ means the density of a pair of body forces. It is obtained from Eq. (6).

$$(P_x 2\xi)\big|_{\xi \to 0} = 4\sigma_\infty \sqrt{b^2 - \eta^2} \tag{9}$$

Therefore,

$$d\sigma_x = \frac{1}{4\pi(y-\eta)^2} \cdot (4\sigma_\infty \sqrt{b^2-\eta^2}) \cdot d\eta \tag{8'}$$

Superposing the stress field of Eq. (8)′ and a uniform stress σ_∞, the stress distribution $\sigma_x(y)$ on the y-axis are obtained as follows[4].

$$\sigma_x(y) = \int_{-b}^{b} \frac{1}{4\pi(y-\eta)^2} \cdot (4\sigma_\infty \sqrt{b^2-\eta^2}) \cdot d\eta + \sigma_\infty \tag{10}$$

when $|y| < b$

$$\sigma_x(y) = \frac{\sigma_\infty}{\pi} \left[\frac{\sqrt{b^2-\eta^2}}{y-\eta} - \sin^{-1}\frac{\eta}{b} + \frac{y}{\sqrt{b^2-y^2}} \ln\left|\frac{b^2-y\eta+\sqrt{(b^2-y^2)(b^2-\eta^2)}}{b(y-\eta)}\right| \right]_{-b}^{b} + \sigma_\infty = 0 \tag{10'}$$

when $|y| > b$

$$\sigma_x(y) = \frac{\sigma_\infty}{\pi} \left[\frac{\sqrt{b^2-\eta^2}}{y-\eta} - \sin^{-1}\frac{\eta}{b} + \frac{|y|}{\sqrt{y^2-b^2}} \sin^{-1}\frac{b^2-y\eta}{b(\eta-y)} \right]_{-b}^{b} + \sigma_\infty \tag{10''}$$

$$= \frac{|y|}{\sqrt{y^2-b^2}} \sigma_\infty \lessapprox \frac{\sigma_\infty \sqrt{\pi b}}{\sqrt{2\pi r}}, \quad r = |y| - b$$

Equations (10)′, (10)″ show that the solution of the infinite plate with a crack can be obtained by superposing the stress field due to a pair of point forces (Fig. 4).

The value of stress at an arbitrary point can be calculated by integrating the stress field due to a pair of body forces $(P_x 2\xi)\big|_{\xi \to 0} d\eta$.

Since the tensile stress which should be canceled along an imaginary crack line is not necessarily uniform, the density of a pair of body forces in general problems cannot be expressed by such a simple form as Eq. (9). Therefore, usually the following equation must be used.

$$(P_x 2\xi)\big|_{\xi \to 0} = 4 f(\eta)\sqrt{b^2-\eta^2} \tag{11}$$

The characteristic of the stress distribution near a crack tip, however, is determined by the value f(b) alone and has no relation with the value f(η) at the point apart from the crack tip. Then, the stress intensity factor K_I becomes

$$K_1 = f(b)\sqrt{\pi b} \qquad (12)$$

2.3 Tension of a Semi-Infinite Plate with a Crack

Although the value of f(η) in Eq. (11) is constant in the case of tension of an infinite plate with a crack $\left\{$Eq. (9)$\right\}$, in general problems f(η) has to be changed with η so as to satisfy the given boundary condition. Taking as an example the case shown in Fig. 5, the numerical procedure for determining the value of f(η) will be explained.

The calculation process in this case is as follows (notations are shown in Fig. 6) ;

Step 1 : The imaginary crack AB in Fig. 6 is divided into MM equal intervals.

Step 2 : The value of f(η) in N-th interval (N:1~MM) is assumed to be constant in the interval. It is denoted by f_N.

Step 3 : The influence coefficient $\tilde{\sigma}_{xM}^{N}$ is calculated, where $\tilde{\sigma}_{xM}^{N}$ is the stress $\tilde{\sigma}_x$ appearing at the midpoint of M-th interval by the pair of body forces (f_N =1) acting on N-th interval.

Step 4 : From the boundary condition at the midpoint of M-th interval (M:1 MM), a set of MM linear equations concerning unknown constant f_N (N:1~MM) is obtained.

$$\sum_{N=1}^{MM} f_N \tilde{\sigma}_{xM}^{N} + \tilde{\sigma}_\infty = 0, \quad (M: 1, 2, \cdots MM) \qquad (13)$$

Solving these equations, the value of f_N is determined.

Step 5 : The values of K_{IA} (MM), K_{IB} (MM) are obtained from $f_N|_{N=1} = f_1$, $f_N|_{N=MM} = f_{MM}$, where K_{IA} (MM), K_{IB} (MM) are the values of K_I at the points A and B respectively in the case when the number of division of AB is equal to MM.

$$K_{IA}(MM) \doteq f_1 \sqrt{\pi b}, \qquad K_{IB}(MM) = f_{MM} \sqrt{\pi b} \tag{14}$$

Step 6 : The real value of K_I is $K_I(\infty)$. $K_I(\infty)$ is obtained by extrapolation

using $K_I(MM)$ (MM: finite), as shown in Fig. 7.

The calculation of $\overline{\sigma}_{xM}^N$ in step 3 are as follows.

The stress due to a point force P acting in the x-direction in a

semi-infinite plate is[2]

$$\sigma_x^P = -\frac{(x-\xi)\{3(x-\xi)^2+(y-\eta)^2\}}{4\pi\{(x-\xi)^2+(y-\eta)^2\}^2}P$$

$$+\frac{1}{4\pi y(A^2+n^2)^3}\left[5A^5+4(n^2-n+1)A^3-n^2(n^2-12n+12)A\right]P \tag{15}$$

$$A=(\xi-x)/y, \qquad n=(\eta+y)/y$$

The first term in Eq. (15) is the stress due to a point force acting in

an infinite plate and the second term is the stress due to the existence of

straight boundary (XX in Fig. 6). Using Eq. (15), the influence coefficient

$\overline{\sigma}_{xM}^N$ is calculated.

$$\overline{\sigma}_{xM}^N = \int_{\eta_{N-1}}^{\eta_N} \frac{\partial \sigma_x^P}{\partial \xi}\Big|_{\substack{P=1 \\ \xi=x=0}} \cdot (4\sqrt{b^2-(\eta-e)^2})\,d\eta$$

$$= \int_{\eta_{N-1}}^{\eta_N}\left[\frac{1}{4\pi(y-\eta)^2}+\frac{1}{4\pi}\left\{-\frac{1}{(y+\eta)^2}+\frac{12y}{(y+\eta)^3}-\frac{12y^2}{(y+\eta)^4}\right\}\right](4\sqrt{b^2-(\eta-e)^2})\,d\eta \tag{16}$$

where η_{N-1}, η_N are the values of η at both ends of N-th interval.

The integral in Eq. (16) is expressed in a closed form.

$$\sigma_{xM}^N = \frac{1}{\pi}\left[\frac{\sqrt{b^2-(\eta-e)^2}}{y-\eta}-\sin^{-1}\frac{\eta-e}{b}+\frac{y-e}{\sqrt{b^2-(y-e)^2}}\ln\left|\frac{b^2-(y-e)(\eta-e)+\sqrt{\{b^2-(y-e)^2\}\{b^2-(\eta-e)^2\}}}{b(y-\eta)}\right|\right]_{\eta_{N-1}}^{\eta_N}$$

$$-\frac{1}{\pi}\left[-\frac{\sqrt{b^2-(\eta-e)^2}}{y+\eta}-\sin^{-1}\frac{\eta-e}{b}+\frac{y+e}{\sqrt{(y+e)^2-b^2}}\sin^{-1}\frac{b^2+(y+e)(\eta-e)}{b(y+\eta)}\right]_{\eta_{N-1}}^{\eta_N}$$

$$+\frac{6y}{\pi}\left[\frac{\sqrt{b^2-(\eta-e)^2}}{(y+\eta)^2}\left\{1-\frac{(y+e)(y+\eta)}{(y+e)^2-b^2}\right\}+\frac{b^2}{\{(y+e)^2-b^2\}^{\frac{3}{2}}}\sin^{-1}\frac{b^2+(y+e)(\eta-e)}{b(y+\eta)}\right]_{\eta_{N-1}}^{\eta_N}$$

$$-\frac{2y^2}{\pi}\left[\frac{\sqrt{b^2-(\eta-e)^2}}{(y+\eta)^3}\left\{-2+\frac{(y+e)(y+\eta)}{(y+e)^2-b^2}+\frac{(2b^2+(y+e)^2)(y+\eta)^2}{((y+e)^2-b^2)^2}\right\}\right.$$

$$\left.+\frac{3b^2(y+e)}{\{(y+e)^2-b^2\}^{\frac{5}{2}}}\sin^{-1}\frac{b^2+(y+e)(\eta-e)}{b(y+\eta)}\right]_{\eta_{N-1}}^{\eta_N} \tag{17}$$

To obtain the value of σ_{xM}^{N} in more complicated problems, it is necessary to use a method of numerical integration.

Table 1 shows the comparison with the value of Isida[5]. As seen from Table 1, the degree of accuracy of this method is satisfactory.

3. Numerical Results

The values of stress concentration factors or stress intensity factors under various geometrical conditions are obtained using the present method. Several examples are shown in charts and tables.

4. Conclusions

A method (body force method) for calculating stress concentration or intensity factors was explained and numerical results were given in the form of charts and tables.

The main cases treated here are the tension of a semi-infinite plate having an elliptic arc notch or V-notch or an inclined edge crack or an elliptic arc notch with a crack emanating from its apex, the tension of an infinite plate having a row of ellipses, the tension of a semi-infinite plate with a row of elliptic arc notches, the tension of a strip with two edge cracks, the tension of a strip with two elliptic arc notches, the tension of a cylinderical bar with a ring-shaped edge crack, the tension of semi-infinite body having a semi-elliptical surface crack and the torsion of a round bar with a semi-elliptical circumferential groove.

References

1) Sih, G. C., ed. : Methods of Analysis and solutions of Crack Problems, (1973), Noordhoff.
2) Nisitani, H. : J. JSME. 71(1967), 627.

3) Timoshenko, S. and Goodier, J. N. : Theory of Elasticity, (1951), McGraw-Hill.

4) Nisitani, H. : Preprint of JSME, No. 69-2(1969), 37.

5) Isida, M. : Trans. ASME, 33(1966), 674.

6) Nisitani, H. : Preprint of JSME, No. 700-2(1970), 105.

7) Nisitani, H. : Preprint of JSME, No 69-5(1969), 41.

8) Nisitani, H. and Hasimoto, K. : Preprint of JSME, No. 720-10(1972), 17.

9) Nisitani, H. : Preprint of JSME, No. 740-2(1974), 17.

10) Nisitani, H. and Saito, K. : Preprint of JSME, No. 700-13(1970), 31.

11) Schultz, K. J. : Advances in Applied Mechanics, (1947), 121. Academic Press.

12) Nisitani, H. and Murakami, Y. : Proc. 1971 Inter. Conf. Mech. Behav. Mat., 2(1972), 346.

13) Nisitani, H. and Suematu, M. : Preprint of JSME, No. 720-2(1972), 17

14) Nisitani, H. and Isida, M. : Trans. JSME, 39(1973), 7.

15) Nisitani, H. : Proc. ICF (München), (1973), I-513.

16) Nisitani, H. : Preprint of JSME, (1974).

17) Isida, M. : Trans. JSME, 19(1953), 5.

18) Isida, M. : Trans. JSME, 25(1959), 1118.

19) Bowie, O. L. : J. Appl. Mech., 31(1964), 208.

20) Tokuda, N. and Yamamoto, Y. : J. Soc. Naval Arch. Japan, 132(1972), 349.

21) Nisitani, H. and Murakami, Y. : To appear in Inter. J. Fracture.

22) Smith, F. W., Emery, A. F., Kobayashi, A. S. : Trans. ASME, 34(1967), 947.

23) Shah, R. C. and Kobayashi, A. S. : ASTM STP. 513. (1972), 3.

24) Murakami, Y. and Nisitani, H. : Preprint of JSME, No. 740-2(1974), 1.

25) Harris, D. O. : Trans. ASME, 89(1967), 49.

26) Nisitani, H. and Hasimoto, K. : Preprint of JSME, No. 730-10(1973), 61.

27) Hamada, M. et al. : Trans. JSME, 33(1967), 1916.

28) Okubo, H. : J. Appl. Mech., 19(1952), 560.

29) Petersen, C. : Forshung, 17(1951), 16.

30) Neuber, H. : Kerbspannungslehre, (1934), Springer.

31) Willers, F. A. : Z. Math. Phys., 55(1907), 227.

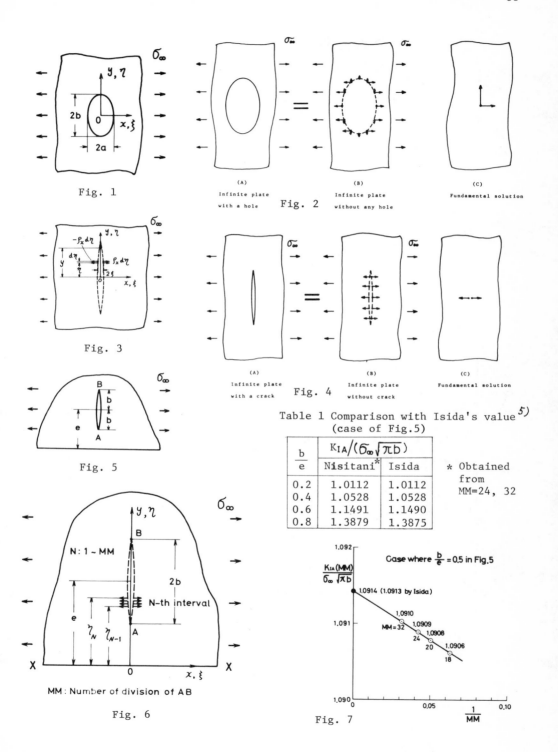

Fig. 1

(A) Infinite plate with a hole   Fig. 2   (B) Infinite plate without any hole   (C) Fundamental solution

Fig. 3

(A) Infinite plate with a crack   Fig. 4   (B) Infinite plate without crack   (C) Fundamental solution

Fig. 5

MM: Number of division of AB

Fig. 6

Table 1 Comparison with Isida's value [5)]
(case of Fig.5)

$\dfrac{b}{e}$	$K_{IA}/(\sigma_\infty\sqrt{\pi b})$		
	Nisitani[*]	Isida	
0.2	1.0112	1.0112	
0.4	1.0528	1.0528	
0.6	1.1491	1.1490	
0.8	1.3879	1.3875	

* Obtained from MM=24, 32

Case where $\dfrac{b}{e}=0.5$ in Fig.5

Fig. 7

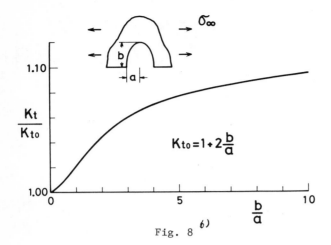

Fig. 8 [6]

Table 2

$\dfrac{b}{a}$	K_t
0.5	2.016
1	3.065
2	5.222
4	9.625
6	14.07
10	22.98

$$K_{to} = 1 + 2\frac{b}{a}$$

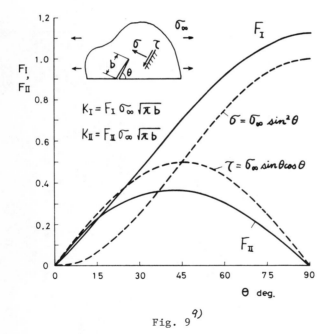

$$K_I = F_I \,\sigma_\infty \sqrt{\pi b}$$
$$K_{II} = F_{II} \,\sigma_\infty \sqrt{\pi b}$$

$$\sigma = \sigma_\infty \sin^2\theta$$
$$\tau = \sigma_\infty \sin\theta\cos\theta$$

Fig. 9 [9]

Table 3

θ deg.	F_I	F_{II}
15	0.225	0.228
30	0.461	0.337
45	0.705	0.364
60	0.920	0.306
75	1.068	0.174
90	1.121	0

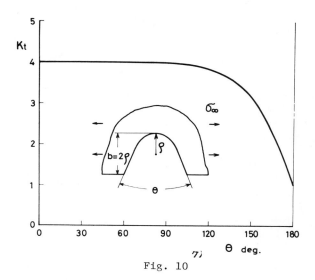

Fig. 10

Table 4

θ deg.	K_t
0	3.997
60	3.995
90	3.976
120	3.839
150	3.145

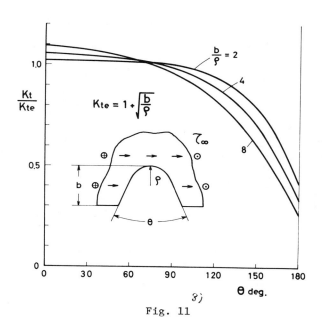

$$K_{te} = 1 + \sqrt{\frac{b}{\rho}}$$

Fig. 11

Table 5

θ deg.	K_t $\frac{b}{\rho} = 2$	4	8
0	2.467	3.165	4.192
30	2.463	3.141	4.117
60	2.449	3.085	3.965
90	2.403	2.957	3.683
120	2.266	2.676	3.178
150	1.908	2.105	2.321

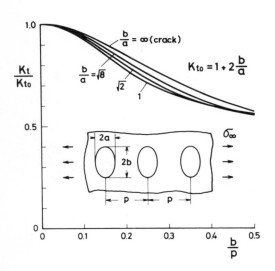

Fig. 12 [10]

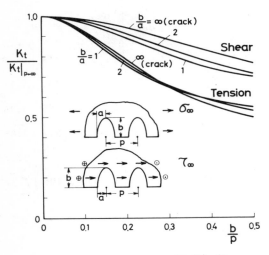

Fig. 13 [10), 12), 13)]

Table 6

$\frac{b}{a}$	$\frac{b}{p}$	K_t
1	0	3.000 (3.000)
	0.1	2.768 (2.768)
	0.2	2.326 (2.326)
	0.333	1.920
	0.5	1.676
$\sqrt{2}$	0	3.828
	0.1	3.565
	0.2	3.034
	0.333	2.476
	0.5	2.114
$\sqrt{8}$	0	6.657
	0.1	6.271
	0.2	5.441
	0.333	4.434
	0.5	3.680

( ) : Schultz[11]

Table 7

$\frac{b}{a}$	$\frac{b}{p}$	K_t or F_I tension	K_t or F_I shear
1	0	3.065	2.000
	0.1	2.791	1.936
	0.2	2.314	1.785
	0.333	1.920	1.564
	0.5	1.676	1.393
2	0	5.222	3.000
	0.1	4.806	2.929
	0.2	4.021	2.748
	0.333	3.254	2.449
	0.5	2.755	2.132
∞	0	1.121	1.000
	0.1	1.039	0.984
	0.2	0.872	0.942
	0.333	0.688	0.863
	0.5	0.558	0.764

$$F_I = \frac{K_I}{\sigma_\infty \sqrt{\pi b}}$$

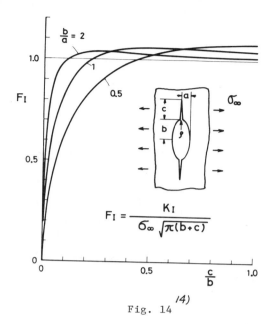

$$F_I = \frac{K_I}{\sigma_\infty \sqrt{\pi(b+c)}}$$

Fig. 14 [4)]

Table 8

$\frac{c}{b}$	$\frac{b}{a}=0.5$	1	2
	F_I		
0.1	0.634	0.839	0.993
0.2	0.823	0.984	1.035
0.4	0.985	1.053	1.031
0.6	1.047	1.057	1.022
0.8	1.071	1.049	1.015
1.6	1.077	1.040	1.011

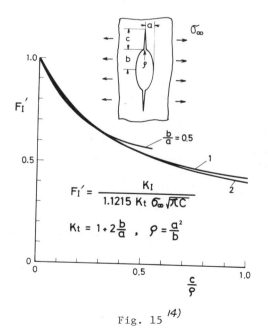

$$F_I' = \frac{K_I}{1.1215\, K_t\, \sigma_\infty \sqrt{\pi C}}$$

$$K_t = 1 + 2\frac{b}{a} \,, \quad \varrho = \frac{a^2}{b}$$

Fig. 15 [4)]

Table 9

$\frac{c}{b}$	$\frac{b}{a}=0.5$	1	2
	F_I'		
0.1	0.939	0.827	0.587
0.2	0.899	0.717	0.452
0.4	0.822	0.586	0.344
0.6	0.762	0.513	0.298
0.8	0.716	0.468	0.272
1.0	0.679	0.438	0.255

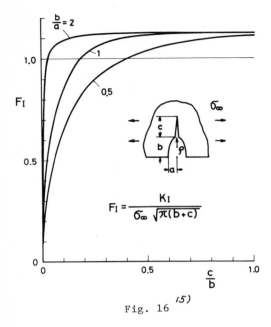

Fig. 16 $^{(5)}$

Table 10

$\frac{c}{b}$	F_I		
	$\frac{b}{a}=0.5$	1	2
0.2	0.834	1.016	1.112
0.4	0.999	1.102	1.122
0.6	1.068	1.118	1.124
0.8	1.097	1.121	1.121
1.0	1.111	1.122	1.121

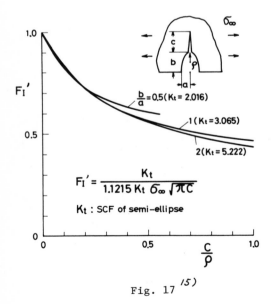

Fig. 17 $^{(5)}$

Table 11

$\frac{c}{b}$	$F_I{}'$		
	$\frac{b}{a}=0.5$	1	2
0.2	0.904	0.724	0.465
0.4	0.827	0.600	0.359
0.6	0.771	0.531	0.314
0.8	0.728	0.489	0.287
1.0	0.695	0.462	0.271

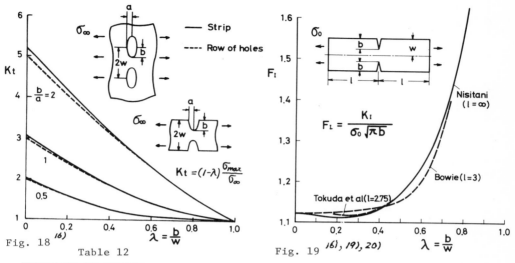

Fig. 18 [16]

Fig. 19 [16], [19], [20]

$$K_t = (1-\lambda)\frac{\sigma_{max}}{\sigma_\infty}$$

$$F_L = \frac{K_I}{\sigma_0 \sqrt{\pi b}}$$

Table 12

$\dfrac{b}{a}$	Case	K_t					
		λ =0	0.1	0.2	0.4	0.6	0.8
0.5	A	2.016	1.803	1.595	1.263	1.102	1.040
	B	2.000	1.791	1.582	1.257	1.102	1.040
1	A	3.065	2.746	2.429	1.864	1.420	1.151
		(3.065)	(2.745)	(2.429)	(1.864)	–	–
	B	3.000	2.700	2.405	1.858	1.419	1.151
		(3.000)	(2.700)	(2.405)	(1.858)	–	–
2	A	5.222	4.679	4.141	3.168	2.316	1.569
	B	5.000	4.511	4.042	3.151	2.321	1.572
5	A	11.85	10.62	9.399	7.183	5.225	3.330
	B	11.00	9.936	8.931	7.037	5.216	3.344

Case A : Strip, Case B : Row of holes ( ):Isida [17], [18]

Table 13

λ	F_I
0	1.121
0.05	1.120
0.1	1.116
0.2	1.111
0.3	1.115
0.4	1.132
0.5	1.169
0.6	1.236
0.7	1.353
0.8	1.574
0.9	2.120

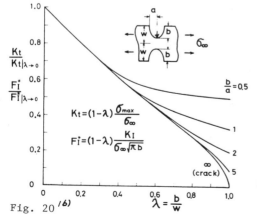

Fig. 20 [16]

$$K_t = (1-\lambda)\frac{\sigma_{max}}{\sigma_\infty}$$

$$F_i^* = (1-\lambda)\frac{K_I}{\sigma_\infty\sqrt{\pi b}}$$

Table 14

λ	$K_t/K_t\|_{\lambda \to 0}$ or $F_I^*/F_I^*\|_{\lambda \to 0}$				
	$\dfrac{b}{a}$=0.5	1	2	5	∞
0	1.000	1.000	1.000	1.000	1.000
0.1	0.893	0.896	0.896	0.896	0.896
0.2	0.791	0.792	0.793	0.793	0.793
0.4	0.626	0.608	0.607	0.606	0.606
0.6	0.547	0.463	0.444	0.441	0.441
0.8	0.516	0.376	0.300	0.281	0.281
1.0	0.496	0.326	0.194	0.084	0.084

68

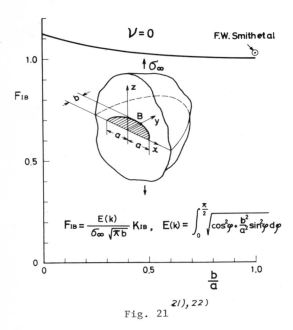

$$F_{IB} = \frac{E(k)}{\sigma_\infty \sqrt{\pi b}} K_{IB}, \quad E(k) = \int_0^{\frac{\pi}{2}} \sqrt{\cos^2\varphi + \frac{b^2}{a^2}\sin^2\varphi}\, d\varphi$$

Fig. 21 [21), 22)]

Table 15

$\frac{b}{a}$	F_{IB}
0	1.121
0.125	1.088
0.25	1.057
0.5	1.022
1.0	0.999
2.0	0.994
4.0	0.998
8.0	0.999

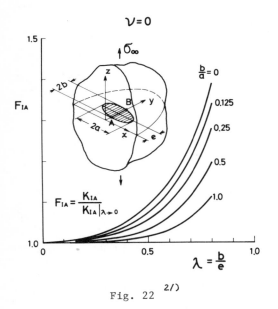

$$F_{IA} = \frac{K_{IA}}{K_{IA}\big|_{\lambda \to 0}}$$

Fig. 22 [2)]

Table 16

$\frac{b}{a}$	F_{IA}		
	$\frac{b}{e}=0.5$	0.625	0.8
0	1.091	1.168	1.388
0.125	1.072	1.141	1.341
0.25	1.055	1.111	1.279
0.5	1.033	1.073	1.197
1.0	1.015	1.037	1.112
0.2	1.061	1.121	1.302
	(1.06)	(1.12)	(1.28)

( ) : Shah & Kobayashi (ν =0.3) [23)]

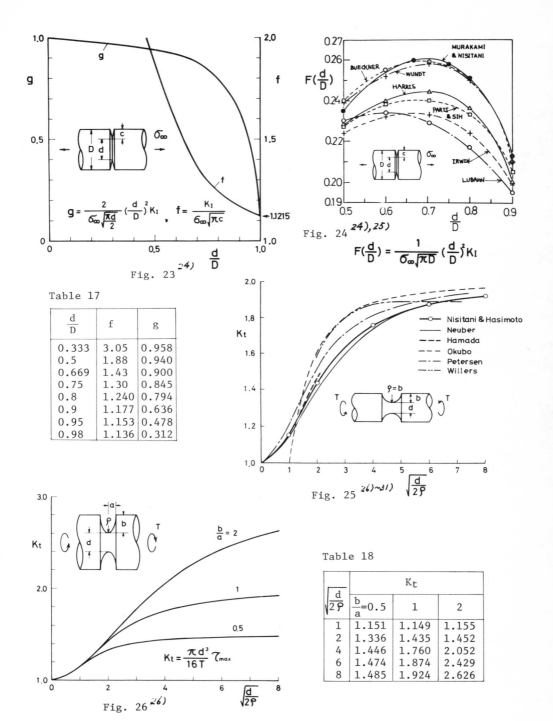

$$g = \frac{2}{\sigma_\infty \sqrt{\frac{\pi d}{2}}} \left(\frac{d}{D}\right)^2 K_I \quad , \quad f = \frac{K_I}{\sigma_\infty \sqrt{\pi c}}$$

Fig. 23 [24)]

$$F\left(\frac{d}{D}\right) = \frac{1}{\sigma_\infty \sqrt{\pi D}} \left(\frac{d}{D}\right)^2 K_I$$

Fig. 24 [24),25)]

Table 17

$\frac{d}{D}$	f	g
0.333	3.05	0.958
0.5	1.88	0.940
0.669	1.43	0.900
0.75	1.30	0.845
0.8	1.240	0.794
0.9	1.177	0.636
0.95	1.153	0.478
0.98	1.136	0.312

Fig. 25 [26)~31)]

$$K_t = \frac{\pi d^3}{16 T} \tau_{max}$$

Fig. 26 [26)]

Table 18

$\sqrt{\frac{d}{2\rho}}$	K_t		
	$\frac{b}{a}=0.5$	1	2
1	1.151	1.149	1.155
2	1.336	1.435	1.452
4	1.446	1.760	2.052
6	1.474	1.874	2.429
8	1.485	1.924	2.626

Highly Accurate Measurement of Residual

Stress on Hardened Steels

by Toru Hayama *

Shinichi Hashimoto **

1. Introduction

In the last decade, remarkable progress has been made in the reliability of X-ray stress measurement, through the activities
of Committee on Mechanical Behavior of Materials, and the X-ray
method has been qualified as the most reliable method for the
measurement of residual stresses. (Ref.1)

Moreover, quantitative as opposed to qualitative evaluation of
the effect of residual stress on the strength and deformation
of machine members and structure is now possible.

In the case of hardened steels, it has been clarified that the
effect of residual stress on the fatigue strength is great and
that the residual stress is one of the major controlling factor
of the fatigue strength.(Ref.2)

However, for the hardended steel, the accuracy of the measurements has not been sufficient because of the broadness of the
diffraction line.

According to the joint research conducted by the Committee on
Mechanical Behavior of Materials, the error of the measurement
of residual stress on hardened steels is $\pm$ 6kg/mm^2 or larger

---

 * Dr., Senior Researcher, Mechanical Engineering
   Research Laboratory, Hitachi Ltd., Tsuchiura, JAPAN.

** Mechanical Engineering Research Laboratory,
   Hitachi Ltd., Tsuchiura, JAPAN.

(standard deviation). (Ref.3, 4)

For a quantitative evaluation of the residual stress, an accuracy of within 2 - 3kg/mm^2 is considered necessary.

In this paper, the newly developed highly accurate fully-automated X-ray stress measuring system with small digital computer is outlined and several experimental results on the accuracy of measurements are presented.

2. Highly accurate fully automated X-ray stress measuring system

For the accurate measurement of stress by X-ray, it is essentially important to obtain accurate profiles of the X-ray diffraction line with little statistical fluctuation. (Ref.4)

Moreover all the processes should be automated to eliminate the influence of individual differences.

In order to satisfy the above conditions the new system, schematically illustrated in Fig.1 was developed.

The system consists of a X-ray stress measuring apparatus (modified Rigaku MSF type), a small digital computer and interfaces.

The specifications of the system are shown in Table 1.

The step scanning method was introduced into the goniometer for the scanning of the counter to increase the sampling time and to eliminate the effect of the statistical fluctuation without increasing the total measuring time.

The parallel beam principle in the geometry of the X-ray diffraction was adopted in order to reduce the error of the measurement due to misfit of the sample. (Ref.1)

The X-ray tube (Toshiba A-30 type), usually used for Japanese stress measuring apparatus makes the possible scanning range of the counter 140 - 170 deg in 2θ.

All the movements of the goniometer were automated and the
output data from the scaler are directly fed to the computer
through interfaces.

For the data processing and the control of the system, a small
digital computer HITAC 10 (12K memory, 16 bits) is used.

3. Soft ware of the system

The soft ware is essentially based on the standard method for
X-ray stress measurement by the Committee on Mechanical Behavior
of Materials.

The most important point in X-ray stress measurement is the
determination of the peak position of the diffraction line so
as to obtain a high degree of accuracy of measurement.

According to the result of an investigation, the half value
breadth method was the most reliable method among the generally
used three methods, center of gravity method, the parabola
method and the half value breadth method. Therefore the half
value breadth method illustrated in Fig.2 was used.

First, the background line AB is calculated as a regression line
of five points at each end of a profile. Then the line A'B' is
determined so that it is parallel to the background and the
distance from it to the point of maximum intensity P and the
distance from it to the background are equal. For each side
of the profile three points X, Y, Z and X', Y', Z', nearest the
line A'B' are selected and the regression lines EF, E'F' of
the respective three points are calculated.

The intersections M and N of the line A'B' and EF, E'F' are
calculated, and then the peak position is determined as the
center of the points M and N.

The flow chart of the computer program for data processing including the program for determining the peak position is shown in Fig.3. After determining the peak positions for all the preset incident angle, the stress value is calculated.

In order to investigate the effect of absorption on the peak position and the calculated residual stress, a computer program including the correction by the absorption factor was used. The correction by the absorption factor was made using the following formula.

$$\Phi_A(x) = \Phi(x) / [1 - \tan(\frac{\pi}{2} + \psi_0 - \frac{x}{2}) \cdot \cot \frac{x}{2}] \quad \dots\dots\dots\dots (1)$$

where

$\Phi_A(x)$:   profile of the diffracted X-ray corrected by the absorption factor

$\Phi(x)$:   measured profile without correction

ψ_0 :   incident angle of X-ray

x: diffraction angle (2θ)

In the case of on-line measurements, the reliability of the measurements is not clear.

For the check of the measurements, the estimated error of the measurement is calculated on the basis of the deviation of the calculated peak position from the regression line in $2\theta - \sin^2\psi$ diagram.

According to the statistical analysis (Ref.4) the limit of confidence for the gradient of the regression line in the $2\theta - \sin^2\psi$ diagram is given by

$$M = t(\alpha, \ n-2) \sqrt{\frac{\Sigma Y_i - (A+MX_i)^2}{(n-2) \Sigma (X-\overline{X})^2}} \quad \dots\dots\dots\dots\dots (2)$$

where

Y_i: obtained peak position

X_i: $\sin^2 \psi$

A: $\overline{Y} - M\overline{X}$

$$M = \frac{\Sigma (X_i - \overline{X})^2}{\Sigma (X_i - \overline{X})^2} (Y_i - \overline{Y})$$

from ΔM the error in measured stress can be calcurated as follows.

$$\Delta \sigma = |K \cdot \Delta M| \quad \ldots\ldots\ldots\ldots\ldots\ldots\ldots\ldots\ldots\ldots\ldots\ldots \quad (3)$$

where

$$K = - \frac{E}{2(1+V)} \cdot \frac{\pi}{180} \cot \theta_0$$

4. Experimental results and Discussion on the accuracy of the measurements

4.1. Fluctuation of measured stress values

Using the system described above, several experiments were conducted.

First, the range of fluctuation or the limit of confidence of the measured stress was investigated.

Specimen used was a carburized S35C (0.35% carbon steel) plate specimen of 3mm thick.

The half value bradth of the diffraction line from the specimen was approximately 6 deg in 2θ.

A loading apparatus as shown in Fig.4 was used to apply bending moment to the specimen.

The stress on the surface of the specimen was measured by X-ray and strain gage at the same time, and these were compared.

The conditions for the measurements were tabulated in Table 2.

As an example of the experimental results, the result of the measurement in the case of FT=10sec, is shown in Fig.5. According to the figure, the deviation of the stress values from the regression line is quite small.

The standard deviation from the regression line is calculated to be 0.27kg/mm^2. In the case of FT=2 - 8sec, the standard deviations were also small, and the maximum value was 1.6kg/mm^2 for FT=2sec.

The error in the X-ray stress measurements is affected by the apparatus, measuring conditions and the materials, but the most fundamental error is considered to come from statistical fluctuation of the diffracted X-ray. The error in the X-ray stress measurements has been theoretically analysed by Shiraiwa and Sakamoto on the basis of the statistical fluctuation (Ref.5).

According to their analysis, the error in measurements is given by the following equation

$$\Delta\sigma = a \cdot \frac{H_w}{I_p} \sqrt{1 + \frac{2}{I_p/I_B} v} + b \quad \dots\dots\dots\dots\dots\dots (4)$$

where

$\Delta\sigma$: error in measurement (kg/mm^2)

H_w: half value breadth (deg.)

I_p: peak intensity excluding background (c.p.s.)

I_B: intensity of background (c.p.s.)

v: scanning speed of counter (deg/min)

a, b: constants

The analysis was applied to the experimental results described above, and the results are illustrated in Fig.6, in terms of

the relation between the error in measurements (standard
deviation) and the parameter related to the statistical fluctuation. Since the measurement were made by step-scanning
method in this investigation, the scanning speed v in Eq.(4)
is substituted by the equivalent scanning speed ($\Delta 2\theta \times \frac{60}{FT}$).
In Fig.6 $\odot$ represents the $\Delta\sigma$ calculated on the basis of the
deviation of the data from the $\sigma_{x-ray}-\sigma_M$ line in Fig.5 and
similar ones and $\bullet$ represents the estimated $\Delta\sigma$ calculated on
the basis of the deviation of the $2\theta - \sin^2\psi$ plots from the
regression line, using Eq.(2) and (3).

The strait line in Fig.6 indicates the regression line for
the estimated $\Delta\sigma$ obtained by the joint research by the
Committee on Mechanical Behavior of Materials.

Since these three kinds of results agree with each other Eq(4)
is proved to be valid for estimating the accuracy of X-ray
stress measurement.

According to the results shown in Fig.6 the accuracy of the
measurements can be expected to be approximately $\pm$ 1kg/mm^2
even for hardened steel by using the new system.

4.2. The effect of absorption on measured stress values

In order to clarity the effect of absorption on the measured
stress value, the measured stress was compared with that corrected using the absorption factor in Eq(1).

The results were illustrated in Fig.7 in terms of the relation
between the stress measured by X-ray and the stress measured
by strain gage.

According to the figure the corrected stress values are much
different from uncorrected values.

However, since these $\sigma_{x-ray}-\sigma_M$ lines in Fig.7 are parallel to each other, the difference of these stress values is independent to the stress level.

Then for the purpose of finding the relation between these stress values measured by X-ray and the real stress value, an experiment on the measurement of powder was conducted.

Two kinds of steel powder were prepared.

One of them, called sample A is relatively soft steel and the other called sample B is hardened steel. The half value breadth of these sample were 2.9 deg and 5.8 deg respectively.

The residual stress of these powder was measured and Tabulated in Table 3.

For the sample A, the corrected stress value is almost zero, hence the measurement with correction using the absorption factor is proved to be correct.

However the effect of the correction is not quite so great in the case of soft material.

On the other hand in the case of sample B, the effect of correction on the stress value is great.

However as shown in Table 3, the measured stress corrected by absorption factor does not necessarily agree with zero.

This is possibly due to the distortion of the profile of the diffraction line caused by high background.

5. Conclusion

(1) For highly accurate measurements of residual stresses and automation of the measurements, a new fully automatic X-ray stress measuring system with step scanning goniometer and with a small digital computer was developed.

Using the system, an accuracy of $\pm$ 1kg/mm^2 (standard deviation) can be expected even for hardened steels.

(2) The fluctuation or error in X-ray stress measurements
depends mainly on the statistical fluctuation of the diffracted X-ray.

(3) As the hardness of the sample increases the effect of the
absorption on the measured stress value become great.
Hence the correction by the absorption factor should be
made especially for hardened steels.

Acknowledgement

The authors would like to express their special thanks to
Mr. K. Ogiso and other members of Rigaku Electric Co. for their
cooperation in the development of the new system.

References

(1) Shuji Taira, "X-ray Diffraction Approach for studies on
Fatigue and Creep", Experimental Mechanics, vol.13, No.11
449 - 463, Nov. 1973.

(2) Toru Hayama, "The Effect of Residual stress on the Fatigue
Strength of Induction-hardened steels", Preprint of the 10th
Symposium on X-ray study on Deformation and Fracture of
Solids, (1972)

(3) Division of X-ray Stress and Strain Measurement, the Committee
for X-ray Study on Deformation and Fracture of Solids, the
Society for Materials Science Japan "Measurement of Residual
Stress in Round Robin Specimens" Journal of the Society of
Material Science, Japan, Vol.20 1251 (1971)

(4) Division of stress Measurement, Committee on Mechanical
Behavior of Materials, the Society of Material Science, Japan,

"Results of Cooperative Work for Standardization of X-ray stress measurement". Preprint of the 10th Symposium on X-ray study on Deformation and Fracture of Solids, 1 (1972)

(5) T. Shiraiwa and Y. Sakamoto, Influence of statistical Fluctuations on the determination of the Peak Position of the X-ray diffraction line", Preprint of the 10th Symposium on X-ray Study on Deformation and Fracture of Solids, 7 (1972).

Table 1. Specifications of automatic X-ray stress measuring system

X-ray generator	Maximum voltage	50 KV
	Maximum current	40 mA
	Stability	Under 0.1 % for input fluctuation
Goniometer	Optical system	Parallel beam
	Scanning range	140°~170°(2θ) preset
	Incident angle	0°~45° arbitrary preset by 5° step
	Step scanning	0.1°, 0.2°, 0.5°, 1.0° step
	Detector	Scintillation counter
Scaler·Ratemeter	Scale range	(10², 10³, 10⁴, 10⁵) x (1, 2, 4, 8)
	Fixed time	(1, 10, 10², 10³) x (1, 2, 4, 8)
Computer	Capacity of memory	12 K
	Number of bit	16 bits

Table 2. Measuring conditions

Table 3. The results of the measurements of powder samples

X-ray	Cr-Kα
Plate voltage	25 KV
Plate current	8 mA
Divergent angle of parallel slit	1.02°
Irradiated area	4 x 4 mm
Filter	V
Incident angle	0°, 15°, 30°, 45°
Scanning range	140°~170°(2θ)
Scanning step	0.5°(2θ)
Fixed time	2, 4, 8, 10 sec
Determination of peak	Half value breadth method (Automatic)

Sample	Half value breadth (deg)	Measured stress (kg/mm²)	
		Uncorrected	Corrected
A	2.9	-4.3	-1.7
B	5.8	-15.5	-3.6

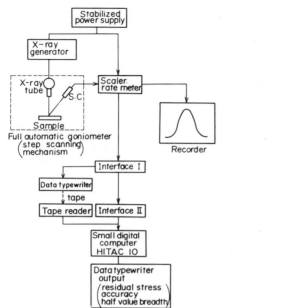

Fig.1. Schematic diagram
of automatic X-ray
stress measuring system

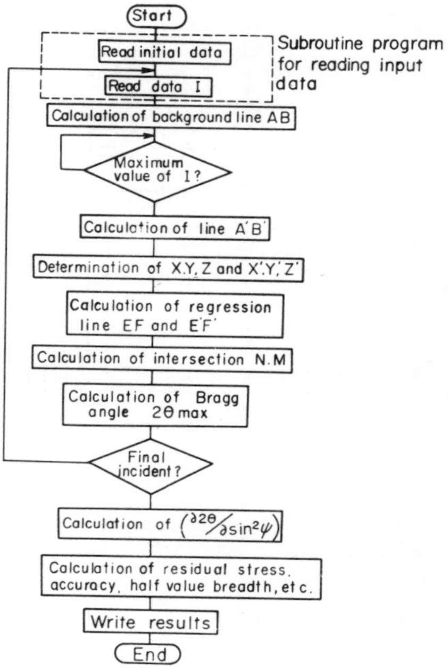

Fig.3. Flow chart of computer
program for data processing

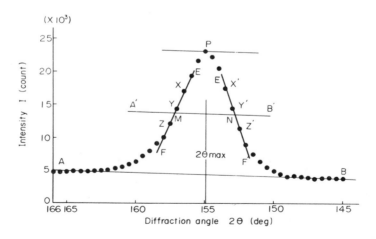

Fig.2. Half value breadth method applied for computer program

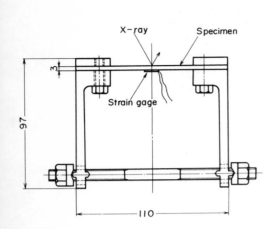

Fig.4. Apparatus for application of bending moment

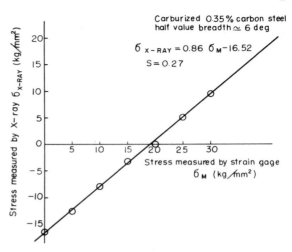

Fig.5. Relation between the stress value measured by X-ray and that measured by strain gage (fixed time: 10 sec.)

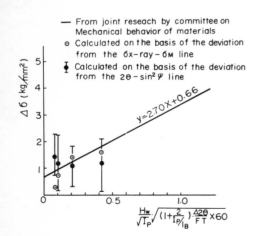

Fig.6. The Error of measurements Carburized 0.35% carbon steel Half value breadth ≈ 6°

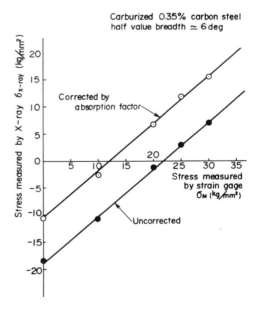

Fig.7. The effect of correction by absorption factor on measured stress value.

SESSION II

Fracture of Metals

Microstructural Aspects of Fracture Toughness – Ductility Relationships

Volker Weiss*

1. THE CONTINUUM MECHANICS BASIS

Fracture toughness is defined as the energy required to create a unit area of new crack surface. In elastic brittle solids this is simply the surface energy. In metals and alloys of technical importance the major resistance to crack extension comes from the associated plastic work. This is schematically illustrated in Fig. 1. As the crack length increases at constant $G = G_c$ the plastic zone is moved ahead. To calculate the work for this process requires a knowledge of the stress and strain distribution inside the plastic zone for a strain hardening solid. Although no general solution is available at present, many researchers are intensively working on the problem. An estimate can be obtained from

$$\Delta W = 2 \int_0^\pi \int_0^{r_p} \bar{\sigma}_y \, d\bar{\varepsilon} \, (r,\theta) \, rd\theta \, dr \tag{1}$$

where $\bar{\sigma}_y$ and $\bar{\varepsilon}$ are the effective flow stresses and strains (1)-(3). Using experimentally supported empirical relationships for the strain distribution (4) a shape factor to describe the plastic zone and an exponential stress-strain law $(\bar{\sigma} = K\bar{\varepsilon}^n)$ one obtains

$$G_R \simeq \frac{S\rho^* \, \sigma_y \, \bar{\varepsilon}_{F,\alpha\beta}}{n} \left(\frac{\bar{\varepsilon}_{F\alpha\beta}}{\bar{\varepsilon}_y} \right)^n \tag{2}$$

where ρ^* is the Neuber micro-support effect constant (4) and $\bar{\varepsilon}_{F\alpha\beta}$ represents the local effective fracture strain for a stress state defined by $\alpha = \sigma_2/\sigma_1$

---

*Professor, Syracuse University, Syracuse, New York

and $\beta = \sigma_3/\sigma_1$. A very simple relationship is obtained by setting $n \to 1$, namely

$$K_R \simeq E\sqrt{\frac{S\rho^*}{\alpha'}}\;\;\bar\epsilon_{F,\alpha\beta} \tag{3}$$

where $\alpha' = 1$ for plane stress and $1 - \nu^2$ for plane strain. Of particular interest is the plane strain case. From slip line field theory we obtain for

$\alpha = \sigma_2/\sigma_1 = (1 + \pi)/(2 + \pi) = 0.806$

and for $\beta = \sigma_3/\sigma_1 = \pi/(2 + \pi) = 0.611$

and

$$K_{Ic} \simeq E\sqrt{\frac{S\rho^*}{1-\nu^2}}\;\;\bar\epsilon_{F\alpha} = 0.081, \;\beta = 0.61 \tag{4}$$

While more exact calculations will undoubtedly lead to much more involved expressions, the character of the expression for K_{Ic}, especially the strong dependence on the fracture ductility, is not likely to change very much. To proceed we must seek an understanding on what affects the "plane strain fracture ductility" and how to determine it (under a tri-axial stress state) from an experimental measurement that can, at best, be made for a bi-axial stress state.

2. EFFECT OF STRESS STATE ON FRACTURE DUCTILITY

The two relationships proposed for the effect of stress state on ductility are illustrated in Fig. 2. The Marin relationship (5) assumes a critical normal stress or a critical shear stress failure criterion. Weiss (1) proposed a critical mean stress (or volume strain) failure criterion, which is also supported by measurements of Bridgemann (6) and recent measurements of Yashima et al (7) and by French and Weinrich (8). The results are best expressed in terms of the ductility ratio, i.e. fracture ductility under a multiaxial stress state $\bar\epsilon_{F,\alpha\beta}$ to fracture ductility under pure (no necking) tension $(\bar\epsilon_{F\;\alpha=o,\beta=o})$. For the critical normal or shear stress criterion Marin obtained

$$\frac{\overline{\epsilon}_{F\alpha\beta}}{\epsilon_{F,o,o}} = [(1 + \alpha + \beta)^2 - 3(\alpha + \beta + \alpha\beta)]^{\frac{1}{2n}} \tag{5}$$

The critical mean stress criterion yields

$$\frac{\overline{\epsilon}_{F\alpha\beta}}{\epsilon_{F,o,o}} = \frac{1}{1 + \alpha + \beta} [(1 + \alpha + \beta)^2 - 3(\alpha + \beta + \alpha\beta)]^{\frac{1}{2}\}\frac{1}{n}} \tag{6}$$

The significant difference between the two relationships is that Equation (5) predicts the tensile ductility for balanced bi-axial tension (bulge test) while Equation (6) predicts a much lower value; at most half the tensile ductility. Except for the extremes, $\alpha=\beta=0$ and $\alpha=\beta=1$ the mean stress criterion predicts lower fracture ductilities for all stress states. Representative experimental results are shown in Fig. 3. These data clearly follow the trend predicted by the mean stress criterion.

One of the problems connected with the determination of fracture ductilities as a function of stress state is the development of a neck prior to failure. Consequently the stress state at fracture is not the same as that at the start of the test. The tri-axial stress state caused by the neck frequently causes premature fracture. This is illustrated in Fig. 4 where the fracture ductilities for D6AC steel heat treated to a hardness of R_c 42.5 are plotted as a function of the nominal (starting) stress state (3). Necking preceeds fracture for the tension and the plane strain tension tests, but not for the bulge test. While the un-corrected data seem to follow the trend predicted by Marin, the corrected data clearly follow the trend predicted by the mean stress failure criterion. It is thus suggested that the bulge test offers the best potential for correlating fracture toughness and ductility. According to the proposed mean stress fracture criterion the ratio of bulge ductility to local fracture ductility ahead of a crack under plane strain condition is given by

$$\bar{\varepsilon}_{F\alpha=0.81,\beta=0.61} = (0.279)^{1/n} \bar{\varepsilon}_{F\alpha=1,\beta=0} \tag{7}$$

A three-dimensional representation of the "ductility index" surface including the tension-compression quadrants is given in Fig. 5. In accordance with the assumption infinite ductility is predicted for $\sigma_1 + \sigma_2 + \sigma_3 = 0$ or $\alpha + \beta = -1$.

From the above and Equation (4) it follows that one might expect a linear relationship between the plane strain fracture toughness and the bulge ductility namely

$$K_{Ic} \approx \frac{0.279}{\sqrt{1-\nu^2}} \ E \ \sqrt{S\rho*} \ \bar{\varepsilon}_{F\alpha=1,\beta=0} \tag{8}$$

Tests of both bulge ductility and plane strain fracture toughness have been performed for a number of high strength steels. The results, shown in Fig. 6, are indeed in good agreement with the predictions of Equation (8).

Models for predicting the specimen thickness effect on fracture toughness and R-curves have also recently been developed on the basis of the observed linear relationship between K_c and the fracture ductility determined for the appropriate stress state (9). These models require assumptions regarding the stress distribution, as a function of thickness and plastic zone size, inside the plastic zone and will undoubtedly be improved when finite element solutions for these stress distributions become available. Examples of the results of these model calculations are shown in Figs. 7 and 8.

3. MICROSTRUCTURAL CONSIDERATIONS

Microscopic examination of the fracture surfaces has given overwhelming evidence that "ductile fracture" is associated with dimples. The formation of these dimples is caused by void nucleation ahead of the crack, void growth, and coalescence. McClintock (10) analyzed a number of 2-dimensional void growth models and obtained for the fracture ductility

$$\left[\overline{\overline{\epsilon_c}}\right]_f = \frac{(1-n) \, \ln \, f^{-1/3}}{\sinh \, [(1-n) \, (\sigma_1 + \sigma_2)/(2\overline{\sigma}/\sqrt{3})]} \qquad (9)$$

where f is the volume fraction of voids. From this relationship one obtains
for the fracture strain ratio

$$\frac{\overline{\epsilon}_{c,\alpha}}{\overline{\epsilon}_{c,o}} = \frac{\sinh \, [(1-n) \, \sqrt{3}/2]}{\sinh \, [(1-n) \, (\sqrt{3}/2) \, (\alpha+1)/\sqrt{\alpha^2 - \alpha+1}} \qquad (10)$$

which is plotted in comparison with the predictions of the volume strain
criterion in Fig. 9. The trends of the two models are the same, however the
McClintock model predicts a much smaller effect of decreasing n-values than
does the mean stress model. Experimental data show a sharper decrease of
fracture ductility than would be possible in accordance with the hole growth
model, even for n → 0.

The hole growth model discussed above presumes the presence of a certain
volume fraction of voids, f. There is little evidence of measured tensile
fracture ductilities that would correspond to a typical void distribution; e.g.
$\ell/2b \simeq 100$ would suggest a tensile ductility of approximately 3.3. A more
realistic assumption is that voids are nucleated during the deformation process
itself. Such void nucleation would occur e.g. at inclusions and grain boundaries.

In a number of recent papers Argon and Co-workers (11)-(13) have examined
the conditions for cavity formation from inclusions. They find that. for non-
interacting stress fields, the interface stress normal to the matrix - inclusion
interface is approximately given by

$$\sigma_{rr} = Y \, (\overline{\epsilon}^p) + \sigma_m \qquad (11)$$

where $Y \, (\overline{\epsilon}^p)$ is the flow stress in the region of the inclusion and σ_m is the
mean stress (or negative pressure component). Experimental determinations of
σ_{rr} (13) yield for

TiC inclusions in unaged maraging steel (Y≈195 ksi) $\sigma_{rr} \approx$ 264 ksi

Cu-Cr inclusions in copper (Y≈70 ksi) $\sigma_{rr} \approx$ 144 ksi

Fe_cC particles in spherodized 1045 steel (Y≈125 ksi) $\sigma_{rr} \approx$ 242 ksi

These values are between 0.008 E and 0.009 E.

Equation 11 indicates that the critical mean stress required for void nucleation varies with temperature as

$$\sigma_m = \sigma_{rr} - Y \, (\bar{\epsilon}, {}^P T) \qquad (12)$$

If σ_{rr} increases only moderately with decreasing temperature (like Young's modulus), much less than the yield strength, then the mean stress required for void nucleation may decrease rapidly in a critical temperature range, thus providing an explanation for the ductility transition phenomenon. For $\bar{\sigma} = K\bar{\epsilon}^n = 3\sigma_m \, f(\alpha,\beta)$ the strain for void nucleation becomes a function of σ_m only, if the stress state (α,β) remains the same. Fig. 10 shows computed values of the void nucleation strain, $\bar{\epsilon}_{i\alpha\beta}$, as a function of test temperature for maraging steel Y(°)= 200 ksi) in the un-aged condition. σ_{rr} was assumed to vary with temperature in the same ratio as Young's modulus. The Y(T) curve for maraging steel was available from a prior study. Since it is only possible to calculate ratios a room temperature value of $\bar{\epsilon}_i$ = 1 was assumed as reference.

An examination of Equation 9 indicates that, for a given material and constant stress state, the strain required for void coalescence is nearly constant and independent of test temperature. The principal influence is the void density f (or 2b/ℓ). For typical materials one might expect a coalescence strain of between 0.01 and 0.5 for the stress state ahead of a crack. These strains are additive, i.e.

$$\bar{\epsilon}_{F\alpha\beta} = \bar{\epsilon}_{i\alpha\beta} + \bar{\epsilon}_{c\alpha\beta}$$

and hence the fracture strain vs. temperature curve will be parallel to the void nucleation strain vs. temperature, displaced upwards by a constant amount, $\bar{\varepsilon}_{c\alpha\beta}$. Finally, becuase of the already discussed linear correlation between fracture strain and fracture toughness, a fracture toughness vs. test temperature curve is obtained which is proportional to the $\varepsilon_{F\alpha\beta}$ – T curve.

Thus, as a first step, a connection between microstructure and fracture toughness of the form

$$K_{c,\alpha\beta} = \{\bar{\varepsilon}_{i\alpha\beta}(\sigma_{rr}-Y(\varepsilon^{P},T,n)) + \bar{\varepsilon}_{c\alpha\beta}\ (f,n)\}\ E\ \sqrt{\frac{S\rho^{*}}{\alpha'}} \tag{13}$$

is proposed. This synthesis suggests that the temperature dependence of ductility and fracture toughness arises in the void nucleation phase. The ease of void nucleation depends primarily on the inclusion – matrix interface strength and the amount by which it exceeds the flow stress. The inclusion density affects primarily the strain required for void coalescence. Both components of the fracture strain, $\bar{\varepsilon}_{i}$ and $\bar{\varepsilon}_{c}$ are strong functions of the stress state (α,β) and decrease with increasing stress triaxility. In fcc materials the yield strength is not a very strong function of temperature, however, the strain hardening rate n is. As n increases with decreasing test temperature the initiation and coalescence strains increase. This might offset to some extent the $(\sigma_{rr}-Y)$ effect and thus produce a much weaker fracture toughness transition with decreasing temperature than observed in bcc materials.

At present the character of the proposed correlations is largely qualitative. However, some quantitative information, such as values of σ_{rr} and the K_{Ic} – vs $\varepsilon_{F\ \alpha=1,\ \beta=0}$ relation for steels, is already available. Further development of reliable quantitative relationships of the type suggested here should be strongly encouraged; they would aid considerably in materials development and also in the development of fracture toughness tests that do not require the difficult and expensive procedures now necessary for typical medium strength high toughness materials.

4.   ACKNOWLEDGEMENT

   This work was supported by the United States Navy Air Systems Command, under Contract Nos. N-62269-73-C-0261 and N-00019-75-C-0065.

5.   <u>REFERENCES</u>

(1)   V. Weiss, "Material Ductility and Fracture Toughness", Proc.
      Intl. Conf. Mech. Behavior of Materials, Kyoto, Japan (Aug.
      15-20, 1971) v. 1, (1972) p. 458.

(2)   V. Weiss and M. Sengupta, "Correlation Between Fracture Tough-
      ness and Material Ductility", Proc. III Intl. Conf. Fracture,
      Munich, Germany, V. 3, (April 1973) p. 341.

(3)   M. Sengupta, "Fracture Resistance of Metals and Alloys" Ph.D.
      Dissertation, Syracuse University (1974).

(4)   V. Weiss, "Notch Analysis of Fracture", in book, <u>Fracture</u> - An
      advanced Treatise in Seven Volumes, (ed.) H. Liebowitz, Academic
      Press, <u>III</u>, (1971) p. 34.

(5)   J. Marin, J.H. Faupel, V.L. Dutton and M.W. Brossman, "Biaxial
      Plastic Stress Strain Relations for 24S-T Aluminum Alloy", Tech.
      Note No. 1536, Nat. Advisory Committee for Aeronautics (May 1948).

(6)   P.W. Bridgeman, <u>Studies in Large Plastic Flow and Fracture,</u> McGraw-
      Hill (1952.

(7)   M. Yajima, M. Ishii and M. Kobayashi, "The Effects of Hydrostatic
      Pressure on the Ductility of Metals and Alloys", Int. J. of
      Fracture Mechanics, v. 6, No. 2, (1970) p. 139.

(8)   I.E. French and P.F. Weinrich, "The Effect of Hydrostatic Pressure
      on the Tensile Fracture of α-brass", Acta Metallurgica, v. 21, No.
      11, (1973) p. 1533.

(9)   V. Weiss and M. Sengupta, "Ductility, Fracture Resistance and R-
      curves", ASTM 8th National Symposium on Fracture Mechanics, Brown
      University, Providence, R. Island (Aug. 26, 1974).

(10) F.A. McClintock, "Plasticity Aspects of Fracture", in book, <u>Fracture</u>
      v. III, (ed) H. Liebowitz, Academic Press (1971) p. 48.

(11) A.S. Argon, J. Im and A, Neelemann, "Distribution of Plastic Strain
      and Negative Pressure in Necked Steel and Copper Bars," Advance
      copy paper -Personal Communication (Sept. 1974).

(12) A.S. Argon, J. Im and R, Safoglu, "Cavity Formation From Inclusions
      in Ductile Fracture," presented in part orally at the Third Intl.
      Conference on Fracture in Munich,Germany (April 1973).

(13) A.S. Argon and J. Im, "Separation of Second Phase Particles in
      Spheriodized 1045 Steel, Cu-0.6% Cr Alloy, and Maraging Steel in
      Plastic Straining", presented orally at the Third Intl. Conference
      on Fracture in Munich, Germany (April 1973).

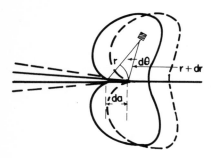

FIG. 1   SCHEMATIC ILLUSTRATION OF THE
         DETERMINATION OF G = δW/δA.

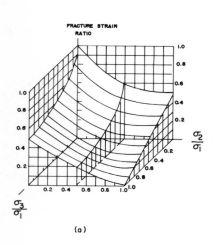

(a)

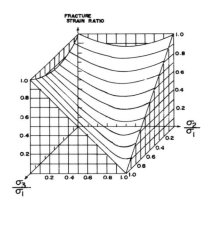

(b)

FIG. 2   DUCTILITY SURFACES FOR THE MEAN STRESS, (a), AND THE MAXIMUM STRESS,
         (b), FAILURE CRITERION.

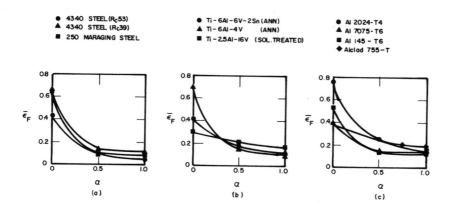

FIG. 3   EFFECT OF STRESS BIAXIALITY ON FRACTURE STRAIN OF (a) STEELS,
(b) TITANIUM ALLOYS AND (c) ALUMINUM ALLOYS.

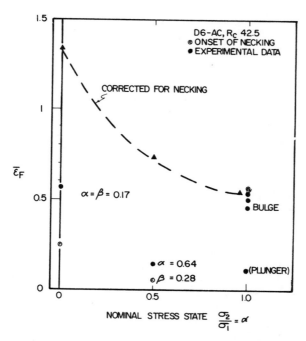

FIG. 4   FRACTURE STRAIN VS. STRESS STATE DATA FOR D6AC
STEEL CORRECTED FOR NECKING

96

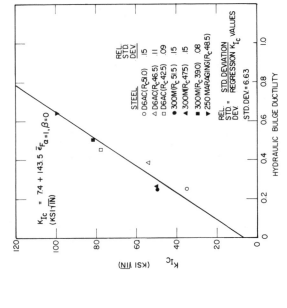

$$K_{Ic} = 74 + 143.5 \ \bar{\epsilon}_{F_{\alpha=1,\beta=0}}$$
(KSI√IN)

STEEL		REL. STD DEV
○ D6AC(R_c50)		.15
△ D6AC(R_c46.5)		.11
□ D6AC(R_c42.5)		.09
● 300M(R_c51.5)		.15
▲ 300M(R_c47.5)		.15
■ 300M(R_c39.0)		.08
▼ 250 MARAGING(R_c48.5)		

REL. STD = STD DEVIATION
DEV. = REGRESSION K_Ic VALUES

STD DEV.=6.63

HYDRAULIC BULGE DUCTILITY

K_{Ic} (KSI √IN)

FIG. 6   CORRELATION BETWEEN PLANE STRAIN FRACTURE
TOUGHNESS AND HYDRAULIC BULGE DUCTILITY
FOR D6AC STEEL, 300M STEEL AND 250 GRADE
MARAGING STEEL.

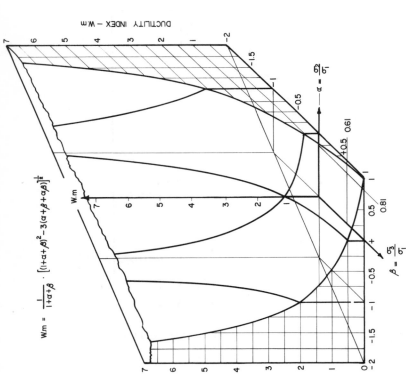

$$W_m = \frac{1}{1+\alpha+\beta} \cdot \left[(1+\alpha+\beta)^2 - 3(\alpha+\beta+\alpha\beta) \right]^{\frac{1}{2}}$$

DUCTILITY INDEX — W_m

$\alpha = \dfrac{\sigma_2}{\sigma_1}$

$\beta = \dfrac{\sigma_3}{\sigma_1}$

FIG. 5   FRACTURE DUCTILITY INDEX SURFACE FOR THE MEAN STRESS

FAILURE CRITERION $\bar{\epsilon}_{F_{\alpha\beta}} = (3\sigma_m \ W_m)^{1/n}$.

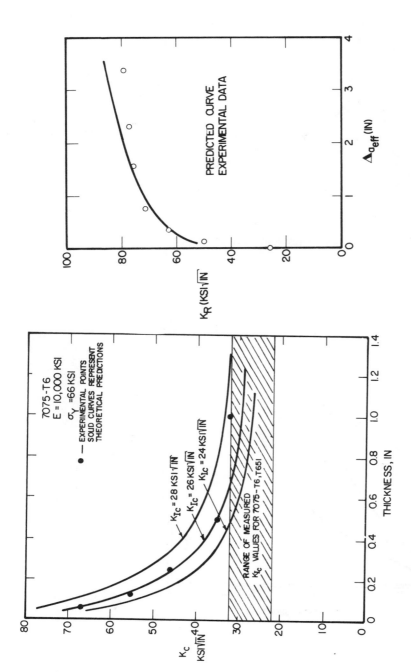

FIG. 8 PREDICTED K_R CURVE FOR ALUMINUM ALLOY 7079 T6.
EXPERIMENTAL DATA SUPERIMPOSED (a_o/W = 0.3805,
t = 0.615 INCH).

FIG. 7 EXPERIMENTAL AND THEORETICAL VALUES OF FRACTURE
TOUGHNESS AS A FUNCTION OF THICKNESS FOR ALUMINUM
ALLOY 7075 T6 AND T651.

98

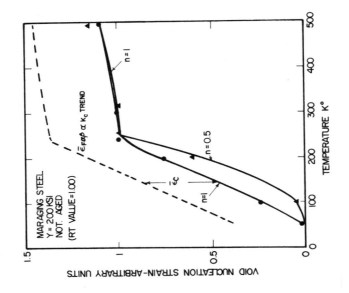

FIG. 10 SCHEMATIC REPRESENTATION OF THE STRAIN REQUIRED FOR VOID NUCLEATION AS A FUNCTION OF TEMPERATURE IN UNAGED MARAGING STEEL.

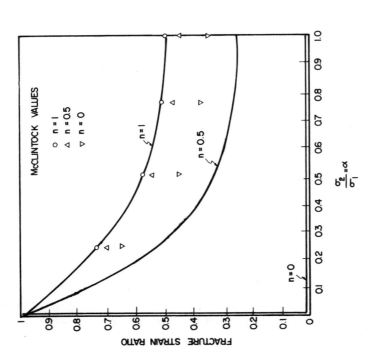

FIG. 9 COMPARISON OF FRACTURE STRAIN RATIOS PREDICTED FROM THE McCLINTOCK HOLE GROWTH MODEL AND THE MEAN STRESS CRITERION (SOLID CURVES).

Investigation of the Basic Mechanisms
of Ductile Fracture in Metals Using
Sintered Copper

by Kiyotsugu OHJI*, Keiji OGURA**,
and Yoshiharu MUTOH***

Abstract

The processes of initiation, growth and coalescence of voids in ductile fracture of copper were investigated.   Growth-and-straightforward-coalescence of pre-existing or initially nucleated voids was rarely observed.   Rather coalescence of the grown-up voids through formation of microvoids was prevailing.   This process of nucleation and unstably rapid growth of these microvoids seems to trigger the final stage of ductile fracture and to determine ductility of industrial materials which contain enough sources for nucleating both the initial voids and the secondary microvoids.

1. Introduction

It has been accepted that ductile fracture of metallic materials occurs through the processes of initiation, growth and coalescence of voids during deformation.   However, the details of the processes are still only poorly known because the processes are often affected sensitively by material struc-tures and mechanical conditions, in addition to the difficulty of direct continuous observations of these processes, which take place inside a specimen. This is particularly true for the process of void coalescence, since this

  * M.S., Dr. Eng., Professor, Department of Mechanical Engineering,
    Osaka University, Suita, Osaka 565
 ** M.S., Dr. Eng., Assistant Professor, Department of Mechanical
    Engineering, Osaka University, Suita, Osaka 565
*** M.S., Graduate Student, Department of Mechanical Engineering,
    Osaka University, Suita, Osaka 565

process is usually accompanied further by mechanical instability.

Naturally the mechanics apporoach to this problem is based on the present state of knowledge on ductile fracture. It has generally been assumed that, except for highly pure metals, the process of void growth may well represent, in a mechanics point of view, the whole processes in ductile fracture. Two of the most representative theories in this line may be McClintock's theory[1] and Thomason's theory[2]. However, the actual behavior of materials often differs considerably from the predicted ones based on these rather oversimplified theories. For further accurate analysis of ductile fracture, an improved theory is needed, which reflects with more fidelity the details of the real processes in ductile fracture.

In this study, therefore, more precise observations of void initiation, growth and coalescence were intended by using tough-pitch copper as well as sintered copper with controlled volume fractions of initial voids. Based on these observations, a mechanical model of ductile fracture of metals with moderate ductility is proposed. Some analytical discussion will be made to show the validity of the proposed model.

2. Preparation of Specimens

The sintered copper investigated was prepared as follows. Atomized copper powder of (-100) mesh was deoxidized at 300°C for 10 hours in hydrogen atmosphere. This deoxidized copper powder was pressed in a die into a piece of 50mm × 14mm × 14mm. This piece was sintered in vacuum (< 1×10^{-4} mmHg) at 900°C for 10 hours. The volume fraction of voids was controlled by changing forming pressure. In order to obtain a low volume fraction piece, the processes of pressing and heating were repeated.

The volume fraction of voids V_f was determined by measuring the density D of sintered copper as,

$$V_f = (8.96 - D)/8.96$$

The procedure specified in JIS Z2505 was followed.

Homogeneity of mechanical properties in sintered pieces was checked by the distribution of indentation hardness values and was found sufficiently uniform.

Photo.1 shows microstructures of the sintered coppers of V_f = 0.04, 0.08 and 0.16. Grains have grown beyond the original powder particle boundaries and voids are randomly located in grains and along grain boundaries.

Tough-pitch copper supplied in a form of round bars in 19mm diameter was also used as a test material.

Hour-glass-shaped specimens shown in Fig.1 were used in this study.

3. Experimental Procedure

The processes of void initiation, growth and coalescence were optically observed by cutting specimens, which were deformed up to various prescribed strains, through the longitudinal specimen axis. Scanning electron fracto-graphical observations were also made. By combining these two types of information, basic mechanisms of ductile fracture were estimated.

Tensile deformation was given by using a Shimadzu Autograph IS-10T, ordinarily at a cross-head speed of 1.0 mm/min, and partly at 0.1 mm/min. The latter speed was used in the instability range preceding the final fracture.

Tension tests under hydrostatic pressures were conducted in a high-pressure testing apparatus[3] attached to the Autograph. This apparatus was a kind of pressure vessel in which a specimen was confined and pulled under a liquid pressure up to 3000 kg/cm^2. Turbine oil #90 was used as a pressure medium. The specimen surface was coated dually by silicone rubber paste over a syn-thetic rubber coating, in order to prevent penetration of the pressure medium into voids inside of the specimen.

4. Experimental Results

4.1 Fracture ductility

Fig.2 shows the relation between fracture ductility and V_f -value for the sintered copper under atmospheric pressure. The solid line in the figure represents the results obtained by Edelson and Baldwin[4]. Good agreement is observed between the present results and the Edelson and Baldwin's results.

By application of hydrostatic pressure, fracture ductility of the sintered copper as well as the tough-pitch copper increased considerably with increase in hydrostatic pressure, as is usually observed[5].

4.2 Behavior of voids under atmospheric pressure

Photos.2 and 3 show micrographs indicating the behavior of voids for a sintered copper (V_f = 0.05) and a tough-pitch copper, respectively. Initiation and growth of voids with increasing strain are evident in these photographs.

In Photo.3, voids have already been initiated in the tough-pitch copper at a strain of ε = 0.3, at which necking was noticed to start. The initiation of voids was observed equally all over the minimum section. Therefore the initiation of voids must have been started during uniform deformation. When the strain is increased to ε = 0.8 [Photo.3(b)], both sizes and number of voids are evidently increased. Clusters of voids are also observed at this strain. At a strain of ε = 1.0 [Photo.3(c)], at which a sharp drop of the true stress started, lateral growth of void is noticeable, probably due to development of the triaxial tension by severe necking. Some voids in a cluster seem to have coalesced by internal necking to form a macrovoid.

For quantitative evaluation of initiation and growth of voids, in what follows the size of a void will be represented by the diameter of a reduced circle, the area of which is equal to that of the void.

Figs. 3 and 4 illustrate the histograms of the void size for sintered coppers of V_f = 0.05 and 0.15, respectively. Fig.5 shows the one for the tough-pitch copper.

In Fig.3, with increase in strain, the number of voids larger than 5 μ (micron) increased, while the number of voids smaller than 5 μ was almost unchanged.   The total number of voids increased by about 15 % during these processes.   It may be noticed that, for a sintered copper with a relatively low value of V_f, void growth occupies the main process in this stage of strain, with slight initiation of voids.

In Fig.4, on the contrary, the number of small voids increased considerably, while the number of large voids was little altered.   It may be said that void initiation is the main process in this stage of a sintered copper with a relatively high value of V_f.

In the case of the tough-pitch copper, it is obvious from Fig.5 that the main process in the stage from $\varepsilon = 0.3$ to 0.8 is void initiation and that in the stage from $\varepsilon = 0.8$ to 1.0 void growth is predominant.   In the latter stage coalescence of large voids by internal necking was also observed.

At strains higher than $\varepsilon = 1.0$, rapid formation of microvoids and their unstable growth occurred.   Because of mechanical instability in this stage, no quantitative measurement was possible in this study.   However, some signs of microvoid initiation were noticed in highly deformed zones between macrovoids.

From the foregoing observations of the sintered coppers and the tough-pitch copper, one may notice that the behavior of voids in the sintered copper with a relatively low value of V_f resembles that of the tough-pitch copper in the stage between $\varepsilon = 0.8$ and 1.0.   Further, the void behavior of the high-V_f sintered copper may correspond to that of the tough-pitch copper at the stage of $\varepsilon > 1.0$.   In this case, nucleation of microvoids may be a common feature to both materials, although it occurs in a limited volume under instability for the tough-pitch copper while it does in a wide-spread manner for the sintered copper.

Because of mechanical instability, precise observations of the final coalescence process in ductile fracture are very difficult.   However, from the foregoing resemblance of the behavior between the tough-pitch copper and the sintered coppers, one may suppose that the final coalescence process of the tough-pitch copper might be estimated from the observations on the sintered copper with a relatively high value of V_f, for which observation of the coalescence process is much easier.   Based on this supposition, the coalescence process of a sintered copper with $V_f = 0.16$ was studied.

Photo.4 shows a scanning electron fractograph and optical micrographs of the longitudinal section in the vicinity of the fractured surface of the sintered copper.   In Photo.4(a), the fractured surface consists of large dimples, which result from initial voids, interlinked by microdimples between them.   By comparing Photo.4(a) with Photos.4(b) and (c), it is evident that the large voids grown-up from the initial voids coalesce with the aid of nucleation and growth of microvoids between them.   This is quite different from the ordinary assumption in mechanical models for ductile fracture that the voids, which are pre-existing or nucleated in a primary stage, grow and coalesce straightforward without any formation of other voids.

Photo.5 shows scanning electron fractographs of the three sintered coppers with $V_f = 0.04$, 0.08 and 0.15.   In general, fractured surfaces consist of grown-up initial voids (IV), ordinary dimples (D) and microdimples (MD) which are about an order smaller than ordinary dimples.   With decrease in the V_f-value, the initial voids (IV) become smaller and the D-region is extended. However the sizes of the microdimples remain almost unchanged in about 3 to 6μ. Further with increase in the D-region and decrease in the MD-region, fracture ductility is increased.

At the bottom of each ordinary or micro-dimple, an impurity particle is often observed.

All the foregoing facts may indicate consistenly a triggering effect of

microvoids nucleated at the final stage of ductile fracture on coalescence
of grown-up initial and nucleated voids.

4.3 Influence of hydrostatic pressure on processes of ductile fracture

Photo.6 shows voids in the tough-pitch copper at various stages of strain
under a hydrostatic pressure of 2000 kg/cm^2.

Initiation of voids has already started at the strain of ε = 0.3, as was
similarly observed in Photo.3 under atmospheric pressure.   No discernible
difference is found at this strain between the results under the pressure of
2000 kg/cm^2 and those under atmospheric pressure, in amount, sizes or shapes
of initiated voids.   Although a micrograph is not shown, the difference was
still not essential until the stage of ε = 0.8; only a slight suppression of
lateral growth was noticed.

However, with further increase in strain, the difference becomes evident;
the lateral growth of the voids is suppressed and stable growth of the voids in
the longitudinal direction continues to a very high strain.

Photo.7 shows comparison of the longitudinal sections in the vicinity of
the fractured surface.   Suppression of both lateral void growth and initiation
of microvoids may be noticed.

Photo.8 shows scanning electron fractographs of a sintered copper of V_f =
0.04 under 3000 kg/cm^2.   By comparing these with Photo.5(a), the suppression
of lateral void growth and microvoid initiation is understood.

This suppression may be a main cause of increase in fracture ductility
under hydrostatic pressure.

5. Discussion

It is said that McClintock's theory predicts fracture ductility larger
than the observed values, in particular for high values of V_f[6]. According to
the present study, this might be attributed to the neglect of both void inter-

action and microvoid initiation. McClintock's theory would be a good approx-

imation for a case of relatively pure metals.

In this respect, discussion will be made on the triggering effect of

microvoids, on the basis of elastic-plastic stress analysis of void interaction,

as combined with some aid from experimental information.

The finite element elastic-plastic analysis[7] was applied to a plate

specimen with two holes, as shown in Fig.6, under the plane strain condition

and a displacement boundary condition of $U_X = -0.5U_Y$.

Bi-linear stress-strain relation with the following characteristic values

was used; Young's modulus E = 12500 kg/mm^2, yield stress σ_y = 23.0 kg/mm^2,

work hardening rate H = 35.7 kg/mm^2 and elastic Poisson's ratio ν = 0.348.

Fig.7 illustrates schematically the analytical distributions of equivalent

strain and triaxiality for three values of D_o/L_o (See Fig.6). With increase

in the D_o/L_o value, the equivalent strain increases while the triaxiality

decreases.

When small holes are introduced between the initial two large holes, the

distribution of either equivalent strain or triaxiality is varied considerably.

Fig.8 illustrates an example of the analysis. The two continuous curves in

the figure represent the values for the plate without small holes. It is

found that the background intensity of the two quantities are raised consider-

ably. This might be a cause of rapid unstable growth of microvoids.

According to the mechanical model proposed in this study, the fracture

strain ε_f may be expressed as the sum of initiation strain ε_{f1} for microvoids

and strain ε_{f2} for their afterward growth until coalescence (fracture). Then

the initiation strain ε_{f1} was computed on the assumption that microvoids may be

initiated when the equivalent strain at the middle point between the initial

two holes (voids) reaches a critical value, which is taken as 0.3. The analy-

tical results are shown in Fig.9 by triangular marks. The initiation strain decreases with increasing initial volume fraction of holes (voids).

For the growth strain ε_{f2}, the finite element analysis is too much time-consuming to conduct it. Therefore a qualitative estimation was made by experiments using polyethylene plate specimens with two kinds of holes as shown in Fig.9. The results are shown in Fig.9 by solid circles. The open circles represent the results for specimens with no small holes. It is found that the fracture strain decreases remarkably when small holes (microvoids) are introduced, with little influence on initial volume fraction of holes (voids). It should be noted that the tendencies of ε_{f1}, ε_{f2} and their sum ε_f resemble that of the sintered copper shown in Fig.2. These may support the proposed mechanical model for ductile fracture.

Some evidence for supporting the proposed mechanism, which emphasizes the role of microvoids at the final stage of ductile fracture, could also found in the literature relating to the fractography of ductile fracture in various industrial materials[8]-[11]. Because of length limitation, further discussion will not be given in this paper.

6. Conclusions

Precise observations of initiation, growth and coalescence of voids in ductile fracture were made using tough-pitch copper as well as sintered coppper. The sintered copper with controlled volume fractions of initial voids was used for simulating the processes of growth and coalescence of voids in ductile fracture of metals.

Growth-and-straightforward-coalescence of pre-existing or initially nucleated voids, which had been taken as the main process in mechanics approach to ductile fracture, was rarely observed. Rather coalescence of grown-up voids through formation of microvoids between them was found predominant. It is proposed that this process of nucleation and unstably rapid growth of these microvoids may trigger the final stage of ductile fracture and, in consequence, may determine ductility of industrial materials, which contain enough sources for nucleating both the initial voids and the secondary microvoids. An elastic-plastic analysis of void interaction, together with experimental results on hole coalescence in plate specimens, supported this triggering effect of micro-voids on the coalescence process. The fractographic data of various industrial materials found in the literature were also compatible with this proposed mechanism of ductile fracture.

References

(1) F.A.McClintock, Trans. ASME, Ser.E, Vol.35 (1968), 363.
(2) P.F.Thomason, J. Inst. of Metals, Vol.96 (1968), 360.
(3) K.Ohji, K.Ogura, and T.Yoshimura, Proc. 16th Jap. Congr. on Materials Research, (1973), 222.
(4) B.I.Edelson and W.M.Baldwin,Jr., Trans. ASM, Vol.55 (1962), 230.
(5) M.Brandes, The Mechanical Behavior of Materials Under Pressure, (H.Ll.D. Pugh ed.), Elsevier, (1970), 236.
(6) F.A.McClintock, Ductility, ASM, (1968), 255.
(7) Y.Yamada, N.Yoshimura, and T.Sakurai, Int. J. Mech. Sci., Vol.10 (1968), 343.
(8) C.D.Calhoun and N.S.Stolloff, Metallurg. Trans., Vol.1 (1970), 977.
(9) I.Kozasu, T.Shimizu, and H.Kubota, Tetsu-To-Hagane, Vol.57 (1971), 2029 (in Japanese).
(10) G.D.Joy and J.Nutting, Effect of Second-phase Particles on the Mechanical Properties of Steel, The Iron and Steel Inst. (1971), 95.
(11) D.Broek, Engng. Frac. Mech., Vol.5 (1973), 55.

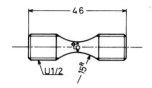

Fig.1 Geometry of specimen.

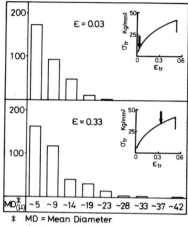

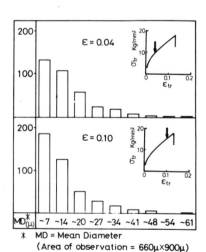

Fig.2 Fracture ductility of
sintered coppers.

✳ MD = Mean Diameter
(Area of observation = 483μ×714μ)

Fig.3 Histogram of void size for
sintered copper with
V_f=0.05.

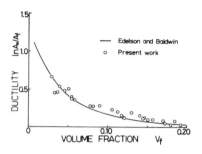

✳ MD = Mean Diameter
(Area of observation = 660μ×900μ)

Fig.4 Histogram of void size for
sintered copper with
V_f=0.15.

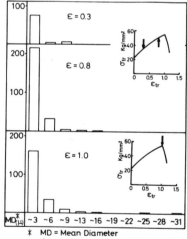

✳ MD = Mean Diameter
(Area of observation = 320μ×334μ)

Fig.5 Histogram of void size for
touph–pitch copper.

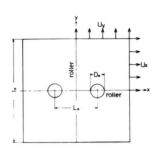

Fig.6 Geometry of the plate
   specimen ($L_0/\ell_0 = 0.18$)
   analysed.

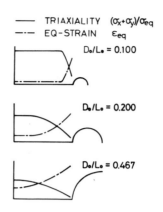

Fig.7 Distributions of
   equivalent strain
   and triaxiality.

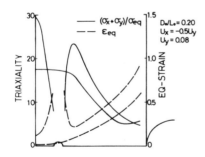

Fig.8 Effect of microvoids on
   distributions of equivalent
   strain and triaxiality.

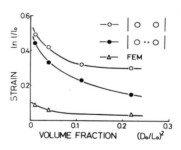

Fig.9 Effect of volume fraction
   of holes on fracture
   strain of a plate.

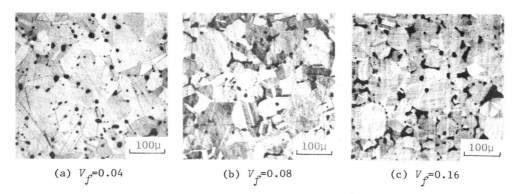

(a) V_f=0.04   (b) V_f=0.08   (c) V_f=0.16

Photo.1 Microstructures of sintered coppers.

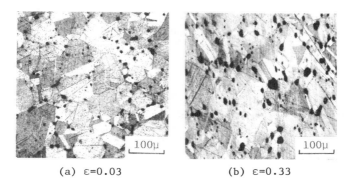

(a) ε=0.03   (b) ε=0.33

Photo.2 Void growth of sintered copper (V_f=0.05).

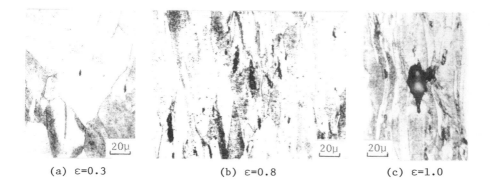

(a) ε=0.3   (b) ε=0.8   (c) ε=1.0

Photo.3 Void initiation and growth of tough-pitch copper.

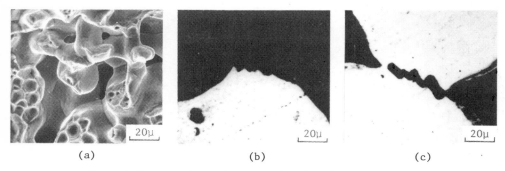

(a)   (b)   (c)

Photo.4 Fractured surface and its longitudinal sections
of sintered copper (V_f =0.16).

(a) V_f=0.04   (b) V_f=0.08   (c) V_f=0.15

Photo.5 Scanning electron fractographs of sintered coppers.

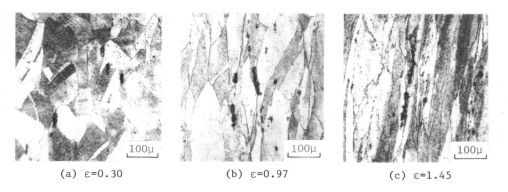

(a) ε=0.30   (b) ε=0.97   (c) ε=1.45

Photo.6 Void initiation and growth of tough-pitch copper
under pressure of 2000 kg/cm^2.

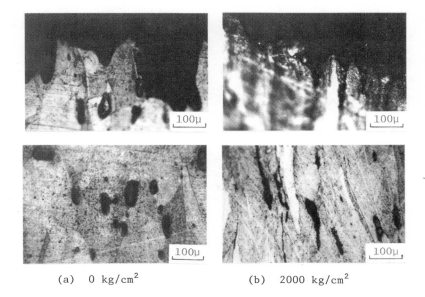

(a)   0 kg/cm^2 (b)   2000 kg/cm^2

Photo.7 Comparison of longitudinal sections in the
vicinity of fracture portion.

Photo.8 Scanning electron fractograghs of sintered
copper (V_f=0.04) under pressure of 3000 kg/cm^2.

Correlation of Plane-Strain and Plane-Stress Fracture Toughness

with Fractographically Derived Stretched Zone Width in High

Strength Steel.

by Hideo KOBAYASHI[*], Hajime NAKAZAWA[**] and Atsushi NAKAJIMA[***]

1. Introduction

The fracture surfaces of fracture toughness specimens comprise three zones;

the fatigue pre-crack, the ductile fracture portion, and a stretched zone be-
  [1]~[5]
tween these two. The general fractographic feature of the stretched zone region

is a fairly smooth or wavy slipped texture associated with plastic deformation

processes as a result of stretching. The stretching is defined as the pro-

duction of new, smooth, featureless surfaces by extremely complex plastic flow

mechanisms acting on too fine a scale to produce characteristic surface traces
  [6]
in present-day replicas.

There has been some speculation as to the significance of the stretched

zone. Several workers consider the stretched zone to be a result of crack tip
  [1]~[5]
blunting. This implies that the dimensions of the stretched zone are a measure

of the critical crack tip opening displacement CTOD at fracture. In the past

several years, work has been done to determine the quantitative correlation

between fracture toughness parameters and the fractographically derived width

of the stretched zone, mainly in valid plane-strain fracture toughness
  [1]~[5]
specimens. It has been observed that the width of the stretched zone increases

with increasing fracture toughness value K_{Ic} and material ductility, and

---

[*]   Dr., Associate Professor, Department of Physical Engineering, Faculty of

   Engineering, Tokyo Institute of Technology, Tokyo

[**] Dr., Professor, Department of Physical Engineering, Faculty of Engineering,

   Tokyo Institute of Technology, Tokyo

[***] Engineer, Engineering Division, Morinaga Milk Industry Co., Ltd., Tokyo

that it exhibits a good correlation with the calculated CTOD value (see open symbols in Fig. 1). More recently, Broek showed that the depth of the stretch-[7] ed zone (not its width) reflects the calculated CTOD value. However, in invalid plane-strain fracture toughness specimens (see solid symbols in Fig. 1 where [8]~[12] data are plotted against the calculated apparent CTOD values) or critical crack opening displacement (COD) test specimens (also see solid symbols in Fig. 1 where data of Elliott and Stuart are plotted against the measured COD [13] values), the stretched zone width shows no apparent dependence on, or correlation with, the calculated CTOD value or the measured COD value at fracture. Moreover, the features of the stretched zone region and of the ductile fracture initiation portion have not yet been clearly defined.

In this work, using single-edge cracked three-point bend (valid and invalid plane-strain fracture toughness) specimens and circumferentially cracked tension (invalid plane-strain fracture toughness) specimens of high strength steel, the fractographic features of the ductile fracture initiation portion and the quantitative correlation of the fracture toughness parameters with the stretched zone width were examined.

2   Material and Experimental Work

2.1   Material

The chemical composition of Ni-Cr-Mo steel (AISI 4340) used in this investigation is listed in Table 1. All test specimens were taken from a single, large 50 mm dia. round bar and were oriented such that the fracture propagation was normal to the rolling direction of the bar. The heat treatment of the specimens consisted of austenitizing for 40 min at 850°C, quenching in oil, tempering for 1 hr at the appropriate temperature (200 - 600°C), and quenching in water. The tensile property data are illustrated in Fig. 3.

2.2   Three-Point Bend Testing

The configuration and dimensions of the single-edge notched specimens are shown in Fig. 2, (a) and (b). The specimen thickness was 20 mm for large type specimens and 10 mm for small type specimens.

The repeated tension fatigue pre-cracking and the three-point bend plane-strain fracture toughness testing were carried out according to the proposed ASTM recommended test procedure.[14] The stress intensity factor was computed from the following relationship,

$$K = \frac{PS}{BW^{3/2}} [2.9(\frac{a}{W})^{1/2} - 4.6(\frac{a}{W})^{3/2} + 21.8(\frac{a}{W})^{5/2} - 37.6(\frac{a}{W})^{7/2} + 38.7(\frac{a}{W})^{9/2}],$$

where B is the specimen thickness, W is the specimen width, P is the applied load determined from the load-displacement record, S is the span length, and "a" is the crack depth. Ferthermore, the following relationship must be completely satisfied before a K_{Ic} result can be classed as "valid".

$$a, (W - a) \text{ and } B \geq 2.5(\frac{K_{Ic}}{\sigma_{Ys}})^2$$

2.3 Tension Testing

The configuration and dimensions of the specimens are shown in Fig. 2 (c). The circumferential fatigue pre-cracking was carried out using the Ono-type rotating bending fatigue machine. The fracture toughness testing was done by static tensile loading. The stress intensity factor was computed from the following relationship,

$$K = \frac{P}{D^{3/2}} (1.72 \frac{D}{d} - 1.27),$$

where P is the applied load at fracture, D is the specimen diameter, and d is the fatigue pre-cracked net section diameter.

2.4 Fractographic Examination

Replicas for fractographic examination were prepared from the fracture surfaces of the specimens by means of a standard two-stage plastic-carbon replication technique in which chromium was used as the replica shadowing material.

For the three-point bend test specimens, the region examined in each specimen
was that at the midthickness of the specimen. This area, presumably of greatest
transverse constraint, should reflect most accurately the influence of a plane-
strain stress state.

3. Results and Discussion

3.1 Tempering Temperature Variation of Fracture Toughness Values

Fig. 3 illustrates the typical tempering temperature variation of the tensile
properties. The tempered martensite embrittlement in the notch tensile strength,
which was obtained by using the circumferentially notched specimens of Fig. 2
(c) without the fatigue pre-cracking, was clearly evident.

Fig. 4 illustrates the typical tempering temperature variation of the apparent invalid plane-strain fracture toughness value K_c in the circumferentially
cracked tension specimens. It is obvious that the tempered martensite embrittlement of the apparent K_c is the same as that of the notch tensile strength.
On the other hand, the valid plane-strain fracture toughness value K_{Ic} in the
single-edge cracked three-point bend specimens did not show the tempered martensite embrittlement, as illustrated in Fig. 5.

3.2 Fractographic Observations

Examination of the features of the stretched zone region provides some evidence regarding the mechanism of the incipient fracture of fatigue pre-cracked
fracture toughness specimens, as well as inferential evidence regarding the
influence of the local stress system. As noted before, the stretched zone
itself is characterized primarily by a "stretching" feature.

Representative fractographs showing the general features of the stretched
zone region are shown in Photos. 1 - 3. In the valid plane-strain fracture
toughness specimens, the features were similar to those shown in Photo. 1,
that is, there was an abrupt transition from the stretched zone to a region
of equiaxed dimple pattern. Also, in view of the nominally low incidence of the

dimples on the surface of the stretched zone in the valid plane-strain fracture toughness specimens, this region presumably forms prior to the incidence of extensive incipient cracking; this represents the creation of a new surface by stretching rather than by an increment of crack extension. The low incidence of dimples results presumably from the absence of triaxiality at this free surface and the fact that this feature is necessary to effect microvoid formation.[4] The abrupt nature of the transition from the stretched zone to the region of equiaxed dimple pattern in the valid plane-strain fracture toughness specimens suggests that the final fast fracture starts below the blunting crack tip surface in the region of higher triaxiality (Fig. 6 (a)).

In the invalid plane-strain fracture toughness specimens, however, the dimples in a narrow band adjacent to the purely stretched zone were extremely small and elongated considerably, as shown in Photos. 2 and 3. Beyond this zone the dimples were large and equiaxial. It is possible that the occurence of the dimples may depend to some extent on the continued plastic deformation process at the blunting crack tip. If the higher triaxiality were not achieved below the blunting crack tip surface, one would anticipate that the further deformation process would take place on the blunting crack tip surface, and the major feature of the surface produced by this process would be that of small elongated tear dimples. This feature exactly coincides with the representative fractographs in the invalid plane-strain fracture toughness specimens, as shown in Photos. 2 and 3. Therefore, this region presumably forms later than the incidence of incipient plastic blunting at the fatigue crack tip, and this represents the increment of crack extension prior to the onset of the final fast fracture rather than the creation of a new surface by stretching (Fig. 6 (b)). Also, in almost all cases where the small elongated tear dimples were observed, their features indicated crack extension away from the fatigue crack tip, because their shapes were an open or parabolic tear in this direction.

In plane-stress fracture toughness (K_c) specimens or critical crack opening displacement (COD) test specimens, it is well known that the slow ductile tear-ing fracture take^s place prior to the onset of the final fast fracture. In such [15] cases, it is possible that the tear dimpled zones adjacent to the purely stretched zones might be present at the ductile fracture initiation portion in those specimens. In this work, fractographic examination of the features of the ductile fracture initiation portion in the circumferentially notched tension specimens also revealed the existance of tear dimpled zones adjacent to the notch root. In this case, the stretching occured on the notch root sur-face, and therefore the stretched zone did not appear on the fracture surface. In view of the above observations, the possibility that the tempered martensite embrittlement of the notch tensile strength and of the apparent fracture toughness value K_c illustrated in Figs. 3 and 4, respectively, is related with the existance of the tear dimpled zones, appears to be a reasonable supposition.

3.3   Correlation of Fracture Toughness with Stretched Zone Width

Fig. 7 illustrates the tempering temperature variation of the stretched zone width in all test specimens examined fractographically. In the valid plane-strain fracture toughness specimens (see open symbols in Fig. 7), the stretched zone width increases greatly with increasing   tempering temperature, and shows no apparent dependence on the specimen ligament. From the result of the tempering temperature variation of the valid plane-strain fracture toughness value K_{Ic}, as shown in Fig. 5, this corresponds to the trends observed by others [1]~[5] that the stretched zone width increases with increasing the K_{Ic} value.

In the invalid plane-strain fracture toughness specimens (see solid symbols in Fig. 7), the stretched zone width increases slightly with increasing tempering temperature, and shows no apparent dependence on the specimen liga-ment, the specimen geometry or the loading type. This suggests that the stretched zone width in the invalid plane-strain fracture toughness specimens is a measure of the critical crack tip opening displacement CTOD at the tearing fracture initiation. It has been shown that, in COD testing, the measured COD value at maximum load varies with specimen size, while the measured COD value for the first appearance of a surface crack in a specimen gives a lower but seemingly more constant value of the critical COD over a range of specimen sizes. Therefore, for [16] COD testing at fracture initiation, the fractographic approach is thought to be very usefull.

The stretched zone width was compared to the crack tip opening displacement CTOD calculated for plane-strain conditions, that is, [17]

$$CTOD = 2V_c = \frac{K_{Ic}^2}{2E\sigma_{Ys}} ,$$

where the quantity $2V_c$ represents the displacement of the faces of the crack in a direction perpendicular to the plane of crack. Correlation of the stretched zone width in the valid plane-strain fracture toughness specimens with the calculated CTOD value is illustrated in Fig. 8. The result reveals the stretched zone width to be numerically nearly equal to the calculated CTOD value. This 1:1 correspondence trend generally confirms previous observation. [1]~[5]

The stretched zone width in the invalid plane-strain fracture toughness specimens was less than that in the valid plane-strain fracture toughness specimens under the same tempering temperature conditions, as shown in Fig. 7. If the constraint varied from a complete plane-strain stress state to a incomplete plane-strain stress state or plane-stress stress state, one would expect that the amount of the small elongated tear dimples on the stretched zone to increase rapidly by the continued plastic deformation process, and that a part of the stretched zone turns into a complete tear dimpled zone. It is thus expected that, for the same material, the more complete the plane-strain stress state, the larger the stretched zone width. [18]. It is also possible that the validity of the plane-strain fracture toughness values K_{Ic}

may be determined by the presence or absence of a tear dimpled zone.
On the other hand, the tear dimpled zone width in the invalid plane-strain
fracture toughness specimens increased greatly with increasing tempering
temperature, and showed apparent dependence on the specimen ligament, the
specimen geometry and the loading type. It should be pointed out that, in COD
testing, the total width of the stretched zone and of the tear dimpled zone reflects the measured COD value up to the maximum load.

4   Conclusion

In this work, using single-edge cracked three-point bend (valid and invalid
plane-strain fracture toughness) specimens and circumferentially cracked
tension (invalid plane-strain fracture toughness) specimens of high strength
steel, fractographic features of the ductile fracture initiation portion and
the quantitative correlation of the fracture toughness parameters with the
stretched zone width were examined.

The results show that, in the valid plane-strain fracture toughness specimens,
the stretched zone width reflects the calculated crack tip opening displacement
CTOD value. In the invalid plane-strain fracture toughness specimens, however,
the stretched zone width does not reflect the apparent calculated CTOD value
at the final fast fracture. For the same material, the stretched zone width in
the valid plane-strain fracture toughness specimens is smaller than that in the
invalid plane-strain fracture toughness specimens, although the apparent
fracture toughness value increases with the transition from valid conditions
to invalid conditions. Additionally, a fourth zone is observed between the
stretched zone and the final fast fracture portion in the invalid plane-strain
fracture toughness specimens. This transition zone is characterized primarily
by a small, elongated dimple pattern, whereas the final fast fracture portion
is characterized by an equiaxed dimple pattern. The stretched zone width shows
no apparent dependence on the specimen ligament, the specimen geometry or the

loading type in the invalid conditions, and incrases slightly with incrasing material ductility. The tear dimpled zone width, however, shows apparent dependence on the above factors, and increases greatly with increasing material ductility.

It is therefore possible that the validity of the plane-strain fracture toughness values K_{Ic} may be determined by the presence or absence of a tear dimpled zone. Also, it is proposed that the stretched zone width in the invalid plane-strain fracture toughness specimens reflects the CTOD value at the slow ductile tearing fracture initiation. Therefore, for crack opening displacement COD testing at the fracture initiation, the fractographic approach is thought to be very useful. It is also expected that, for COD testing at the maximum load, the total width of the stretched zone and of the tear dimpled zone reflects the measured COD value up to the maximum load.

References
(1) Spitzig, W.A., Pellissiev, G.E., Beachem, C.D., Brothers, A.J., Hill, M. and Warke, W.R.: ASTM STP, 436 (1968), 17.
(2) Spitzig, W.A.: Trans. ASM, 61 (1968), 344.
(3) Spitzig, W.A.: ASTM STP, 453 (1969), 90
(4) Brothers, A.J., Hill, M., Parker, M.T., Spitzig, W.A., Wiebe, W. and Wolff, U.E.: ASTM STP, 493 (1971), 3.
(5) Wolff, U.E.: ASTM STP, 493 (1971), 20.
(6) Beachem, C.D. and Meyn, D.A.: ASTM STP, 436 (1968), 59.
(7) Broek, D.: Engineering Fracture Mechanics, 6 (1974), 173.
(8) Bates, R.C., Clark, W.G. and Moon, D.M.: ASTM STP, 453 (1969), 192.
(9) Bates, R.C. and Clark, W.G.: Trans. ASM, 62 (1969), 380.
(10) Gerberich, W.W. and Hemmingo, P.L.: Trans. ASM, 62 (1969), 540.
(11) Kobayashi, H., Nakazawa, H. and Miyazawa, T.: J. Japan Soc. Materials Science, 10 (1973), 67.
(12) Kobayashi, H., Nakazawa, H. and Kawamura, H.: Preprint of Japan Soc. Mechanical Engineers, No. 740-1 (1974), 103.
(13) Elliott, D.E. and Sturt, H.: BISRA Open Report, MG/C/49/70 (1970).
(14) Brown, W.F., Jr. and Srawley, J.E.: ASTM STP, 410 (1967).
(15) Knott, J.F.: "Fundamentals of Fracture Mechanics", 1973, Butterworths.
(16) Harrison, T.C. and Fearnehough, G.D.: Int. J. Fracture Mechanics, 5 (1969), 348.
(17) Hahn, G.T. and Rosenfield, A.R.: ASTM STP, 432 (1968), 5.
(18) Kobayashi, H., Nakazawa, H. and Naito, K.: Preprint of Japan Soc. Mechanical Engineers, to be published.

124

Table 1   Chemical composition of AISI 4340 steel.

C	Si	Mn	P	S	Ni	Cr	Mo	Cu
0.40	0.27	0.84	0.010	0.005	1.66	0.78	0.16	0.03

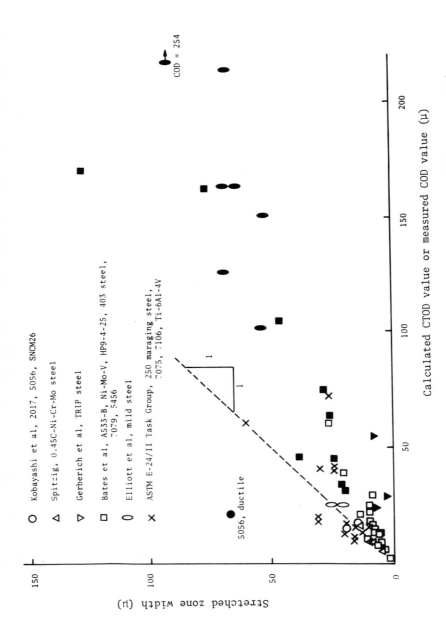

Fig. 1   Correlation of stretched zone width with calculated CTOD or measured COD value (Refs. $(1) \sim (5)$ and $(8) \sim (13)$).

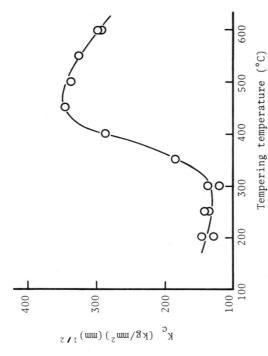

Fig. 4　Tempering temperature variation of apparent fracture toughness value K_c in circumferentially cracked tension specimens.

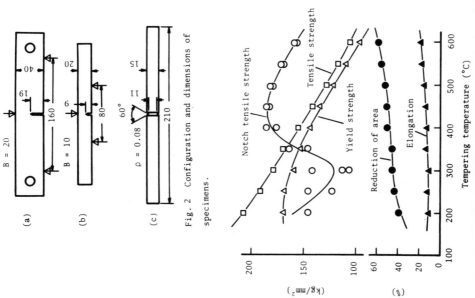

Fig. 2　Configuration and dimensions of specimens.

Fig. 3　Tempering temperature variation of tensile properties.

126

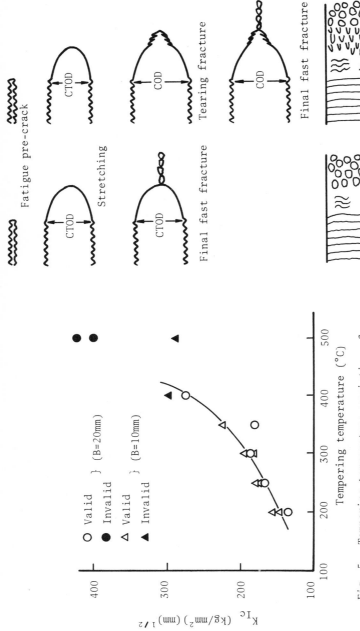

Fatigue pre-crack

Stretching

CTOD

CTOD

CTOD

Final fast fracture

CTOD

CTOD

COD

Tearing fracture

COD

Final fast fracture

Fracture surface

(a) Valid condition   (b) Invalid condition

Fig. 6   Schematic crack profiles at fracture initiation site and fracture surfaces.

K_{Ic} (kg/mm^2)(mm)$^{1/2}$

Tempering temperature (°C)

○ Valid   } (B=20mm)
● Invalid   }
△ Valid   } (B=10mm)
▲ Invalid   }

Fig. 5   Tempering temperature variation of plane-strain fracture toughness value K_{Ic} in single-edge cracked three-point bend specimens.

127

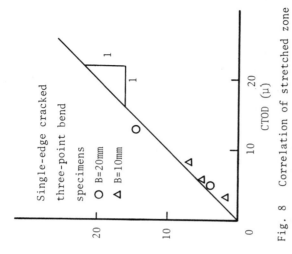

Fig. 8 Correlation of stretched zone width in valid plane-strain fracture toughness specimens with calculated CTOD value.

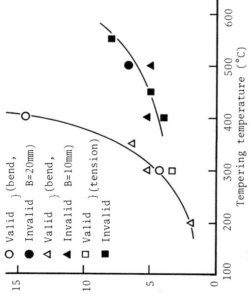

Fig. 7 Tempering temperature variation of stretched zone width.

128

Photo. 1   Fractograph showing an abrupt transition from the stretched zone to a region of equiaxed dimple pattern (bend, B=20mm, T.T. 300°C, valid).

5μ

Fracture direction

Photo. 2   Fractograph showing a tear dimpled zone between the stretched zone and the region of equiaxed dimple pattern (bend, B=10mm, T.T. 400°C, invalid).

Photo. 3   Fractograph showing a tear dimpled zone between the stretched zone and the region of equiaxed dimple pattern (tension, T.T. 550°C, invalid).

Probability of Flaw Detection for Use
In Fracture Control Plans

P.F. Packman [1]
J.K. Malpani [2]
F.M. Wells [3]

1. INTRODUCTION

High performance structures such as critical components
of advanced aircraft require the use of high strength-to-density
materials. These materials are usually designed by procedures
that limit the operating applied stress levels to some values
below the uni-axial tensile yield stress. However, it is known
that brittle failure occurs at stresses considerably below the
uni-axial tensile yield for materials that contain crack like
defects. The introduction of fracture mechanics design concepts
into the design phase of such critical components can minimize
the possibility of premature failure of these components. The
fracture mechanics analysis relates to the ability of the
material to withstand the presence of small crack like defects
and can be used to predict the growth of these flaws in loaded
structures if the loading and material properties are known. [1]

In order to use these fracture concepts in a design process,
knowledge of the size, shape and orientation of these crack like
flaws must be obtained for the part by nondestructive testing of
the component so that it may be used in the actual structure.

[1] Professor, [2] Graduate Student, [3] Assn't Professor
Vanderbilt University, Nashville, Tenn. 37235

The present state of the art of Nondestructive Inspection for
flaw quantification is such that it is difficult to fully
characterize the size of defects except under simple laboratory
conditions.[2]   Thus the utilization of a Fracture Mechanics/
NDT design process is limited by the ability of the NDT to
measure flaws in production processes.

This difficulty can be modified slightly if one considers
the NDT as a quantitative screening process. Since the fracture
formulation presumes an inverse square root relationship
between flaw size and applied stress, only large flaws need be
considered. Missing the presence of a small flaw in a part
would not be as critical as missing the presence of a large flaw
in the same part. Thus one is tempted to determine the size of
the largest flaw which can be overlooked in a component, and
base the initial design process on that flaw size, assuming that
flaws more critical, i.e., larger than this size would be
eliminated by the NDT inspection, while smaller flaws would not
be critical to the component within the design lifetime. It is
the purpose of this paper to present some available data regarding the detection of surface fatigue cracks by advanced NDI.

2.   MATERIALS AND PROCEDURES

The materials used in this investigation consisted of
plates of 7075-T6511 aluminum, Ti6Al-4V and high strength
D6AC steel heat treated to 220-240 tensile strength. Machined
plates approximately 4" x 12" x 0.50" were fatigued in a three
point bend fixture to produce a fatigue crack grown from an
initial laser or weld spot. The specimens were then machined to

a final size of 4" x 8" x approximately 0.50", with the flaw randomly located on the surface.

The inspection procedures to be used in the production inspection were developed in the laboratory on additional specimens prepared with identical procedures. These laboratory specimens were not used in the production inspections.[3]

Once the inspection procedures had been developed and approved, the flawed specimens and controls were sent to the production inspection lines in groups of approximately 25-30. Each group contained about 50% dummy samples (not containing flaws) the actual flawed specimens contained randomly selected flaws ranging in size from 0.008" to 0.25". Each inspection record was examined, and the samples cleaned and at some later time resubmitted for second and third inspections until a sufficient number of inspections had been made in each flaw size range. The results were then tabulated and examined to see if sufficient inspections had been made of flawed samples within each flaw size range. It should be noted that in all cases the flaw size reported is the surface flaw size.

3. DETECTION PROBABILITIES

The probability of detection of a crack in a specimen is defined to be the probability that a trained inspector will verify the presence of a crack during a given inspection process. The inspection process is the total sequence of steps used in verifying the presence or absence of the defect. The probability of detection at a given confidence level guarantees

with the stated degree of confidence that a crack will be
detected with the stated probability.

Let S_n be the number of verifications of the presence of a
crack out of a total of n cracks truly present in the parts.
A natural estimate for the probability of detection P is the
ratio S_n/n. This sample proportion is an unbiased estimate for
the true probability of detection P. Since there are only two
outcomes of the inspection, (a) a flaw is indicated, and (b)
a flaw is not indicated, S_n has a binomial distribution with
mean nP and variance nP(1-P).

There is appreciable probability that the unbiased estimate
of the detection capability (S_n/n) is not exactly equal to the
true probability of detection, P, of the NDI process under
consideration. The probability that the estimate differs from
the true value gets smaller if more flaws (n) are verified by
the NDI process. The magnitude of the error, (S_n-nP), tends
to zero as n is increased. Obviously, for economic reasons,
n cannot be increased indefinitely. Usually, n is chosen such
that magnitude of the error is small. The smaller the error,
the greater the degree of confidence in the estimate.

If the value of P is close to unity (or zero), the number
of observations required for a given magnitude of error and
confidence level is less than if the value of P is close to
0.5. The value of n should be chosen to keep the magnitude
of the error constant. This is not practicable because:

(1) n is a function of the unknown P which varies as
a function of flaw size;

(2) it is usually impossible to get the required number of observations of flaws in certain size ranges because of the error in producing the flaws.

In practice, therefore, the magnitude of the error has not been kept constant. Irrespective of the magnitude of the error, a lower bound probability of detection P' is calculated such that true probability of detection P is at least equal to this value P', with a higher probability, called confidence level (usually specified at 90%, 95% and 99%). The true probability of detection is at least equal to or greater than the lower bound P' which is obtained from the estimate S_n/n.

In other words, the probability that $P \geqslant P'$ (S_n, n, α) is equal to $1 - \alpha$, where α is called level of significance and $100(1-\alpha)$ is called the confidence level. When the total number of observations is small, say 25 or less, the exact binomial distribution should be used to find the lower bound probability of detection P'. For a given S_n, n, α, we find P' or 1-P' from the cumulative binomial table [4] such that:

$$\alpha = \sum_{X=Sn}^{n} b(X,n,P') \tag{1}$$

$$\alpha = \sum_{X=Sn}^{n} \binom{n}{Sn} P'^{Sn} (1-P')^{n-Sn} \tag{2}$$

For example, if $n = 20$, $S_n = 14$ and $100(1-\alpha) = 95$ or $\alpha = 0.05$, then $P' \approx 0.49$. For large n and very large or small P (usually $P > 0.90$ or $P < 0.10$), the binomial distribution can be approximated by a poisson distribution. Tables of cumulative poisson distribution [5] may be used in such cases to find P' or $1 - P'$.

The Poisson parameter given by $\lambda = nP'$ or $n(1 - P')$ such that:

$$\alpha = \sum_{\substack{X=Sn \\ \text{or} \\ X=n-Sn}}^{\infty} e^{-\lambda} \lambda^{Sn} \cdot \frac{1}{Sn!} \qquad (3)$$

The Poisson approximation, when suitably applicable, is very useful, since it is tabulated for Sn or $n - Sn$, whichever is lower and is not dependent on $\underline{n}$.

For example, if $n = 45$, $Sn = 43$, $\alpha = 0.05$, we have $n - Sn = 2$, from the cumulative poisson distribution table. The upper bound for $n(1 -P)$ is given as 4.75, hence $1 - P' = 0.106$ or $P' = .904$.

Had we used exact binomial distribution, the value of P' would have been .863 (against .904).

For large $\underline{n}$ and $.10 \leq P \leq .90$, the normal approximation to the binomial distribution is used. The standardized binomial variable $Z = \frac{Sn - np}{\sqrt{nP(1-p)}}$ is approximately distributed as a standard normal variable . Thus:

$$P\left[Sn > nP - Z_\alpha \sqrt{nP(1-P)}\right] = 1 - \alpha. \qquad (4)$$

where the probability that a standard normal random variable is greater than or equal to Z_α is $1-\alpha$. Solving the inequality $Sn > nP - Z_\alpha \sqrt{nP(1-P)}$ for P, one can obtain the lower bound for P. For example, if $n = 215$ tests for crack sizes of size 0.0323 inches or larger, $Sn = 203$, and $1-\alpha = .99$, then the probability of detection is .91 or greater with confidence 99%.

4. PRESENTATION AND INTERPRETATION OF DATA

There are two distinctly different methods of presenting
and analysing the data. These are:

4.1 Cumulative Detection Probabilities: This develops the
probability of detection of flaw sizes equal to or greater than
the given flaw size. Thus a cumulative probability of detection
of 90/95 for a flaw 0.050" means that there is that probability
of detection of all flaws 0.050" or larger. This biases the
detection capabilities in favor of the large flaws, provided
that no large flaw is missed. If a large flaw is missed, and
smaller flaws are not missed, it penalizes all of the smaller
flaw observations. This appears to be most in accordance with
the fracture mechanics analysis, since the fracture design is
an extreme value flaw criticality concept.

4.2 Range Detection Probabilities: This measures the probability of detection of flaws within a given flaw size range.
Thus a range detection probability of 90/95 for a flaw 0.050"
means that there is that probability of detection of all flaws
within a range of 0.025 to 0.050". This does not bias the
results, if sufficient inspections are made within each flaw size
range. Hence 30 correct observations must be made within each
range to guarantee a 90/95. A miss of a flaw within one range
does not influence the results within another range. This
was the procedure used for most of the Air Force inspection
demonstration programs for use on the fracture critical parts.
The difficulty here is compounded by the fact that it is not

known beforehand how many of each flaw size will be inspected, and hence certain critical ranges may not have sufficient numbers of inspections.

The results of the three NDT production inspections are presented in Tables I through VI. Tables I and II list the cumulative probability of detection and range probability of detection of the titanium penetrant system used. The results are given in terms of surface flaw size and a function of the degree of confidence. In all cases, as the degree of confidence increases, the probability of detection at a given flaw size decreases.

It should be noted that in Table I that the range probability of detection increases as the flaw size increases. This indicates that the range probability of detection approaches 95% at a 95% confidence level if the flaw is greater than 0.060 inches in length. For flaw sizes greater than 0.040 inches, the proba-bility of detection of the flaw is greater than 90% at 95% confidence level.

Selected portions of the cumulative probability for titanium penetrant data is tabulated in Table II. In contrast to the range probability, it should be noted that the cumulative detection probability increases to a maximum and then decreases. This maximum corresponds to the flaw size for which no further misses were found. The resulting decrease in probability of detection is not necessarily due to a decrease in the "sensi-tivity" of the inspection technique, but is due in part to the fact that there is a decrease in the cumulative number of observations.

Tables III and IV show the cumulative and range probabilities of detection as a function of flaw size for different confidence levels for the aluminum penetrant inspection. It should be noted that the particular aluminum penetrant system used is less sensitive to small flaws in the 0.020 to 0.040 and the 0.040 to 0.060 inch ranges than the penetrant system used for the titanium. The probabilities of detection for the aluminum inspection are 51% and 93%, as opposed to 80% and 96% for the titanium inspection system for thesame flaw size range. This is due in part to the increased sensitivity of the particular penetrant system used for the titanium. The sensitivity of this penetrant is higher than that used on the aluminum penetrant system. It should be noted that for flaws greater than 0.060 inches, the detection probabilities are all greater than 90% at the 95% confidence level.

Tables V and VI show the cumulative and range probabilities of detection as a function of the flaw size for different confidence levels for the magnetic particle inspection appeared to produce results that were lower than that of the titanium penetrant system and aluminum penetrant systems. This is somewhat different from the results obtained previously by Packman, et al [6] where the point estimates S_n/n for the magnetic particle systems were higher than the point estimates of the penetrant systems for the same flaw size ranges. It should be noted, however, that for the penetrant inspections in the present study, all specimens were subjected to a half mil etch prior to inspection, while no etch was used in the earlier work. It appears that the use of the etch is critical to increasing the detection sensitivity of the penetrant system. It should be noted that no etch was used in the magnetic particle inspection process.

For the magnetic particle inspections, the flaw size must be about 0.125 inches in length before the detection probabilities are greater than 95%. This is higher than the results found for the titanium penetrant (0.06 to 0.08 inches) and of the same order of magnitude for that found for the aluminum penetrant system.

The cumulative detection probability drops off rapidly for flaw sizes greater than 0.08" and decreases below 90% at the 95% confidence level for flaws greater than 0.160". This is due to two factors: first, the decreasing sample size; and second, a decrease in sensitivity. Here two flaws were missed whose surface length was greater than 0.125".

Because of these misses in the larger flaw size ranges, the cumulative detection probabilities now penalize the complete flaw size range, while the range probabilities only penalize that particular range.

5. SUMMARY AND CONCLUSIONS

Some of the statistical concepts and analysis techniques used to examine the probability of detection of small surface flaws by advanced aerospace NDT production inspection procedures have been developed and examined. It has been shown that high probabilities of detection with high confidence levels can be developed if sufficient preparation and laboratory examination has been conducted prior to introduction of the inspection instructions and controlled procedures into the production inspection.

The difference between range and cumulative detection probabilities has been introduced and data presented which shows that cumulative detection probabilities, although significantly higher than the range detection sensitivities, exhibit a characteristic drop in probability of detection at the larger flaw sizes. This may be due to two effects: (1) a decrease in number of observations as was observed in the aluminum and titanium inspection procedures, and (2) an apparent decrease in sensitivity at the higher flaw size ranges as was seen in the magnetic particle inspection process.

The range detection probabilities showed that all three procedures were capable of high probabilities of detection of flaws whose size is greater than 0.080 inches in length. For flaw sizes in the range 0.070 to 0.080" only the high resolution titanium penetrant system was able to show detection probabilities greater than 95% at all three confidence levels. The relative ranking of the inspection procedures used for the respective base materials is:

      1. Titanium penetrant

    2.   Aluminum penetrant

    3.   Magnetic particle on high strength steel.

This does not imply that the aluminum penetrant, if applied to the titanium or steel specimens, would produce the same results.

6.   <u>ACKNOWLEDGEMENTS</u>

The authors wish to thank many of the people involved in the design and conduct of this program. These include: Mr. Ed Caustin, Mr. Bob Gray of Rockwell International, Dr. Goodman, B-1 SPO, C.F. Tiffany, H. Wood, N. Tupper, AF ASD, and Mr. T. Cooper, W. Trapp and Dr. H. Burte of AFML. Portions of the data analysis and laboratory work were conducted while Dr. Packman was a National Academy of Sciences Senior Fellow at the Air Force Systems Command, under AFOSR Contract #F44620-73-C-0073, under the direction of Mr. W. Walker

REFERENCES

1. H. Wood, <u>The Role of Applied Fracture Mechanics</u> in the Air Force Structural Integrity Program, AFFDL-TM-70-5 (June, 1970).

2. <u>Proceedings</u> of the Interdisciplinary Workshop in Non-Destructive Testing - Materials Characterization, AFML-TR-73-69, (April, 1973).

3. T. McCann and Ed Caustin, B-1 Demo Program, presented WESTEC (March, 1973).

4. William H. Beyer, <u>CRC Handbook of Tables for Probability and Statistics</u>, 2nd Edition, The Chemical Rubber Company, Cleveland, Ohio (1968).

5. Irwin Miller and John E. Freund, <u>Probability and Statistics for Engineers</u>, Prentice-Hall, Inc., Englewood Cliffs, New Jersey (1965).

6. P.F. Packman, et al, "The Applicability of a Fracture Mechanics-Nondestructive Testing Design Criterion," Technical Report AFML-TR-68-32 (May, 1968).

TABLE I

CUMULATIVE PROBABILITY OF DETECTION, TITANIUM PENETRANT

Flaw Size Greater Than In Inches	Cumulative No. of Observations	Cumulative No. of Misses	Probability of Detection at Confidence		
			90%	95%	99%
0.02	410	7	.974	.971	.965
0.04	368	7	.971	.967	.961
0.06	295	0	.992	.990	.985
0.08	196	0	.989	.985	.977
0.125	93	0	.976	.969	.952

TABLE II

RANGE PROBABILITY OF DETECTION, TITANIUM PENETRANT

Flaw Size Range in Inches	Number of Observations	Number of Misses	Probability of Detection at Confidence		
			90%	95%	99%
0.00-0.02	19	7	.447	.369	.237
0.02-0.04	55	7	.809	.782	.736
0.04-0.06	60	0	.962	.951	.926
0.06-0.08	99	0	.977	.971	.955
0.08-0.125	103	0	.978	.972	.957
0.235 ...	93	0	.976	.969	.952

TABLE III

CUMULATIVE PROBABILITY OF DETECTION, ALUMINUM PENETRANT

Flaw Size Greater Than In Inches	Cumulative No. of Observations	Cumulative No. of Misses	Probability of Detection at Confidence		
			90%	95%	99%
0.04	243	2	.982	.980	.976
0.06	200	0	.989	.985	.977
0.08	174	0	.987	.983	.974
0.235	74	0	.969	.960	.940

TABLE IV

RANGE PROBABILITY OF DETECTION, ALUMINUM PENETRANT

Flaw Size Range in Inches	Number of Observations	Number of Misses	Probability of Detection at Confidence		
			90%	95%	99%
0.00-0.04	9	2	.51	.45	.34
0.04-0.06	34	0	.935	.916	.873
0.06-0.08	25	0	.915	.891	.840
0.98-0.125	100	0	.977	.971	.955
0.125-0.20	74	0	.969	.960	.940

TABLE V

CUMULATIVE PROBABILITY OF DETECTION, STEEL MAGNETIC PARTICLE

Flaw Size Greater Than In Inches	Cumulative No. of Observations	Cumulative No. of Misses	Probability of Detection at Confidence		
			90%	95%	99%
0.04	215	12	.924	.918	.908
0.06	203	8	.943	.939	.929
0.08	177	2	.979	.976	.970
0.125	91	2	.957	.948	.927

TABLE VI

RANGE PROBABILITY OF DETECTION, STEEL MAGNETIC PARTICLE

Flaw Size Range in Inches	Number of Observations	Number of Misses	Probability of Detection at Confidence		
			90%	95%	99%
0.00-0.04	5	3	.415	.341	.219
0.04-0.06	7	1	.535	.477	.355
0.06-0.08	26	6	.643	.583	.500
0.08-0.125	86	0	.973	.965	.947
0.125-0.20	91	2	.957	.948	.927

AN INVESTIGATION ON FRACTURE MECHANISM
OF NOTCHED STEEL SPECIMEN

Muneyoshi MORI*, Eisuke NAKANISHI** and Kazuo TAGUCHI***

1. Introduction

When a notched steel specimen is fractured at low temperature, cleavage fracture sometimes occurs after a ductile crack is formed at the tip of a notch. This precedence of a ductile crack prior to cleavage fracture is thought to have a great influence on the low temperature toughness of steels, and the nonlinear fracture mechanics is applied to the analysis of brittle fracture preceded by ductile crack.

Recently, as one of the nonlinear test methods, COD[1] (crack opening displacement) test has often been used for determining the criteria of brittle fracture of steels which usually exhibit nonlinear behavior prior to unstable fracture. It has generally been said in the COD test that brittle fracture occurs at the critical COD which is usually determined by the test using a small size specimen and is not affected by the yielding phenomena of steels.[2] This utility of COD test is sufficiently reconized when a small size specimen is fractured without yielding across its load-bearing cross-section. However, when the fibrous fracture begins at the region just beneath the notch, some considerations should be paid to the utilization of COD test.

From the point of view, the authors have tried in this investigation to observe the initiation and propagation process of ductile cracks by using an electron microscope, a scanning electron microscope and the X-ray microbeam

---

*   Researcher.
**   Dr., Senior Researcher.
*** Manager, Material Laboratory, Technical Research Institute, KOMATSU, LTD.
     2597 Shinomiya, Hiratsuka, Japan.

diffraction technique. On the basis of these observations, the low temperature toughness of steels is also discussed in connection with the initiation and propagation of ductile cracks.

2. Experimental Procedures

Materials employed in this investigation were mild steels, high strength steels and low temperature service steels. The chemical compositions and mechanical properties of the steels are listed in Table I. Specimens were machined from the materials to the dimensions shown in Fig. 1 (a). There are two steps of notch at the center of each specimen and the final narrow deep notch was machined by using a thin grinder (0.1 mm in thickness).

Bending tests were conducted in a cold bath and crack opening displacement was measured by a ring gauge which was fitted in the first step of the notch as shown in Fig. 1 (a). Load was applied at three points indicated by A, B and C in Fig. 1 (a). The proportional method [1] was employed as the method for calculating of COD at the bottom of a notch. Each specimen was bent to various values of COD. After unloading, the specimens were cut by a thin grinder to observe the notch tip behavior as illustrated in Fig. 1 (b). The cross-section surface was polished by emery paper and powdered alumina, and then lengths of ductile cracks which were generated at the tip of a notch were measured by using an optical microscope and a scanning electron microscope.

The substructures of the plastically deformed region at the tip of a notch were observed by a scanning electron microscope after electropolishing and etching by nital. The observations by thin foil by an electron micro-scope were also carried out.

Determination of the plastic zone size and the excess dislocation density were made by analyzing the microbeam X-ray diffraction patterns. Taira's method [3] was applied to calculate the excess dislocation density. The conditions of microbeam X-ray diffraction are listed in Table II.

3. Results and Discussions

When cleavage fracture occurs after a notched specimen generally yields across its load-bearing cross-section, the fracture process of steel is often consists of the three processes including the initiation, the propagation of a ductile crack and the transition from ductile to cleavage fracture. The processes of the fracture are illustrated in Fig. 2 in which the initiation and propagation behaviors of a ductile crack are shown by scanning electron micrographs.

COD increases with an increase in load as the edge of a notch is blunted by the applied stress. After the edge of a notch is blunted, one of the several tears generated in the blunting zone at COD of 0.18 mm developed into a main ductile crack. A main ductile crack continues to grow to a length sufficient for ductile-cleavage transition to occur as shown in the figure.

The value of COD at which one of the tears is developed to a main ductile crack is thought to be a criterion of the initiation of macroscopic ductile fracture. Fig. 3 shows the result of COD test of the steels at $-60^{\circ}C$. It should be noticed in the figure that all the main ductile cracks are initiated at the COD value of 0.3 mm and the critical COD (δ_{r}) of ductile crack initiation is almost independent of steel grade or strain rate. This result is in good agreement with Terry's[4] and Fearnehough's[5]. These authors have indicated that a constant COD is observed at the initiation of ductile cracks.

Furthermore, the observations by an electron microscope have revealed that subgrains of about 6 microns in diameter are formed just beneath the notch even at COD of 0.07 mm while no substructures are observed in a region at a

distance of 100 microns from the edge of a notch. This means that there
exists a high strain zone around the tip of a notch induced by a strain
concentration. Some microcracks along subgrain walls were also found out
in the high strain zone at COD of 0.07 mm. These microcracks are seemed to
grow into the V-shaped tears by propagating to the free surface of a notch,
as shown in Fig. 4. G. T. Hahn[6] has pointed out that the initiation and
propagation of cracks occurs within the heavily strained light etching zone
similar to high strain zone in this study.

Since cracks are initiated and propagated within the high strain zone
accompanied with subgrain formation, properties of the high strain zone are
thought to have a great influence on the low temperature toughness of steels.
In order to clarify the effect of the high strain zone, changes in the distribution of excess dislocation density just beneath a notch were examined at
each COD by using the microbeam X-ray diffraction technique. In this observation, the high strain zone was defined as the region where the excess dislocation density was beyond $3 \times 10^{10} \text{cm}^{-2}$ because the region was approximately
equivalent to the area where the microcracks were observed by a scanning
electron microscope. Thus, the region of $3 \times 10^{10} \text{cm}^{-2}$ in the excess dislocation density is thought to have the potential of initiating cracks.
According to the definition, a good linear relationship was obtained between
the high strain zone size and COD as indicated in Fig. 5.

Fig. 6 shows the relation between the length of ductile cracks and COD.
The lengths of ductile cracks increase with an increase in COD. Since the
rate of COD ($d\delta/dt$) is independent of steel grades or temperatures after
general yielding occured across the load-bearing cross-section[7], the
slope of curves appears to indicate the growth rate of a ductile crack.
It is, therefore, concluded that the critical length of a ductile crack at
the initiation of cleavage fracture (l_c) is approximately equal to the critical

COD (δ_c) though the rate of a ductile crack propagation increases with a decrease in temperature.

The ductile crack appears to propagate utilizing the interaction between cracks and microvoids because the fracture surface created by a ductile crack is composed of a dimple pattern. Small propagation rate in a ductile crack at higher temperature seems to depend on the microvoid-coalecence. It is, however, noticed that even at higher temperatures the critical COD is nearly equal to the critical ductile crack length that is the length at the initiation of cleavage fracture[7]. As the results, the length of a ductile crack never exceeds the high strain zone size through its initiation and propagation process. Cleavage fracture occurs when a ductile crack length becomes equal to the critical COD. The void formation during the process will occur within the high strain zone at the tip of a ductile crack[8]. Thus, the value of COD becomes larger at higher temperature since it includes the increment due to the interaction between a ductile crack and microvoids.

Therefore, when cleavage fracture occurs before a ductile crack formation, COD is nearly equal to the high strain zone size through the deformation process, and COD theory[9][10] is appropriately applied to the analysis of the brittle fracture. However, when there exists a ductile crack formation prior to cleavage fracture, some considerations should be paid to the application of COD theory because the COD includes the increment due to the interation between a ductile crack and microvoids.

Fig. 8 shows the relation between the critical COD and temperature of SM41B steel. In the figure, the critical COD (δ_c) seems to have a nearly constant value below the temperature of $-70°C$, and the brittle fracture is thought to occur before a ductile crack initiation in this region. At the higher temperature, however, the critical COD increases remarkably with temperature because of formation of a ductile crack. This increment of the critical COD will be represented by the equation $\varDelta = \delta_c - \delta_I$. Since the fracture of this mode is significantly influenced by the propagation behavior of a ductile crack, the low temperature toughness of steels can be estimated by introducing the concept of $\varDelta$ when a ductile crack is formed prior to cleavage fracture.

4. Conclusions

Conclusions obtained from the investigation are as follows:

(1) In the plastic deformation area at the tip of a notch there is a high strain zone where the dislocation density is beyond a certain limit and the ductile crack is initiated and is propagated within this zone.

(2) The size of the high strain zone is approximately equal to COD (crack opening displacement).

(3) The ductile crack is initiated when COD reaches a definite value (δ_I) independent of steels and strain rate.

(4) The cleavage fracture starts when the length of the ductile crack becomes equal to the critical COD.

(5) The low temperature toughness of steels can be estimated by the equation $\varDelta = \delta_c - \delta_I$ even when a ductile crack is formed prior to cleavage fracture.

5. References

(1) Nichols, R. W., et al. "The use of critical opening displacement technique for the selection of fracture resistant materials", CODA panel of the Navy Department Advisory Committee on Structural Steels, Kyoto (1970).

(2) Koshiga, F., and Ishihara, K., "A Preliminary Study of the COD Concept for Brittle Fracture Initiation", Journal of the Society of Naval Architects of Japan, Vol. 24, 244 (1968).

(3) Taira, S., and Hayashi, K., "X-ray Investigation on Fatigue of Notched

Specimens", JSMS, Vol. 33, 1 (1967).

(4) Terry, P., and Barnby, J. T., "Determining Critical Crack Opening Displacement for the Onset of Slow Tearing in Steels", Met. Const. & British Welding Journal, Vol. 3, 343 (1971).

(5) Fearnehough, G. D. et al. "The role of stable ductile crack growth in the failure of structures", Inst. of Mech. Eng., 28 (1971).

(6) Hahn, G. T., et al. "Local Yielding Attending Fatigue Crack Growth", Metallur. Trans., Vol. 3, 1189 (1972).

(7) Mori, M. et al. "An Investigation on Brittle Fracture of Notched Specimens of Mild Steels at Low Temerature", Proc. Sym. Mech. Behav. Mat. Japan, 122 (1974).

(8) Mori, M., unpublished.

(9) Bilby, B. A., et al. "The Spread of Plastic Yield from a Notch", Proc. Roy. Soc., A272, 304 (1963).

(10) Wells, A. A., "Application of Fracture Mechanics at and beyond General Yielding", Brit. Weld. J., Vol. 10, 563 (1963).

Table I Chemical compositions and mechanical properties of steels

Steels	C	Si	Mn	P	S	Ni	Cr	No	V	Nb	B	Cu	Al	O_2 ppm	N_2	σ_y kg/mm²	σ_y kg/mm²	δ (%)	φ (%)
SS41P	0.21	0.03	0.86	0.013	0.019	-	-	-	-	-	-	0.04	-	125.5	-	28.4	47.2	38.7	69.1
SM41B	0.18	0.03	1.13	0.013	0.019	-	-	-	-	-	-	0.03	-	106.8	-	26.4	45.5	34.2	63.6
SM50A	0.18	0.42	1.39	0.008	0.013	0.13	<0.05	0.02	-	-	-	0.05	0.04	37.6	-	-	-	-	-
HT60(a.r.)	0.15	0.49	1.45	0.015	0.010	<0.05	<0.05	<0.01	0.05	0.04	5×10^{-4}	0.08	0.05	19.1	-	49.8	64.7	31.0	74.4
HT80(Q.T.)	0.15	0.37	0.92	0.017	0.017	<0.05	0.74	0.05	<0.03	<0.01	21×10^{-4}	0.28	0.09	9.9	-	82.7	88.4	21.1	70.4
Low Temperature Service Steels L-A	0.07	0.28	1.28	0.015	0.015	-	-	-	-	-	-	0.05	0.04	16.2	0.006	35.8	47.3	42.5	85.0
L-B	0.01	0.30	1.50	0.014	0.014	0.21	-	-	-	-	-	0.20	0.03	13.7	0.009	36.5	50.9	36.4	74.6
L-C	0.08	0.23	1.32	0.012	0.012	-	-	-	-	-	-	<0.03	0.04	18.3	0.007	33.5	46.7	38.7	83.5

Table II. Diffraction conditions of microbeam X-ray

X-ray condition

Micro focus X-ray generator	Hivax
Tube current	1.1 mA
Tube voltage	50 KV
Target	Cr
Diffraction plane	(211)
Filament current	5.2 A
Slit	50μ single pinhole
Distance between focus & pinhole	83 mm
Distance between film & specimn	33 mm
Effective focus size	90×105μ
Divergence	5.0×10^{-4}
Radiated area	110μ

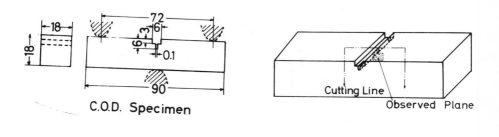

C.O.D. Specimen

(a)   (b)

Fig. 1   The specimen of COD test

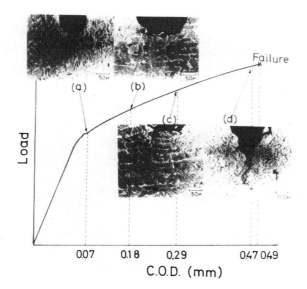

Fig.2 Ductile cracks initiated and
propagated from the tip of a notch at
various values of COD.

154

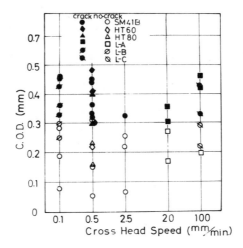

Fig.3 Influence of crosshead speed
for macroscopic ductile crac initiation
in various types of steels at -60 C.

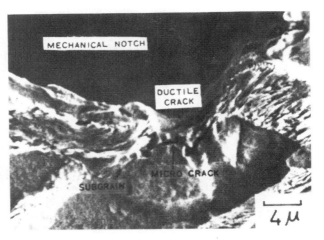

Fig. 4 V-shaped tears generated at the
tip of a notch (-60°C, δ = 0.3 mm SM41B)

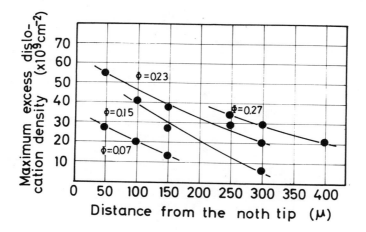

Fig. 5 Distributions of excess dislocation density near the tip at each COD (-60°C, SM41B)

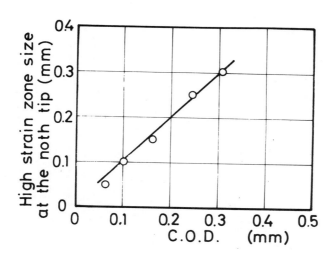

Fig. 6 The relation between COD and total high strain zone size generated at the tip of a notch (-60°C, SM41B)

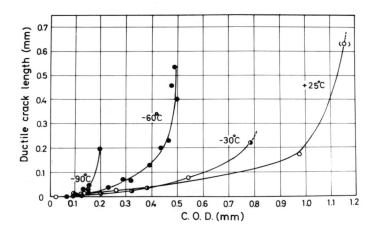

Fig. 7 Influence of testing temperature on
growth of ductile crack in SM41B steel.

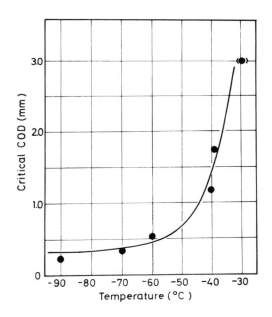

Fig. 8 Variation of critical COD at each
temperature (SM41B –60°C)

Predicting Ductile Fracture Initiation In Large Parts

J. A. Joyce* and F. A. McClintock**

1. INTRODUCTION

Large-sectioned parts of microscopically ductile materials fracture in an elastic fashion when the plastic zone is small compared to the geometrical dimensions of the part. The elastic crack singularity controls because it sets the boundary conditions on the constrained plastic region. Obtaining the critical stress intensity factor for such ductile materials. however, uneconomical when the specimens required are too large to handle in the laboratory.

It is thus desirable to predict the elastic fracture initiation expected in large parts from laboratory-sized specimens whose behavior is elastic-plastic or fully plastic.

A method has recently been proposed by Begley and Landes (1) which does that, subject of course to some restrictions. The work of Hutchinson (2) and Rice and Rosengren (3) shows that the stress and strain concentrations around a sharp crack in a strain hardening material can be characterized by a collection of singularities for the plane strain and plane stress cases. The plastic stress and strain singularities can be expressed in terms of the path independent contour integral of Eshelby (4), Cherepanov (5), and Rice (6). This J_I integral contains the dependence of the stress and strain singularities on the specific crack geometry and loading in direct analogy to K_I in the elastic singularity of Irwin and Williams.

*Assistant Professor of Mechanical Engineering
U.S. Naval Academy, Annapolis, MD   21402

**Professor of Mechanical Engineering
MIT, Cambridge, MA   02139

Begley and Landes (1) noted this analogy and proposed that fracture would occur when J_I reached a critical value J_{IC} for the material. McClintock (7) pointed out that such a single scaler fracture parameter was impossible in the limit of a nonhardening material, where the governing equations were hyperbolic in nature and the crack tip stress and strain fields were not unique. In this presentation a tentative criterion for a valid J_{IC} test is suggested and compared with experimental work by Landes and Begley (8) Zaverl (9), Andresen (10), and the present authors.

2. DISCUSSION OF THE J_I INTEGRAL AS A FRACTURE CRITERION

In the Williams (11) Irwin (12) elastic solution for the stress field around a sharp tensile crack, the magnitude of the stress components is proportional to a stress intensity factor K_I. The stress intensity factor is determined by the loads, the general shape of the body, and the path of the crack. If in a cracked body the plastic zone is an order of magnitude smaller than any characteristic crack or ligament dimension, an annular region exists around the crack tip in which the elastic singularity is valid. The elastic singularity sets the boundary conditions on the plastic zone. Crack initiation occurs when the stress intensity factor reaches the critical stress intensity factor for the material, even though the fracture initiates inside the plastic zone.

The stress intensity factor fails as a fracture criterion when the elastic singularity fails to dominate the stress field in some annular region around the crack tip. This occurs when the plastic zone radius, $r_Y = K_I^2/(2\pi Y^2)$, approaches 1/3 to 1/2 the radius to the next most distant geometrical feature in the cracked body.

For an incompressible, deformation theory plastic material with a uniaxial stress-strain relation

$$\sigma = \bar{\sigma}_1 \, (\bar{\varepsilon}_p)^n,$$ (1)

the Hutchinson (2), Rice - Rosengren (3) singularity (the HRR singularity) gives the stress, strain, and displacement components near a sharp crack in terms of the path independent J_I integral of Rice (6) as:

$$\frac{\sigma_{ij}\,(r,\theta)}{\bar{\sigma}_1} = (\frac{J_I}{\bar{\sigma}_1 I_n r})^{n/n+1} \, \tilde{\sigma}_{ij}(\theta)$$

$$\varepsilon_{ij}(r,\,\theta) = (\frac{J_I}{\bar{\sigma}_1 I_n r})^{1/n+1} \, \tilde{\varepsilon}_{ij}(\theta)$$ (2)

$$\frac{u_{ij}\,(\theta)}{r} = (\frac{J_I}{\bar{\sigma}_1 I_n r})^{1/n+1} \, \tilde{u}_i\,(\theta),$$

where $\tilde{\sigma}_{ij}(\theta)$, $\tilde{\varepsilon}_{ij}(\theta)$, $\tilde{u}_i\,(\theta)$, and I_n are functions of θ and n which have been evaluated numerically by Hutchinson (2).

Only J_I in Eq. 2 depends on the applied boundary conditions or specimen geometry, J_I thus sets the intensity of the stress, strain, and displacement fields in a manner completely analogous to K_I in the Williams, Irwin elastic solution. Carrying the analogy one step further suggests that crack initiation occurs when J_I reaches a critical value, J_{IC} which is a material property. A further justification for this analogy is that for n = 1, Eq. 2 reduces to the Williams' elastic solution. Rice has demonstrated that for this case

$$J_I = \frac{(1 - \nu^2)K_I^2}{E}$$ (3)

J_I can be relied on as a valid fracture criterion only as long as the singularity of Eq. 2 is valid in some annular region around the crack tip. In Fig. 1 two circles are drawn around the crack tip. The smaller radius r_p is termed here the fracture process zone and corresponds to the region very close to the crack tip where strains are greater than unity and large shape changes are taking place, or where microcracks or voids are forming. Inside this radius the HRR singularity is not valid. The larger radius r_{HRR} is taken to be that available to convert the stress and strain distributions from those of the HRR singularity to the non-hardening values controlled by the next nearest boundary. We test the idea that once plane strain is assured, this conversion requires a given ratio between the stress at r_{HRR} and that at r_p, calculated according to the HRR singularity:

$$\frac{\sigma_p}{\sigma_{HRR}} = \left(\frac{r_{HRR}}{r_p}\right)^{n/n+1} = \left(\frac{\sigma_p}{\sigma_{HRR}}\right)_C . \tag{4}$$

If the crack mechanism inside r_p is limited by blunting, take the process zone to be half the crack tip opening displacement (the radius at which the maximum displacement is equal to the radius and the maximum equivalent strain is unity in the HRR singularity)

$$r_p = \frac{J_{IC}}{\bar{\sigma}_1 I_n} . \tag{5}$$

Then if the critical stress ratio is known, the smallest valid specimen size can be determined from the critical J value and the conversion radius r_{HRR} which will be taken to be half the ligament dimension which must be traversed by a non-hardening flow field:

$$b/2 > r_{HRR} = \left(\frac{\sigma_p}{\sigma_{HRR}}\right)^{(n+1)/n} \frac{J_{IC}}{I_n \bar{\sigma}_1} . \tag{6}$$

Equation 3 allows expressing the result in terms of the critical stress intensity factor K_{IC}:

$$b/2 \; > \; r_{HRR} \; = \; \left(\frac{\sigma_p}{\sigma_{HRR}} \right)^{(n+1)/n} \frac{(1-\nu^2)K_{IC}^2}{I_n \, \bar{\sigma}_1 \, E} \; . \qquad (7)$$

Note that the above criterion takes no account of the differences between the
HRR and the various fully plastic flow fields, both in regard to the presence
or absence of a fan of distributed strain, and the normal stress acting on it.

The only known data to check the proposed criterion are the tension and
bending tests of Landes and Begley (8) on Ni.-Cr.-Mo.-V. steel, the bending
tests of Zaverl (9) on two heat treatments of 4340 steel, and the tension
tests by Andresen (10) on single and double grooved AISI 1117 steel specimens,
presented here in a corrected form. The results, summarized in Table 1 ,
require considerable comment.

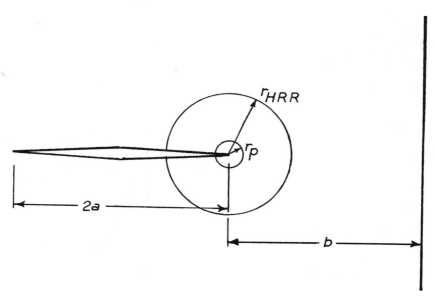

Fig. 1   Geometry of the Annular Region Around the Crack Tip.

Table 1 Table of Specimen Geometries

Ref. Material	Landes and Begley (8) Ni Cr Mo V		Zaverl, (9) 4340 113 ksi	4340 174 ksi
Groove				
type	external	internal	external	external
number	single	single	single	single
equation	w=a+b	w=a+2b	w=a+b	w=a+b
a,mm	7-8.0	7.0	19.4	22.4
b,mm	5-4.0	9.2	6.0	3.0
Loading	bending	tension	bending	bending
Non-hardening				
fan?	no	no	no	no
$\sigma_{nmax}/2k$	1.54	0.5	1.54	1.54
$r_{HRRC} = b/2$,mm	2.5-2.0	4.6	3.0	1.5
Tensile properties				
E,MN/m^2	207,000.	207,000.	207,000.	207,000.
σ_y,MN/m^2	930.	930.	780.	1200.
TS,MN/m^2	1030.	1030.	876.	1300.
ε_u	0.056	0.056	0.056	0.0418
σ_{fnom},MN/m^2	1630.	1630.	1580.	1864.
ε_f	0.77	0.77	0.98	0.71
Calculated power law fit and critical radius ratio				
J_{IC}MN/m	0.17	0.38	0.17	0.084
from σ_y:n	0.0806	0.0806	0.0955	0.0633
$\bar{\sigma}_1$MN/m^2	1438.	1438.	1327.	1662.
$r_{HRRC}/(J_o/\bar{\sigma}_1 I_n)$	93.	76.	105.	127.
$\sigma_{HRR}/\bar{\sigma}_1$	0.713	0.724	0.667	0.750
$\sigma_y/\bar{\sigma}_1$	0.647	0.647	0.588	0.722
from TS:n	0.1064	0.1064	0.1452	0.0658
$\bar{\sigma}_1$,MN/m^2	1454.	1454.	1336.	1627.
$r_{HRRC}/(J_c/\bar{\sigma}_1 I_n)$	97.	80.	112.	125.
$\sigma_{HRR}/\bar{\sigma}_1$	0.644	0.656	0.550	0.742
$\sigma_y/\bar{\sigma}_1$	0.640	0.640	0.584	0.738

Table 1 Continued

Ref. Material	Andresen (10) 1117 CF		Joyce (17) 7075-T651	

Groove
  type
  number
  equation
  a,mm
  b,mm
	external
	single
	w=a+b
	9.2
	>3.2
		external
		double
		w=2a+b
		9.9
		>2.5
			external
			single
			w=a+b
			9.7
			<2.7
				external
				double
				w=2a+b
				9.7
				<2.8

Loading Non-hardening fan? $\sigma_{nmax}/2k$ $r_{HRRC} = b/2$,mm	tension no 0.5 >1.6	tension yes 2.07 >1.25	tension no 0.5 <1.35	tension yes 2.07 <1.40

Tensile properties
  E,MN/m^2
  σ_y,MN/m^2
 	
  TS,MN/m^2
  ε_u
  σ_{fnom},MN/m^2
  ε_f
	
	200,000.
	510.
	
	550.
	0.080
	867
	0.761
		
		200,000.
		510.
		
		550.
		0.080.
		867.
		0.761
			
			71700.
			500.
			
			556.
			0.149
			774
			0.434
				
				71,700
				500.
				
				556.
				0.149
				774.
				0.434

Calculated power law fit and critical radius ratio				
J_{IC}MN/m	>0.18	>0.17	0.026	0.026
from σ_y:n	0.0675	0.0675	0.0874	0.0874
$\bar{\sigma}_1$,MN/m^2	764.	764.	770.	770.
$r_{HRRC}/(J_c/\bar{\sigma}_1 I_n)$	29.	24.	177.	183.
$\sigma_{HRR}/\bar{\sigma}_1$	<0.808	<0.818	0.660	0.658
$\sigma_y/\bar{\sigma}_1$	0.668	0.668	0.649	0.649
from TS:n σ_1,MN/m^2	0.1055 777.	0.1055 777.	0.106 785.	0.106 785.
$r_{HRRC}/(J_o/\bar{\sigma}_1 I_n)$	31.	26.	185.	192.
$\sigma_{HRR}/\bar{\sigma}_1$	<0.720	<0.733	0.606	0.604
$\sigma_y/\bar{\sigma}_1$	0.656	0.656	0.637	0.637

Table 1 Concluded

Ref. Material	Joyce (18) 2024–T4		Joyce (18) 6061–T6	
Groove type number equation a,mm b,mm	external single w=a+b 10.1 <2.3	external double w=2a+b 9.7 <2.7	external single w=a+b 10.3 <2.1	external double w=2a+b 9.8 <2.5
Loading Non-hardening fan? $\sigma_{nmax}/2k$ $r_{HRRC} = b/2$,mm	no 0.5 <1.15	yes 2.07 <1.35	no 0.5 <1.05	yes 2.07 <1.25
Tensile properties E,MN/m^2 σ_y,MN/m^2 TS,MN/m^2 ε_u σ_{fnom},MN/m^2 ε_f	71700. 390. 495. 0.161 670 . 0.351	71700. 390. 495. 0.161 670. 0.351	71700. 281. 324. 0.120 497. 0.71	71700. 281. 324. 0.120 497. 0.71
Calculated power law fit and critical radius ratio				
J_{IC} MN/m from σ_y:n $\overline{\sigma}_1$,MN/m $r_{HRRC}/(J_o/\sigma_1 I_n)$ $\sigma_{HRR}/\overline{\sigma}_1$ $\sigma_y/\overline{\sigma}_1$	0.042 0.117 717. 90. 0.624 0.544	0.038 0.117 717. 117. 0.607 0.544	0.061 0.0837 447. 34. 0.762 0.629	0.079 0.0837 447. 31. 0.767 0.629
from TS:n $\overline{\sigma}_1$,MN/m^2 $r_{HRRC}/(J_c/\overline{\sigma}_1 I_n)$ $\sigma_{HRR}/\overline{\sigma}_1$ $\sigma_y/\overline{\sigma}_1$	0.120 720. 91. 0.617 0.542	0.120 720. 118. 0.600 0.542	0.1002 451. 35. 0.723 0.623	0.1002 451. 32. 0.728 0.623

The specimens were all in the plane strain regime, judging from the data
of Zaverl (9) and lower bounds to the limit load (McClintock, 13). Under
plane strain we may regard the cracks as being grooves perpendicular to the
face of a plate. The geometry is defined by the number of grooves (single or
double), by whether they are internal or external, and by the equation relating the groove length(s) a and the ligament thickness(es) b to the total
thickness w of the plate. The ligament and groove dimensions in Table 1.1 are
those at the limiting values for constancy of the J integral. These dimensions, along with the loading, define the slip line field that would hold for
a non-hardening material (loc. cit.). The non-hardening plane strain slip line
field and the specimen geometry, along with the strain hardening, should define the critical radius r_{HRRC} required for conversion to the HHR strain singularity. One might expect that the influence of the slip line field would
make itself felt through two quantities. First, are there fans of shear
strain singularity above and below the crack tip as found in the HRR singularity and the double external groove tension specimen, or does the slip line
field at the tip of the groove consist of shear on a single pair of planes?
Second, what is the ratio of maximum normal stress on a slip plane to the
plane strain yield strength 2k? This ratio is .5 in plane strain tension and
.5 + $\pi/2$ for the HRR strain singularity and the external, double groove tension
test. For this preliminary study, we test the hypothesis that these two quantities, fan and normal stress, are not important, and r_{HRRC} can be simply
taken to be half the ligament thickness that would have to be traversed by the
non-hardening slip line field.

The tensile properties are those given by the respective authors except
for the cold finished 1117 steel of Andresen, for which data were interpolated
on the basis of hardness information from the properties for annealed and cold
rolled steel in McClintock and Argon (13, p. 327). The uniform strain was

taken to be that at the maximum load. The true stress at fracture was corrected by the Bridgman factor for the necked shape. The ratio of cross section radius to profile radius, a/R, was taken from the results for a wide
variety of materials given in McClintock and Argon (13, p. 324) to be the
amount by which the equivalent strain exceeds the uniform strain:

$$a/R = \epsilon_f - \epsilon_u \quad . \tag{8}$$

The HRR singularity is exact for the Ilyushin (14) form of the stress-
strain relation of Eq. 1, which is based on the total strain (Goldman and
Hutchinson, 15), provided that in the region of interest the stresses exceed
the yield strength, and the strains are large enough so that incompressibility
is a good approximation. We therefore, fitted the above stress-strain data
through the origin and the fracture point, and either the yield strength and
strain or the tensile strength and the uniform strain ϵ_u at which it is attained. The fitting was done iteratively, using a programmable calculator,
starting with an assumed n of .1. Since for the Ilyushin stress-strain relation the strain ϵ_u at the tensile strength is the exponent n, Eq. 8 and the
Bridgman correction give for the true stress at fracture:

$$\sigma_f = \sigma_{from} \, / [(1 + 2/(\epsilon_f - n)) \ln \, (1 + .5(\epsilon_f - n))] \quad . \tag{9}$$

An improved value for the exponent n is now found from

$$n = \frac{\ln(\sigma_f/\sigma_y)}{\ln(\epsilon_f/(\sigma_y/E))} \quad . \tag{10}$$

Equations 9 and 10 are iterated until convergence, which takes only 5 to 10
cycles.

The procedure for fitting an Ilyushin equation through the tensile
strength T.S. is similar except that one must find the true stress at the
tensile strength from the area ratio, which is in turn given in terms of the
logarithmic strength:

$$\frac{A_u}{A_o} = e^{-\epsilon_u} \quad . \tag{11}$$

The true stress is now

$$\sigma_u = T.S. \ A_o/A_u = T.S.e^{\varepsilon u} \tag{12}$$

An improved value for the exponent n is now found from

$$n = \frac{\ln(\sigma_f/\sigma_u)}{\ln(\varepsilon_f/\varepsilon_u)} \tag{13}$$

n is substituted into (9) and the process again converges quickly.

On convergence, the stress at unit strain is found by substituting the fracture stress and strain into the Ilyushin equation (1):

$$\bar{\sigma}_1 = \sigma_f/\varepsilon_f^n \ . \tag{14}$$

The crack tip opening displacement can be estimated from the J integral, $\bar{\sigma}_1$, and the numerical function of n, I_n according to Eq. 1 - 5. Values of I_n are given by Shih (16) and can be approximated within 2% over n = 0 to 1 by the empirical equation

$$I_n = 10.3 \ \sqrt{.13+n} \ -4.8n \ . \tag{15}$$

The available radius ratio, r_{HRR}/r_p is now found, and from it and the exponent n the stress ratio is found according to the HRR singularity, Eq. 4.

The results of the above calculations are shown in Table 1. For all of Andresen's specimens the J integral was varying, so that a valid result had not been obtained. The stress ratios therefore represent inequalities. When the method of Paris, Rice and Merkel (17) for single specimen determination of J is applied to the internally notched specimens of Landis and Begley, it turns out that varying values of J were encountered in those tests too. Furthermore, it has been pointed out to us that there was an error in their calculations so that the J values for the bending and internally grooved tension specimens are different from each other. This leaves the only clearly valid results to be those of Zaverl (9). His results show that the concept of a critical stress ratio is not valid. Insofar as one can draw any conclusions from all these data, the best hypothesis is that the stress ratio

from the point of unit strain out to the mid-ligament, calculated according to
the HRR singularity, must be no more than 10% above the ratio of yield
strength to stress at unit strain.

$$\frac{\sigma_{HRR}}{\sigma_1} < 1 \text{ to } 1.1 \frac{\sigma_y}{\sigma_1} \quad . \tag{16}$$

3.   DESCRIPTION OF THE EXPERIMENTS

3.1 Materials

     The materials used in this study were the aluminum alloys 6061-T6,
2024-T4, and 7075-T651 and a cold worked 1117 steel, fractured in earlier
work by Andresen (10). Mechanical properties are included in Table 1.
Chemical compositions are available in Joyce (18) and Andresen (10).

3.2 Specimen Design

     All of the specimens were machined from 25 mm. diameter bar stock and
loaded axially. The single and double grooved specimen geometries used are
drawn in Fig. 2. The ratio of specimen ligament b after pre-cracking in fatigue to specimen width B was chosen so that b/B = 10 to guarantee plane
strain conditions at least near the centerline of the specimen even if the
specimens became fully plastic before fracture (see McClintock, 13, for a
limit analysis). The ratio of crack length a to ligament b was chosen as
a/b = 4.5 so that flow would not break through to the shoulder unless the
specimen were very nearly fully plastic. An additional requirement to assure
plane strain conditions is that the fatigue pre-crack be straight across the
width of the specimen. This requirement was the most difficult to achieve
and required a modification of the double grooved specimen to include relief
grooves to control the stress intensity at the quarter points of the specimen.
This modification is shown in Fig. 2 and discussed further in Joyce (18).

3.3 Method of the Tests

     The test specimens were first machined and then the final 1.3 mm was cut with a special saw blade which had been ground to a sharp tip with an included angle of 30°. The specimens were then compressed to sharpen the groove tips and to introduce tensile residual stresses which facilitate the initiation of fatigue cracks. Fatigue cracks 1.3 mm deep were then grown from the root of each groove by fully reversed bending in a four point bending fixture. Deflections of the specimen centerline were measured using a calibrated cathetometer so that the specimen compliance could be monitored continuously. An analytical expression relating specimen compliance to crack length was obtained for each specimen geometry by the method outlined by Bucci et al. (19) from Castigliano's Theorem. For the double grooved specimen in four point bending for example

$$\frac{\delta_{TOTAL}}{M} = \frac{36s}{BEw^3} + \frac{48(1-\nu^2)}{BE\ \pi} (\frac{1}{b^2} - \frac{1}{w^2}) \quad , \tag{17}$$

where s is the specimen length between the rigid grips.

Measured compliance changes were then related directly to crack growth and specimens with close to the desired length fatigue cracks were readily obtained. To obtain a final crack tip opening displacement less than a tenth of the critical crack tip opening displacement, the applied moment was reduced for the final 5% of crack growth.

On the double grooved specimens the mean load was varied to obtain equal growth of both cracks. It was possible to detect any unequal crack growth from the compliance measurements because the cracks closed upon themselves when in compression and a distinctly smaller compliance in one direction resulted when the longer crack was in compression.

After the specimen compliance has increased by the desired amount, the specimen was removed from the bending fixture and loaded in tension in an Instron testing machine.

For each specimen a load versus cross-head displacement curve was plotted. After the specimens had fractured, precise ligament measurements were taken from the fracture surface using a calibrated cathetometer.

4.   DISCUSSION OF THE J_I INTEGRAL RESULTS

An important experimental consideration in obtaining a valid value for J_{IC} is the determination of the point on the load displacement curves where crack growth initiates. One double grooved specimen of each of the 2024-T4, 7075-T651, and 6061-T6 aluminum alloys was loaded in tension to maximum load, as determined by prior test specimens, and then broken in bending fatigue. Scanning microscope studies of the resulting fracture surfaces showed that crack initiation in these materials did in fact occur close to maximum load.

From each load-displacement curve, $J_{I\ MAX}$ was determined using the single load displacement curve method reported by Rice, Paris, and Merkel (17). In this method the total J_I is evaluated from two components, $J_{I\ ELASTIC}$ and $J_{I\ PLASTIC}$. The elastic component is equal to the Griffith energy release rate and can be evaluated in terms of the elastic stress intensity factor K_I by

$$J_{I\ ELASTIC} = \mathcal{G} = \frac{(1-\nu^2)K_I^2}{E} . \tag{18}$$

The $J_{I\ PLASTIC}$ can be evaluated from the load displacement curve in terms of the specimen ligament thickness b, plate width B, and two areas shown in Fig. 3:

$$J_{I\ PLASTIC} = \frac{1}{bB} [2\ Area_{ODBCO} - Area_{OABC}]. \tag{19}$$

Cancelling overlapping areas of Eq. 19 gives a simple relation in terms of the cross-hatched area ODBO shown on Fig. 3.

$$J_{I\ PLASTIC} = \frac{2}{bB} Area_{ODEO} . \tag{20}$$

The total J_I is given by

$$J_I = \frac{2}{bB} Area_{ODEO} + \mathcal{G}. \tag{21}$$

The values of $J_{I\ MAX}$ for each specimen are summarized in Table 2 and are plotted versus ligament size b in Figs. 4 - 7. Because the values for $J_{I\ MAX}$ reported in Andresen's thesis for double grooved specimens were incorrectly calculated, corrected values for all of his tests are plotted versus

ligament size in Fig. 7. Also included in these figures and in Table 2 are

estimates of the critical conversion radius r_{HRR} and the fracture process

radii r_p. The conversion radius r_{HRR} was obtained from Eq. 4, 5, and 16, with

a constant c taken to be either 1 or 1.1:

$$\frac{b}{2} > r_{HRR} = \frac{J_{IC}}{\bar{\sigma}_1 I_n} \left(\frac{\bar{\sigma}_1}{c\sigma_y}\right)^{(n+1)/n} \tag{22}$$

Equation 22 was evaluated for Table 2 using all combinations of c = 1 or 1.1,

the average J_{IMAX} for single or double grooved specimens, and the stress-strain

parameters n and $\bar{\sigma}_1$ obtained by fitting through the yield or the tensile

strengths. The calculated fracture process zone r_{pc} was found from Eq. 5.

The results show a very great uncertainty in the conversion radius, a-

rising primarily from different exponents due to fitting the Ilyushin curve

to tensile test data. Experience may show which fit to use. Otherwise more

exact stress and strain distributions must be found to provide assurance of

validity of the HRR singularity. Here, only the 7075-T651 specimens are

clearly in the range of valid J_{IC} tests; the 2024-T4 specimens are at best

marginal, and the 6061-T6 and 1117 steel specimens cannot be counted on.

The observed fracture process zone radius, r_{po}, was estimated from

stereo pairs of scanning electron micrographs of the fracture surface. For

the 2024-T4 and 7075-T651 alloys, there was almost no blunting so the observed

fracture process zone radius, r_{po}, was taken to be the average dimple diameter

at the transition line from fatigue growth to monotonic growth of the crack.

For the 6061-T6 alloy, r_{po} was taken as the magnitude of the crack tip blunt-

ing, which in these specimens was approximately .20 mm (about five times the

average dimple diameter).

The J_{IC} tests for the three aluminum alloys were converted to K_{IC} values

using Eq. 8. These K_{IC} values should, if the J_{IC} tests are valid, agree

closely with K_{IC} values obtained for these materials using large enough specimens for standard linear fracture mechanics tests to be valid. Unfortunately
very little information on K_{IC} for these materials is available. Included in
Table 2 are the best values which are available, obtained from a fracture
toughness compilation of Matthews (20), for plate material in orientations
indicated by the normals to the crack and its leading edge. Surprisingly,
values obtained here lie above Matthew's values, even though finally, for
small specimens, geometrical scaling makes J_{IMAX} fall off linearly with size
as shown in Fig. 7 for 1117 steel, and also by Zaverl (9). A possible reason
for the difference is that the handbook specimens were plate specimens in the
orientations shown on Table 2 while the J_{IC} specimens were 25 mm diameter bar
specimens. For H-11 steel, where K_{IC} values for both geometries were available in the handbook, the K_{IC} for bar specimens is 32% greater than for the
plate specimens. The 2024-T4 specimens used to obtain the handbook value of
K_{IC} were loaded in a transverse WR orientation, and for most materials K_{IC}
is 20 to 50% smaller in the transverse specimen orientation.

Another possible reason that the K_{IC} obtained from J_{IC} is larger than
the handbook values of K_{IC} could be that fracture initiated before the point
of maximum load which has been used in this work as the point of initial
crack growth. A scanning electron microscope fractography study was done on
specimens which were pre-cracked, loaded near to maximum load, and then broken in fatigue. The fracture surface topology shows that fracture occurs
close to maximum load in these materials. No attempt was made, however, to
use ultrasonic or other techniques to determine the point of crack initiation
to a high degree of accuracy, and it is possible that crack initiation could
occur slightly before maximum load. If this is the case a smaller value of
J_{IC} and thus of K_{IC} would result. The fact that the single and double
grooved specimen J_{IC}'s correspond so well, though, is some evidence that

Table 2 Critical conversion radius and fracture process radii

Material	7075-T651		2024-T4		6061-T6		1117CF	
σ_y MN/m^2	500.		390.		281.		510.	
Fit through	Y	TS	Y	TS	Y	TS	Y	TS
n	0.0874	0.106	0.117	0.120	0.0837	0.1002	0.0675	0.1055
$\bar{\sigma}_1$ MN/m^2	770.	785.	717.	720.	447.	451.	764.	777.
I_n	4.43	4.55	4.61	4.62	4.41	4.51	4.30	4.54
Single								
$J_{I\ MAX}$	0.026		0.042		0.061		0.18	
r_{HRR}, c = 1., mm	1.6	0.81	4.3	3.9	12.6	5.4	32.7	4.2
r_{HRR} c = 1.1, mm	0.50	0.30	1.7	1.6	3.7	1.9	7.2	1.55
b/2 ,mm	1.4		1.2		1.0		1.6	
$J/\bar{\sigma}_1 I_n$, mm	0.008	0.007	0.013	0.013	0.031	0.030	0.055	0.051
Double								
$J_{I\ MAX}$	0.026		0.038		0.079		0.17	
r_{HRR} c = 1., mm	1.6	0.81	3.9	3.5	16.3	7.0	30.9	4.0
r_{HRR} c = 1.1, mm	0.50	0.30	1.6	1.4	4.8	2.5	6.8	1.5
b/2 ,mm	1.4		1.4		1.2		1.2	
$J/\bar{\sigma}_1 I_n$,mm	0.008	0.007	0.012	0.011	0.040	0.039	0.052	0.048
r_p observed mm	0.012		0.02		0.20		0.25	
K_{ICC} ksi$\sqrt{in.}$	44.		52.		72.		–	
K_{IC} exp. ksi$\sqrt{in.}$	28 RW		31 WR		30 RW		–	

these J_{IC} values and the corresponding K_{IC} values are valid for these materials. Andresen's results, on the other hand, show surprisingly little effect of single vs. double grooving, even in an invalid regime.

A detailed fractographic study was completed on these same specimens using a scanning electron microscope. A complete discussion of the observations is too lengthly to be included here and will be published elsewhere. A general observation bearing on the above conclusions is that the initial fracture topography in the single and double grooved specimens of 7075-T651 and 2024-T4 were very similar. It was impossible to tell from stereo microphotographs of the initial stages of crack growth whether the specimen had a single or double grooved geometry.

In 6061-T6 aluminum and 1117 steel, for which valid J_{IC} results were not obtained, the initial fracture surface topographies were noticeably different with the single grooved specimens demonstrating sliding-off phenomena and the double grooved specimens demonstrating delamination and more crack tip blunting and hole growth.

The calculation of r_{HRR} from Eq. 22 presented in Table 2 correlate these fractographic results if the stress-strain curve, Eq. 1, is fitted through the tensile strength and the constant c is taken to be 1.1, so $\sigma_{HRR} < 1.1 \sigma_y$.

5. CONCLUSIONS

A test of minimum ligament size b for valid J_I integral results is proposed which is based on the Ilyushin stress-strain curve with exponent n, stress at unit strain $\bar{\sigma}_1$, HRR constant I_n, yield strength σ_y, and a numerical constant c in the form

$$b/2 > r_{HRR} = \frac{J_{IC}}{I_n \bar{\sigma}_1} \left(\frac{\bar{\sigma}_1}{c\sigma_y}\right)^{(n+1)/n} . \tag{22}$$

For the steels except the higher strength 4340 of Zaverl (9), the criterion appears to give valid results if the stress-strain relation is fitted through the yield strength and the constant c is taken to be 1.1. For the higher strength steel, c must be 1.05. For the aluminum alloys fractographic evidence for uniform fracture mechanisms at the crack tip in both single and double groove specimens is correlated by fitting the stress-strain curve through the tensile strength and taking the constant c to be 1.1. This correlation is based on the idea of a certain stress ratio being required to develop the HRR singularity from a fully plastic stress field, but is otherwise empirical in nature and does not take differing stress states and strain distributions in the non-hardening slip line field into account. Further experiments and also numerical calculations of the stress and strain fields for strain hardening exponents of the order of 0.1 would be very desirable.

Specimens of 2024-T4 and 7075-T651 aluminum alloys with ligaments as small as 1.91 mm. gave apparently valid J_{IC} values of 0.026 and 0.036 MN.-m/m^2. Values of K_{IC} calculated from these J_{IC} values were found to be larger by 30 to 50% than handbook K_{IC} values for these materials. This difference might have resulted from the different forming processes of the specimens. Any difference due to crack initiation occurring before maximum load in these materials, is likely to be less than 20%, judging from fatigued-loaded-fatigued specimens.

For the 6061-T6 specimens the ligament sizes tested were greater than the r_{HRR} critical validity radius required for valid J_{IC} tests. The J_I at maximum load, $J_{I\ MAX}$, for the one specimen with ligament b less than the average was .75 of the average of the other $J_{I\ MAX}$ values. J_{IC} was 0.071 MN.-m/m^2, which gives a value of K_{IC} almost 100% over the handbook value of K_{IC} for this material, but that is comparing a 3 in. thick plate with a 1 in. round bar, so differences would be expected.

For the specimens of 1117 steel the critical validity radius r_{HRR} calculated from the fracture process zone r_p and Eq. 4 was too large compared to the specimen ligament for valid J_{IC} values to be obtained.

Fractographic observations showed less delamination and more sliding-off in single grooved specimens than in double grooved ones when a valid J_{IC} criterion was not satisfied.

ACKNOWLEDGEMENT

The support of the Army Research Office through Contract DAHG04 - 72 - C 0043 in the preparation of this paper is gratefully acknowledged.

6.   REFERENCES

(1)   J. A. Begley, and J. D. Landes, "The J Integral as a Fracture Criterion,"
       Fracture Toughness, ASTM STP 514, Amer. Soc. for Testing and Mat.,
       Philadelphia, (1972), pp. 1-20.

(2)   J. W. Hutchinson, "Singular Behavior at the End of a Tensile Crack in a
       Hardening Material," J. Mech. Phys. Solids 16, (1968), pp. 13-31.

(3)   J. R. Rice, and G. F. Rosengren,"Plane Strain Deformation Near a Crack
       Tip in a Power Law Hardening Material." J. Mech. Phys. Solids 16, (1968),
       pp. 1-12.

(4)   J. D. Eshelby, "The Force on an Elastic Singularity," Phil. Trans. Roy.
       Soc. London, Ser. A 244, (1951), pp. 87-112.

(5)   G. P. Cherepanov, "Crack Propagation in Continuous Media," Appl. Math.
       and Mech. (Prik. Matem. i Mekh.) 31 (3), (1967), pp. 503-512.

(6)   J. R. Rice, "A Path Independent Integral and the Approximate Analysis
       of Strain Concentration by Cracks and Notches," J. Appl. Mech. 35,
       (1968), pp. 379-386.

(7)   F. A. McClintock, "Plasticity Aspects of Fracture," Fracture 3,
       H. Liebowtz ed., Academic Press, N. Y., (1971), pp. 47-225.

(8)   J. D. Landes, and J. A. Begley,"The Effect of Specimen Geometry on
       J_{IC}," Fracture Toughness, ASTM STP 514, Amer. Soc. for Testing and Mat.,
       Philadelphia, (1972), pp. 24-39.

(9)   F. Zaverl, Jr., "The Influence of Specimen Dimensions on a J_c Fracture
       Toughness Test," M.S. Thesis, U. of Ill., Theo. and Appl. Mech. Dept.,
       Urbana, Ill., (1974).

(10)   J. A. Andresen, "The Range of Validity of the J Integral for Predicting
       Fracture in Steel," S. M. Thesis, Dept. of Mech. Engr., MIT, Cambridge,
       MA, (1973).

(11)   M. L. Williams, "On the Stress Distribution at the Base of a Stationary
       Crack," J. Appl. Mech. 24, (1957), pp. 109-114.

(12)   G. R. Irwin, "Analysis of Stresses and Strains Near the End of a Crack
       Traversing a Plate," J. Appl. Mech. 24, (1957), pp. 361-364.

(13)   F. A. McClintock, and A. S. Argon, Mechanical Behavior of Materials,
       Addison Wesley, (1966), 770 pp.

(14)   A. A. Ilyushin, "The Theory of Small Elastic-Plastic Deformations,"
       Prik. Matem. i Mekh., P.M.M. 10, (1946), pp. 347-356.

(15)   N. L. Goldman, and J. W. Hutchinson, "Fully Plastic Crack Problems: The
       Center-Cracked Strip Under Plane Strain," Int. J. Sol. Struct. 11,
       (Feb. 1975), pp. 575-591.

178

(16) C. F. Shih, "Small Scale Yielding Analysis of Mixed Mode Plane-Strain Crack Problems," Fracture Analysis, ASTM STP 560, Amer. Soc. for Testing and Mat., Philadelphia, (1974), Fig. 4, p. 196.

(17) J. R. Rice, P. C. Paris, and J. G. Merkle, "Some Further Results of J-Integral Analysis and Estimates," Progress in Flow Growth and Fracture Toughness Testing, ASTM STP 536, (1973), pp. 231-245.

(18) J. A. Joyce, "On the Mechanisms and Mechanics of Plastic Flow and Fracture," Sc.D. Thesis, Dept. of Mech. Engr., MIT, Cambridge, Ma., (1974).

(19) R. J. Bucci, P. C. Paris, J. D. Landes, J. R. Rice, "J Integral Estimation Procedures," Fracture Toughness, Proceedings of the 1971 National Symposium on Fracture Mechanics, Part II, ASTM STP 514, (1972), pp. 40-69.

(20) W. T. Matthews, "Plane Strain Fracture Toughness (K_{IC}) Data Handbook for Metals," Army Mat. and Mech. Res. Center, Watertown, Ma., AMMRC. MS. (1973), pp. 73-76.

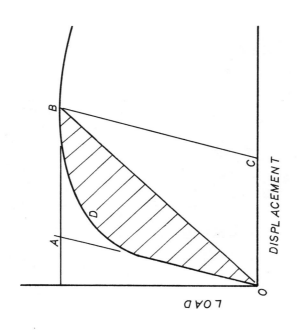

FIG. 3 DESCRIPTION OF AREAS USED TO EVALUATE J_I PLASTIC

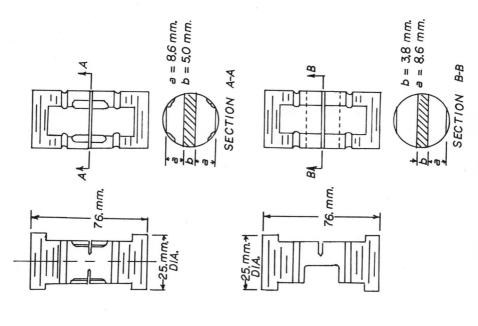

FIG. 2 DOUBLE AND SINGLE GROOVED SPECIMEN GEOMETRIES

180

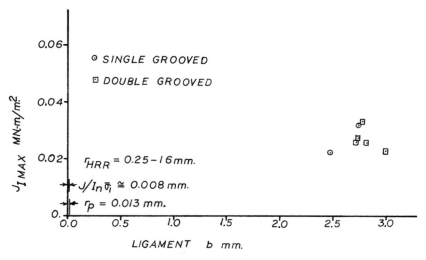

FIG. 4 J_{Ic} VERSUS SPECIMEN LIGAMENT FOR THE 7075–T651 SPECIMENS

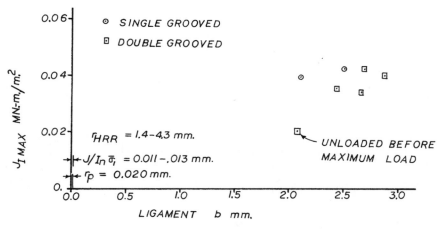

FIG. 5 J_{Ic} VERSUS SPECIMEN LIGAMENT FOR THE 2024–T4 SPECIMENS

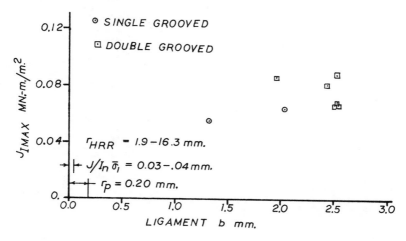

FIG. 6 J_{Ic} VERSUS LIGAMENT OF THE 6061-T6 SPECIMENS

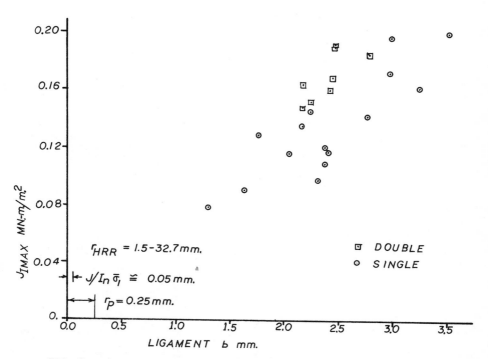

FIG. 7 $J_{I\ MAX}$ VERSUS SPECIMEN LIGAMENT FOT THE 1117 STEEL SPECIMENS

SESSION III

Effects of Elevated Temperature

Fatigue Life Predictions at Elevated Temperature

L. F. Coffin, Jr.*

1. THE PLASTIC ZONE AND ITS RELATIONSHIP TO LOW-CYCLE FATIGUE

The various stages of the fatigue process as they are
currently viewed are represented in Fig. 1. Important in this
model is (a) an analytical knowledge of the deformation state
within the plastic zone and (b) the establishment of a failure
criterion for crack initiation and early growth in terms of the
strains so produced in this zone. To be as quantitative as
possible it is desirable to have a complete elasto-plastic
solution for stresses and strains in the vicinity of the notch.
Although rigorous constitutive equations are lacking for such
calculations, a recent approach by Mowbray and McConnelee appears
attractive in attacking this problem. They utilize a finite
element analysis where constitutive equations are derived from
families of "iso-cycle" stress-strain curves converted to
effective stress-strain curves to construct a complete stress-
strain representation within the plastic zone. It would appear
that the analytical tools for treating complex geometries are
developing more rapidly than concomitant material laws.

The fatigue literature contains numerous references in which
the stabilized or steady-state, stress-strain curve is similarly
utilized to establish the local strain distribution at the notch
root. Notch root strains are commonly and simply determined by
the Stowell or Neuber formulation The applicability of these

* Corporate Research & Development, General Electric Co.,
Schenectady, New York 12301

methods to high-temperature problems will be discussed later.

As a result of repeated cyclic plastic strain in the plastic zone, microcracks nucleate and grow to some observable size at the free surface. One way to simulate this situation is to imagine a low-cycle fatigue specimen located as in Fig. 1 such that its minimum cross section is some fraction of the plastic zone. Provided suitable corrections are introduced to account for the different stress states, fatigue failure of this specimen can be made to approximate the nucleation and early growth of the component of Fig. 1. There is, however, a size and strain gradient problem since, in the test specimen, the crack is nucleated and grows in a uniform strain field to some fraction of the cross-sectional area, while in the plastic zone this same process occurs in a strain gradient. Hence the concept of "smooth specimen simulation"--as Morrow et al. have name it--is sound, provided the plastic zone is sufficiently large relative to the minimum diameter of the standard low-cycle fatigue test specimen (~0.25 inch) and the strain gradient is acceptably small. For smaller plastic zones, either proportionally smaller fatigue specimens or earlier indications of failure in standard size specimens are required. By these arguments, and with the assumption that the plastic zone of our structure (Fig. 1) is large relative to the test specimen, we can insert the massive volume of low-cycle fatigue information as a critical link in the chain of events leading to structural fatigue failure.

2. HIGH TEMPERATURE BEHAVIOR

When one performs low cycle fatigue tests on a given material at increasingly higher temperatures, it is commonly observed in equation (1) that the exponent α increases, approaching unity and the fatigue ductility ε_f' correlates less well with the tensile ductility. Concurrently it is found that the mode of cracking becomes more intergranular. The view was present that because of the high exponent for β in the Coffin-Manson equation

$$N_f^{\alpha} \Delta\varepsilon_p = \text{Const} \tag{1}$$

reported for many materials at elevated temperature, extrapolation to low lives could lead to plastic strain ranges exceeding the tensile ductility. Since this situation is highly unlikely, there must be a break in the slope of $\Delta\varepsilon_p$ vs N_f so that the curve will more nearly approach the tensile ductility value at $N_f = \frac{1}{4}$. This break was attributed to a change in the mode of crack propagation to a more ductile mode with increasing plastic strain and shorter lives. Several examples were given to show this two-slope behaviour of which Fig. 2 is typical.

A model was presented to describe this break in slope, based primarily on environmental considerations. This is shown in Fig. 3. Note in Fig. 3b the observed behaviour at low and high temperature at a specific plastic strain range. First, the view is presented that low cycle fatigue is principally a crack propagation process; this has been already discussed. Next it is assumed that the mode of crack propagation affects the fatigue life. Given a particular plastic strain range as is

shown in Fig. 3a, an intergranular crack propagates more rapidly
through the structure than a transgranular one since less distortion is involved. Hence fracture will occur sooner. At low
temperature, where the fracture is entirely transgranular, the
normal single-loop line is produced, while at high temperature,
where the fracture mode has become intergranular, the life is
reduced.

To explain the intergranular failure mode, the steeper
slope and the break at O', the effect of the environment is
introduced. Because the cracking process continually exposes
fresh active metal at the crack tip, oxidation processes take
place which "damage" the material at the crack tip surface.
Exposed grain boundaries may be more aggressively attacked than
grains themselves because of their chemical complexity (including
carbides and other precipitates) and these are prime fracture
sites. Assuming the period in the cycle remains constant, the
effect that this oxidation damage has on the fracture mode will
depend on the strain amplitude. At low strains, where the
crack growth is small, the oxide-damaged region constitutes a
large fraction of the cracked surface and the environmental
effect is large. At high strains where the crack growth is
large, the oxide-damaged fraction of the crack advance is too
small for much environmental attack to occur. Increasing the
period allows more time for these processes to occur and shifts
the fatigue line to the left; this is observed in Fig. 2. Thus
an environmental model appears attractive in accounting for
deviations in the Coffin-Manson relationship.

To further confirm the importance of environment on high-temperature low cycle fatigue in the frequency range normally employed for tests of this type, the same materials can be similarly tested in high vacuum.

The results of these experiments are striking. Fig. 4 is typical. It shows the comparative behaviour in air and vacuum of A286 at 593°C, together with room-temperature vacuum test results. Note the significant effect of frequency of cycling when testing in air which is not found in vacuum. Also note the large difference in lifetimes between the highest frequency air test and the vacuum results. It is also interesting to observe that the effect of the test temperature is minimal in a vacuum environment. Additionally it was found that crack propagation was transgranular. Tests of this type have been performed on several materials at temperatures where significant degradation in fatigue resistance occurs, with results similar to those obtained for A286.

To summarize these findings, plastic strain range-life data for several materials have been obtained and plotted in Fig. 5. Included here are low cycle fatigue test results reported many years ago for room-temperature air (the open point), together with a number of high temperature tests run either in high vacuum or a highly purified argon atmosphere (the solid points). In Fig. 5 the test results are normalized to the plastic strain range for failure at one cycle, in order toestablish a common base for comparison of the slope by eliminating ductility differences. Lines corresponding to various exponents β in equation (1) are also drawn, showing that the test results fit within a scatter band $0.45 < \beta < 0.60$, and the bulk of the data fit between $0 \cdot 50 < \mathrm{R} < 0 \cdot 55$. Note that temperatures as high as 1150°C are employed in some of the experiments. Frequency of cycling for all these data is $0 \cdot 2$ cycle/min or greater. These findings provide further endorsement for the Coffin-Manson relationship, but more importantly, they show the significance of the environment in elevated temperature low cycle fatigue. An additional feature of these several tests is that all fractures are transgranular.

3. PHENOMENOLOGICAL RELATIONSHIPS

In the subject discussed above it is clear that, when dealing with high-temperature fatigue, time dependency must be introduced into the Coffin-Manson relationship. It has been argued that the strain rate should be considered, since high-temperature deformation processes are commonly represented by this quantity. However, from the position described above, namely, that environmental "damage" would depend on the period of the cycle, the frequency or period of cycle would seem more meaningful. Stemming from the work of Eckel and that of Coles et al. for hold time results, fatigue tests can be predicted for a specific plastic strain by the relationship

$$\nu^k t_f = \text{const} = f(\Delta \varepsilon_p) \ldots \tag{2}$$

where ν is the frequency (expressed here as cycles/min), t_f is the time to failure and k is regarded as a constant dependent only on the temperature. Equation (2) can be rewritten as

$$\nu^k t_f = \nu^k \frac{N_f}{\nu} = N_f \nu^{k-1}$$

a quantity which has been called the frequency modified fatigue life. Note that k serves as a measure of time dependency; when k = 1, equation (3) is independent of frequency. It has been found convenient to use $N_f \, \nu^{k-1}$ as a parameter for combining frequency and life in the Coffin-Manson equation. Fig. 6 is such an example for AISI 304 stainless steel, taken from the work of Berling and Slot. Note that this figure displays the characteristic two-slope features of the high-temperature fatigue behaviour described above as well as test results in high vacuum at 816°C.

That part of the fatigue curve to the right of the convergence point can be represented by substituting equation (3) for N_f in equation (1) such that the high temperature form of equation (1) becomes

$$(N_f \nu^{k-1})^\beta \Delta \varepsilon_p = C_2 \tag{4}$$

Equation (4) provides a way to account for time dependency when dealing with plastic strains. It is attractive to also apply this approach when considering the elastic strain component. The Basquin relationship can be applied to relate the stress range (or elastic strain) and the life. A high temperature form which includes the frequency of cycling can be shown to describe the elastic strain. Thus

$$\Delta \varepsilon_e = \frac{\Delta \sigma}{E} = \frac{A'}{E} N_f^{-\beta'} \nu^{k_1'} \tag{5}$$

Equations (4) and (5) can be combined to eliminate N_f. This gives the cyclic stress-strain relationship

$$\Delta\sigma = A\Delta\varepsilon_p^{n'}\nu^{k_1} \tag{6}$$

where $n' = \beta'/\beta$, $A = A'C_2^{n'}$ and $k_1 = k'_1 + \beta'(k-1)$. Equation (6) can also be represented in terms of the strain rate rather than the frequency or

$$\Delta\sigma = B'\Delta\varepsilon_p^{n_1'}\dot{\varepsilon}_p^{m} \tag{7}$$

where $B' = A/2k_1$, $n'_1 = n'-k_1$ and $m = k_1$. Equation (7) has as its counterpart the monotonic relationship between stress, strain and strain rate.

Application of equations (4-7) to several materials has been reported recently. One example is given in Fig. 7. For this purpose equations (4) and (5) are rewritten as

$$\Delta\varepsilon_p \nu^{\beta(k-1)} = C_2 N_f^{-\beta} \tag{8}$$

and

$$\Delta\varepsilon_e \nu^{-k'_1} = \frac{A'}{E} N_f^{-\beta} \tag{9}$$

The left hand terms of equations (8) and (9) are called the frequency-modified elastic and plastic strains. The material here is AISI 304 stainless steel. Three temperatures are represented. Note the increasing slope of the plastic strain-life relationship, the decreasing magnitude of the elastic strain

with increasing temperature for a given life, and the decreasing
transition fatigue life with increasing temperature. Values of
the coefficients for equations (4)-(9) are given in Table 1 for
the temperatures indicated.

4. FREQUENCY REGIMES

The question arises as to whether the time dependency
approach described by equations (2)-(7) applies over a broad
frequency range. We refer here to the work of Solomon and
Solomon and Coffin on high strain fatigue crack growth for clues
to this question. Crack growth measurements are useful for
determining effects at very low frequencies since only a few
cycles and a reasonable time period are required to accumulate
test data. Tests were performed as described earlier on
prenotched bars of A286 at 593°C subjected to controlled cyclic
plastic strain over a broad range of frequencies in air and in
vacuums of 10^{-8} torr. Over a frequency range of $0 \cdot 05 < \nu < 90$
cycle/min a crack growth law of the form

$$\frac{dC}{dN} = \phi C (\Delta \epsilon_p)^{\alpha} \nu^{k-1} \tag{10}$$

was found to apply. Integration of equation (10) between C_o,
the initial notch, to a final crack length C_f identified as N_f
yields equation (4) where $\alpha = 1/8$ and the two k quantities are
equivalent (k = 0·55). For frequencies less than 0·05 cycle/min,
however, a more or less abrupt change in crack growth occurred,
corresponding to k = 0. This is equivalent, in a low cycle
fatigue test described by equation (4), to

$$t_f^{\,\beta} \Delta\varepsilon_p = \text{const} \tag{11}$$

since, when $k = 0$, $N_f \nu^{k-1} = N_f/\nu = t_f$. Fig. 8 shows the frequency time to failure results for frequencies ranging from $0 \cdot 002$ to 90 cycle/min corresponding to a plastic strain of $0 \cdot 002$. As the frequency is decreased there is a progressive change from a transgranular to intergranular fracture until, below $0 \cdot 05$ cycle/min, the crack is entirely intergranular in nature. Although not covered by specific tests, it was suggested that at very high frequencies a regime existed in which the crack growth process would be frequency independent because there would be insufficient time for chemical effects to act. In such instances $k = 1$. Consequently over a broad range of frequencies, for A286 at 1100°F three regimes could be considered. At very high frequencies, crack growth dC/dN is independent of frequency and $k = 1$. At intermediate frequencies, $0 \cdot 05 < \nu < 90$ cycle/min, $k = 0 \cdot 55$, and crack growth is frequency dependent. At low frequencies, $\nu < 0 \cdot 05$ cycle/min, crack growth varies inversely with the frequency, that is, the total time for failure is constant, but independent of the frequency.

Solomon and Coffin have reported results of similar tests for A286 at 593°C conducted in a vacuum of 10^{-8} torr. In contrast to the air results, no effect of cyclic frequency was observed above 1 cycle/min ($k = 1$). On the other hand, at frequencies less than $0 \cdot 01$ cycle/min the vacuum crack propagation is highly time dependent ($k \approx 0$) and undergoes intergranular fracture. It was concluded earlier that the enhanced crack

growth observed in the previous air tests must be largely due
to the influence of the air environment. This, of course, is
consistent with the low cycle fatigue results shown in Figs. 4
and 5. Note, however, that at low frequency in vacuum, crack
growth was found to be intergranular, suggesting some change in
the basic material response at low frequencies. Possibilities
here include creep rupture, or metallurgical changes due to the
formation of some brittle grain boundary phase or to some strain-
induced ageing process that preferentially strengthens the
grains and enhances deformation and fracture at the grain
boundaries.

More recent unreported work has been performed on nickel A
at 550°C and 304 stainless steel at 816°C subjected to high
strain crack growth in air and high vacuum at low frequencies.
It was determined that there was no evidence of a frequency
effect for nickel between 1 and 0·001 cycle/min and for stainless
steel between 1 and 0·01 cycle/min. Neither crack growth rate
nor mode of cracking changes significantly; in all cases
cracking was transgranular. On the other hand, air tests on
nickel A revealed a progressive increase in growth rate and
transition to intergranular cracking as the frequency was lowered.
This indicates that the pure time-dependent regime may depend on
the specific alloy, and will be more likely found for those
metals alloyed for enhanced creep strengthening.

Fig. 9 is therefore proposed to represent the low cycle
fatigue behaviour of smooth bar specimens, based on the above
observations for crack growth and by applying equation (2) for
tests conducted in air and in vacuum. We note that there are

three possible regimes as described above, and two critical
frequencies, ν_e, the subscript representing the environment, and
ν_m, the subscript representing material effects. The lower
critical frequency may not exist at all or may occur at too low
a frequency to be of practical value. The model is incomplete
with respect to changes in plastic strain, although according
to equation (3) a decrease in frequency would cause a parallel
shift of the curves drawn. It is not clear how ν_e and ν_m would
shift with different plastic strain. Admittedly Fig. 9 is only
a model, and as such is subject to further scrutiny. Note that
there is a similarity between this view and that recently
suggested by Skelton.

5. WAVE SHAPES

Up to this point we have reviewed the state of understanding
of high-temperature fatigue as it pertains to test specimens
subjected to fully reversed straining and most commonly employing
a triangular wave shape (Fig. 10a). Compared to the large body
of possible wave shapes that one can now investigate with
closed loop testing equipment, this particular form would appear
to be highly specialized. Largely because of the large body of
information that has been accumulated, however, it is common to
look upon the triangular wave shape test as a standard from which
to predict other wave shapes and to predict service behaviour.

Other wave shapes are now easily obtainable for study and
consideration as a base for material evaluation and life prediction. Some of these possibilities are shown in Fig. 10. One
family involves a strain hold period, either in tension,

compression or both, and introduces a stress relaxation during the hold. Fig. 10b illustrates the tensile hold. Note that the hysteresis loop may show a mean stress, necessary in order to correct for the difference in the time spent in tension vs the compression part of the cycle. This mean stress effect is particularly important for lives greater than the transition fatigue life. It was shown that substantially longer lives obtain for tension holds over compression holds for cast Rene 80 at 870°C. This surprising effect is in contrast to the work of Conway et al. who found that tension hold periods were more damaging than compression hold periods for AISI 304 stainless steel at 650°C. Here, however, conditions were such that the lives were less than the transition fatigue life. We will return to this point later.

In Fig. 10c is shown a wave shape due to continuous stress cycling. Here a mean stress exists and ratcheting may take place, depending on the transition fatigue life. This form of ratcheting can be time independent, and will occur at all temperatures, depending only on the magnitude of the mean stress and the plastic strain range. Additionally time dependent ratcheting can be expected at elevated temperature due to creep effects.

Another family of wave shapes also involves stress control, Fig. 10d. When the stress is fully reversed, but hold periods are introduced, ratcheting may occur to increase the strain during each cycle. The ratcheting here is the result of creep during the hold period and is associated with the difference in the time period for tensile vs compressive stress.

A wave shape now receiving much attention is that described in Fig. 10e. Here a stress hold period is introduced, either in tension, compression or both, not unlike the strain hold situation. In order to balance the unequal time periods occurring during tension only or compression only hold periods so as to maintain a stable loop without ratcheting, a mean stress may be necessary as shown. Wave shapes in this category have been identified as CP, or PC, depending on whether the hold period is in tension or compression, or CC if the hold period is both tension and compression. The wave shape of Fig. 10a is identified as PP when cycling is sufficiently rapid that no creep deformation occurs.

Another category of wave shapes can be produced when temperature cycling is combined with mechanical cycling, Fig. 10f. Techniques for producing these conditions are now possible and the variations are indeed large. Here we show the situation where the mechanical strain and thermal strain are in phase. Note the rather unusual hysteresis loop produced under these conditions. With this wave shape we can reintroduce laboratory simulated thermal fatigue, an approach which first enmeshed me in the fatigue problem over twenty years ago.

6. THE STATE OF LIFE PREDICTION

First, we must assume a common basis for agreement. This includes the following points:

(a) that the fatigue process is one of nucleation and propagation and that these processes are involved in all of the testing methods considered above;

(b)   that elastic and plastic strains can be separated;

(c)   that we can identify low and high cycle fatigue regimes, the transition between the two being defined by the transition fatigue life;

(d)   that each regime exhibits distinct characteristics, the low cycle regime being characterized best by the plastic strain range, the high cycle regime by linear elastic fracture mechanics.

If we accept these concepts, then some of the complexity of the strain hold time experiments may be removed. For example there seems to be good agreement between the frequency-modified continuous cycling approach and some of the hold time test results (those of Conway et al.) as shown earlier. At least two of the papers given at a recent conference suggest modifications to this approach, indicating that the assumption that equivalence of frequency and of plastic strain range may not be sufficient in predicting life from one wave shape to the other. Part of the problem may be that the specifics of these two wave shapes (those of Figs. 10a and 10b) have to be accounted for more carefully than heretofore in applying this equivalence. This is considered elsewhere. It is also important to keep in mind that when strain hold time tests are conducted above the transition fatigue life, a crack growth approach rather than a plastic strain approach applies. By adopting this position, it is possible to explain the otherwise confusing results found from hold time experiments on cast Rene 80 at 870°C. There it was observed that compressive strain hold periods were more damaging than tensile strain holds. It was further observed that concurrently with the compression hold, a loop shift

developed so as to produce a tensile mean stress, while, with

tensile holds, a compressive mean stress was found. Since the

lives were above the transition fatigue life such that failure

could be predicted by elastic crack growth, the differences in

life could be accounted for by the role of the mean stress on

crack growth. Below the transition fatigue life these mean

stress effects are washed out, and tensile strain hold periods

become more damaging. Identification of regimes of applicability

should be helpful in sorting out some of the differences found

when comparing wave shapes.

     There is another very important topic not yet discussed.

This is the concept of strain range partitioning. What Manson

and his associates have found is that hysteresis loops

consisting of creep deformation at constant stress in one

direction and high strain rate in reverse, cause additional

degradation over and above what has been discussed to date. An

example is the wave shape shown in Fig. 10e. Represented in

terms of inelastic strain range vs life, they compare four wave

shapes:

(a) a rapid, strain controlled, fully reversed cycle identified

as $\Delta\varepsilon_{pp}$;

(b) a constant tensile stress creep followed by a rapid reversal

to a constant compressive stress creep while maintaining a

constant strain range, identified as $\Delta\varepsilon_{cc}$;

(c) a constant tensile stress creep followed by a rapid strain

reversal, for which the strain range is identified as $\Delta\varepsilon_{cp}$;

(d)   a constant compressive stress creep followed by a rapid strain reverse, the strain range for which is identified as $\Delta\varepsilon_{pc}$.

It develops that for the same inelastic strain and at temperature ranges where the creep effects are sufficiently rapid to perform these tests, the $\Delta\varepsilon_{pc}$ tests give shorter lives than the $\Delta\varepsilon_{pp}$, with the $\Delta\varepsilon_{cc}$ and $\Delta\varepsilon_{pc}$ intermediate. It is clear that the frequency of cycles is unaccounted for, but even correcting for the frequency, the $\Delta\varepsilon_{cp}$ form of test is more damaging. It would appear from these tests that a wave shape effect not heretofore considered must be accounted for; that is, a possible ratcheting effect. Thus in Fig. 10e all the tensile-going deformation is by a time-dependent mechanism, while all compressive-going deformation is presumably by a time-independent deformation process. If the microstructural features of these two distinct modes differ, as, for example, if all the time-dependent deformation is by grain boundary sliding and the time-independent were by deformation within the grain, internal damage processes could be expected to be greater than for the same deformation mode in both directions. For example, in crack propagation, crack advance by grain boundary sliding followed by plastic crack closure can be pictured as contributing to greater crack growth than crack advance by grain boundary sliding followed by crack closure by the same mechanism.

From what has been said above, it would appear that this wave shape effect is important under the following circumstances:

(a)   after frequency effects have been accounted for;

(b)   at temperatures and strain rates where grain boundary

sliding is an important deformation mechanism;

(c) for unbalanced loops, that is $\Delta \varepsilon_{cp}$ and $\Delta \varepsilon_{pc}$, since balanced loops, i.e. $\Delta \varepsilon_{cc}$, can be predicted by low frequency continuous cycling;

(d) for thermal cycling. Here the effect can be accentuated by changing the temperature such that high temperature occurs during the creep cycle, and low temperature during the fast reversal. This situation is shown in Fig. 10f.

How can we bring together all of the important considerations discussed above into some reasonable predictive method? There is no single answer at the present, but some of the possibilities include:

(a) extending the frequency-modified fatigue equations to include wave shape effects including thermal cycling and further refining the frequency regimes;

(b) extending the strain range partitioning approach to include elastic effects and better separate the actual material effects including environment and modes of inelastic deformation;

(c) develop approaches which consider frequency-time to failure relationships such as suggested by Skelton following possibly analogous creep-rupture models;

(d) extend high strain crack growth laws to include frequency, environment, wave shape and thermal cycling.

7. REFERENCES

Coffin, L. F. Jr. "Fatigue at High Temperature-Prediction and
Interpretation" James Clayton Memorial Lecture, Inst. Mech. Eng.
1974. Also General Electric Company Report 74CRD066.

Coffin, L. F. Jr. "Fatigue at High Temperature," Symp. on
Fatigue at Elevated Temperature 1972, A.S.T.M. S.T.P. 520, 5-34.

TABLE I

Coefficients for AISI 304 Stainless Steel

Constant	Temperature, °C		
	430	650	816
A'	529,000	226,000	68,000
n'	0.200	0.187	0.094
k_1'	-0.02	0.089	0.071
A	944,000	214,100	63,260
n'	0.486	0.258	0.105
k_1	-0.035	0.053	0.053
C_2	0.300	1.108	1.72
k	0.93	0.81	0.81
β	0.41	0.707	0.87
$E \times 10^{-6}$	23.4	21.6	18.8

204

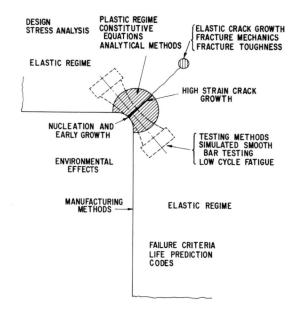

FIG. 1   Schematic view of high-temperature fatigue
         problem showing physical stages in failure
         process and relevant disciplines.

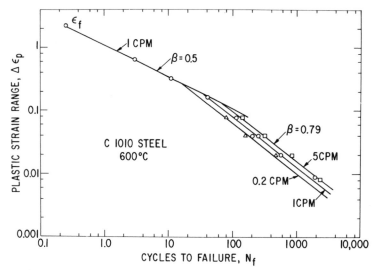

FIG. 2   Effect of frequency of cycling on plastic strain range
         vs fatigue life for AISI–C1010 carbon steel at 600°C.

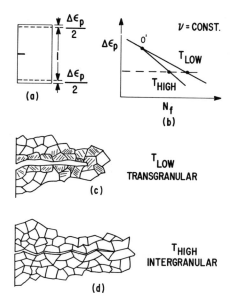

FIG. 3   Schematic representation of fracture
mode change with temperature and
plastic strain range.

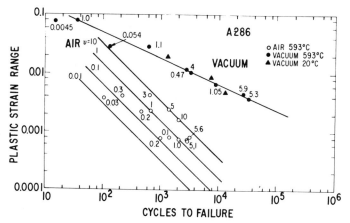

FIG. 4   Plastic strain range vs fatigue life for A286
in air and vacuum at 593°C.   Numbers adjacent
to test points indicate frequency in cycle/min.
Solid lines are regression analysis of eq. (4).

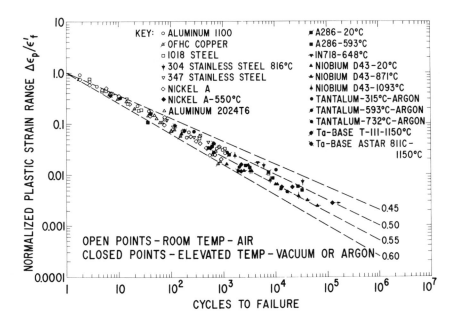

FIG. 5   Summary plot of plastic strain range vs cycles to failure for
several metals in room-temperature air or high-temperature
vacuum or argon. Plastic strain range normalized to fatigue
ductility.

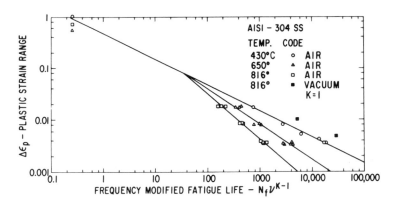

FIG. 6   Representation of data of Berling and Slot for AISI 304 stainless steel by equation (7), showing plastic strain range vs
frequency-modified fatigue life at several temperatures in air.
Vacuum data at 816°C.

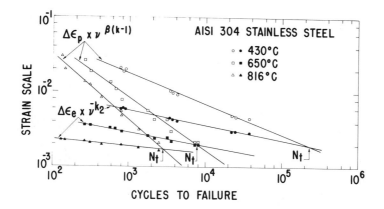

FIG. 7 Representation of data of Berling and Slot for AISI 304
stainless steel by equations (8) and (9), showing frequency-modified elastic and plastic strain range at
several temperatures in air. Note transition fatigue
life.

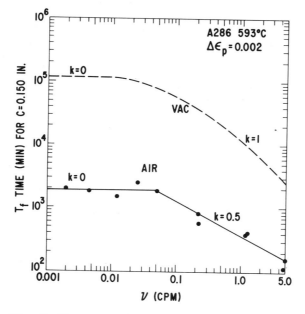

Fig. 8 Time to produce a 0.15 in. crack in
crack growth specimen vs frequency.

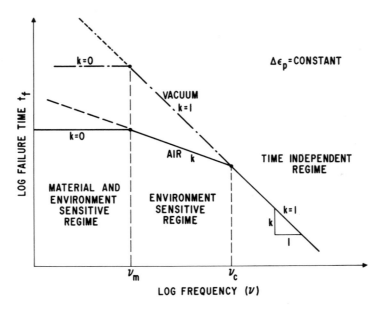

FIG. 9 Schematic representation of frequency vs time to failure at high temperature, identifying three frequency regimes.

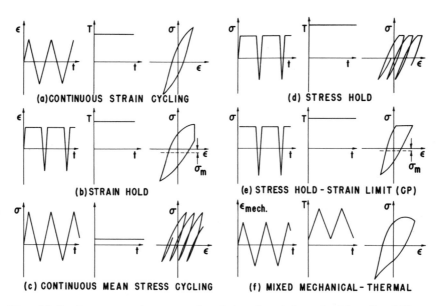

Fig. 10 Various wave shapes produced in closed loop testing for life prediction

Effects of Temperature Level and Temperature Cycling on
Crack Growth during Low Cycle Fatigue of Steels

by Shuji Taira and Motoaki Fujino

1. Introduction

Comparably large thermal strain can be induced in the structural
component of chemical plants and power generators during start-stop
period. Recent advanced projects require a more precise assesment
of the resistance of materials to thermal fatigue.

Studies by the authors have been focused on the relation between
thermal fatigue and isothermal fatigue, concerning the effect of the
synchronized changes of temperature and strain on crack initiation
and crack growth these three years.

Prior to the main theme of the paper "the effects of temperature
level and temperature cycling on crack growth during low cycle
fatigue of steels", thermal and isothermal fatigue strength tested
with smooth specimen are described.

2. Thermal and Isothermal Fatigue Strength of Steels

Thermal and isothermal fatigue tests were conducted on two
steels, 0.16% carbon steel (S15C) and 18Cr-12Ni-Nb austenite steel
(SUS 347). Both inner and outer surfaces of thin-walled cylindrical
specimen, of which gage section is 15 mm long and 1.5 mm thick, were
carefully polished to be as smooth as 0.6 μm in ten point height of
irregularity. Test machine for thermal fatigue was a servo-
controlled push-pull type with a high frequency generator together
with control-record part. Utilizing either attenuated or amplified

Professor, and assistant researcher, respectively, Department
of Mechanical Engineering, Kyoto University, Kyoto, Japan

electric signal from thermocouple wire brought a perfect synchroni-
zation of temperature and strain, even when the temperature wave
was distorted from a triangular shape. Temperature difference
along the gage section was less than ±10°C. Axial elongation and
shrinkage of the gage section was picked up by touching the end of
a pair of glass-bar of a extensometer. Specimens were regarded as
failed when the stress response dropped to three- quarters of the
maximum value (Ref. 1).

Relations between the plastic strain range $\Delta\varepsilon_p$ and the number
of cycles to failure N_f are shown in Figs. 1 and 2, where experimental data points are seperated into five groups according to their
temperature conditions. In the Fig. 1, mean temperature and the
maximum cycle temperature rises by steps of 100°C in the order of
○◑●, and the cycle temperature range becomes larger by steps of
100°C in the order of ◑▲◪. The thermal fatigue life became shorter
as the maximum cycle temperature rose, or as the cycle temperature
range became larger.

Isothermal fatigue data are also shown with arrows and thin
lines. Thermal fatigue lives are fairly shorter than isothermal ones
in the case that the maximum cycle temperature is high and the cycle
temperature range is large. This tendency is developed at lower
strain range in the carbon steel (Ref. 2).

Partially fatigued specimens were cut along the axis and
polished. The newly created surfaces containing cracks were observed
through an optical microscope. A geometrical mean value of the total
crack depth and the deepest crack depth, effective crack depth d*,
represents the feature of cracks. The change of d* during the fatigue
tests is shown in Fig. 3. The slope of the line in $\log(d*/2R)$ -
$\log(N)$ plots for the isothermal fatigue test at room temperature is

steep. This means that cracks initiated in the later stage of life and propagated rather rapidly. Cracks at 500°C initiated in the early stage of life and propagated rather slowly. Cracks in the thermal fatigue test (200-500°C) behaved more radically that those in isothermal fatigue at 500°C (Refs. 3, 4 and 5).

3. Crack Propagation during Thermal and Isothermal Fatigue in a Notched Specimen

Test machine and materials are same to those described in Chapter 2. Specimen used in crack propagation test is such that a drilled hole of 1.0 mm diameter is at the mid-position of the gage section of the thin-walled cylindrical specimen. Distance between crack tips measured through microscope of 17x was corrected to true crack length $2l$ by taking curvature into consideration. Crack is defined as the alternatively opening and closing region which begins from a notch hole (Ref. 6).

Isothermal fatigue tests with notched specimens were performed at six temperature levels from R.T. to 600°C, where 1% and 5 cpm were chosen to be the basic condition of strain range and frequency respectively. Other test conditions of strain range from 0.3% to 1.5%, of frequencies from 0.1 cpm to 50cpm. were added suitably.

In thermal fatigue tests, the twelve test conditions were arranged from a set of two kinds of phase-relation, three mean temperatures, and two strain ranges. The two phase-relations were in-phase and out-of-phase. Under the in-phase condition, tensile strain was induced during the heating half-cycle. The three temperature conditions are 100-300°C, 300-500°C and 400-600°C. Through the studies on isothermal fatigue of the low carbon steel, it is found that fatigue failure process above 350°C is different from one below 350°C. The second temperature condition (300-500°C) corresponds to the case where temperature cycle spanns two characteristic temperature regions.

Logarithmic half crack length log l vs. number of cycles N at six temperature levels are plotted in Fig. 4. Half crack length l is equal to the mean distance from the center of the notch hole to crack tip measured over the specimen surface. As has been reported by H.D. Solomon, logarithmic half crack length increases linearly with number of cycles (Ref. 7). The characteristics of the figure, the steepness of the slope and the starting position of the plots, give us two kinds of information as follows: -(i) the gradient of plots devided by 2.3, "logarithmic rate of crack propagation (1/l) (dl/dN)", which is the representative of test condition, usually independent of crack length. (ii) Starting position of the plots represented by the number of cycles when the size of induced crack becomes 0.1 mm in length (that corresponds to 0.6 mm in half-crack-length), which is defined now "the number of cycles to crack in-itiation $N_{C.I.}$".

As shown in Fig. 4, the logarithmic rates of crack propagation obtained by the slopes of the plots depend complexly on temperature level, as they are largest in the middle temperature range of 300°C and 400°C, smallest in higher temperature range of 500°C and 600°C. The numbers of cycles to crack initiation decrease monotonously, as the temperature rises.

Logarithmic half crack length vs. number of cycles during thermal fatigue under three temperature conditions are plotted in Fig. 5 (Ref. 8). Under the out-of-phase condition (compressive strain is induced during heating half cycle), the slopes become steeper in order of 400-600°C, 100-300°C, 300°-500°C. Under the in-phase condition, slopes become steeper in order of 100-300°C, 300-500°C, 400-600°C.

It can be said that the rate of crack propagation becomes larger as the test temperature (especially maximum cycle temperature T_2) rises under the in-phase condition. On the other hand, it becomes rather small as the test temperature (especially minimum cycle temperature T_1) rises into higher temperature range under the out-of-phase condition.

Studies on crack propagation during isothermal fatigue have confirmed the relation between strain range and the rate of crack propagation as shown in Eq. (1). Assuming that the relation in Eq. (1) holds in thermal fatigue, such exponential values as α and α_p can be derived. α in isothermal fatigue decreases from 2.0 to 1.5 as the temperature rises from 300°C to 500°C, α in thermal fatigue also decreases from 2.4 to 1.4 as the maximum cycle temperature rises from 500°C to 600°C.

$$\frac{1}{l} \frac{dl}{dN} = K \ (T, \nu) \cdot \Delta \varepsilon^{\alpha} = K_p (T, \ \nu) \cdot \Delta \varepsilon_p^{\alpha_p} \ . \tag{1}$$

Temperature dependence of the rate of crack propagation during thermal and isothermal fatigue tests are summarized in Fig. 6. Under the out-of-phase condition, the rate remains at a high value as the maximum cycle temperature T_2 rises up to 500°C, then drops at above 500°C. On the other hand, it increases as the maximum cycle temperature T_2 under the in-phase condition. Compared with isothermal fatigue crack, thermal fatigue crack propagates at the equal to the maximum of isothermal fatigue crack propagation rates in the temperatures between T_1 and T_2.

Temperature dependence of the numbers of cycles to crack initiation in thermal fatigue tests are compared with those in isothermal fatigue tests in Fig. 7. Thermal fatigue crack initiates earlier at higher maximum cycle temperature. Temperature dependence

of crack initiation period in thermal fatigue is the same as that
in isothermal fatigue at temperature levels equal to the maximum
cycle temperature of thermal fatigue. The ratio, $N_{C.I.}(T_1-T_2)/$
$N_{C.I.}(T_2)$ is nearly 0.8.

4. Analysis of Temperature Dependece of the Rate of Crack Propagation

V. Weiss proposed that increase in crack length in a strain
cycle is equal to the width of the region of strain distribution
curve at crack tip where induced strain is larger than fracture
strain ε_f (Ref. 9).   Strain distribution and the resultant rate of
crack propagation are given in Eqs. (2) and (3).

$$\Delta\varepsilon^*(x) = (\frac{l}{x-l+\rho/4})^\delta \cdot \Delta\varepsilon, \quad \delta = \frac{1}{n+1}, \tag{2}$$

$$\frac{dl}{dN} = \frac{1}{2} (\frac{\Delta\varepsilon}{\varepsilon_f})^{n+1} l, \tag{3}$$

where x is distance from the center of notch hole, ρ; radius of
notch tip, n; strain hardening coefficient. As this theoretical
equation (3) has a good coincidence with experimental equation (1)
both in the indices of strain range and crack legnth, regardless of
the numerical coefficient, the analysis found on the principles
given by Eqs. (2) and (3) (Refs. 7 and 9).

V. Weiss made no more comment on fracture strain ε_f than to be
a materials constants. Tensile ductility ε_{f_o} ($ln\{1/(1-R.A.)\}$) may
be used as a fracture strain ε_f when structure changes do not occur.
Both the logarithmic rate of crack propagation $(1/l)(dl/dN)$ and
tensil ductility ε_{f_o} are shown at various temperature levels in
Fig. 8.   The rate-changes with temperature show a reverse inclination
to those the ductility-changes. On the other hand, the rate-change
with strain-rate shows a same inclination to that of the ductility-
change. These contraversy may be brought about by the discrepancy

between the vulky fracture strain ε_{f_o} and the local fracture strain ε_f* as will be discussed in the following section.

4.1 Effect of sub-microcracks on the rate of crack propagation

Photo 1 shows micrographs near the fatigue crack tip grown to about 2 mm. In both specimens fatigued at 300°C and thermal fatigued of 100-300°C, only surface detriorations are found. On the other hand, both in isothermal and thermal fatigued specimens tested above 400°C, many sub-microcracks are found. Fig. 9 shows the effect of precycling on crack propagation rate. The crack propagates more rapidly as more strain cycling have been applied prior to making a notch hole on the smooth thin-walled cylindrical specimen. This acceleration of the propagation rate may results from the reduction of ductility near the crack tip.

Fracture strain at the crack tip ε_f* can be given by the modified tensile ductility, taking subcracks into consideration.

$$\varepsilon_f* = \varepsilon_{f_o} \cdot f \ (d*) \simeq \varepsilon_{f_o} \cdot (1 - B \ \frac{d*^2}{2\pi Rt})^X \qquad (4)$$

where $d*$ is effective crack depth, R; specimen radius, t; specimen thickness, B, X; constants. Substituting Eq. (4) into Eq. (3), the rate of crack propagation surrounded by subcracks is obtained

$$\frac{dl}{dN} \simeq K_o \ (\frac{\Delta\varepsilon}{\varepsilon_{f_o}})^\alpha \ (1 + \frac{BXd*^2\alpha}{2\pi Rt}) \ l \ = \frac{dl}{dN}\bigg|_{d*=0} \ (1+Ad*^2) \qquad (5)$$

where K_o is determined by the experiment at R.T. Quantitative treatment of Eq. (5) starts from the evaluation of microcrack formation in the next section.

4.2 Evaluation of microcracks formed at the main crack tip

As having described in chapter 2, microscopic observation of cracks in partially fatigued specimen gave the sum total area of micro cracks contained in a unit gage section $d*^2$. Empirical equation of the rate of crack formation is given below.

$$d*^2/N = C (T, \nu) \cdot \Delta\varepsilon^\gamma = C_o \exp [- \frac{Q}{T}] \cdot \nu^{-m} \cdot \Delta\varepsilon^\gamma ,$$

$$C_o = 4.2 \times 10^{-3} \text{ mm}^2\text{cycle}^{-1}, \quad m=0.5, \quad \gamma=3, \quad Q=2820. \tag{6}$$

Eq. (6) does not express the rate of crack formation in a notched specimen, but that in a smooth specimen. In the notched specimen, the rate of crack formation at the point apart x from the main crack tip depends on both intensified strain range $\varepsilon^*(x)$ and the cycles from the beginning of the test to the time when the main crack reaches to the point. The relation between microcrack increment, $\Delta(d*^2)$, and the strain cycle increment, ΔN, is formulated as,

$$\Delta(d*^2) = C \cdot \Delta\varepsilon*^3 \cdot \Delta N = C \cdot \Delta\varepsilon*^3 \cdot \Delta l (dl/dN)^{-1} \tag{7}$$

Substituting Eq. (5) into Eq. (7), then integrating with l, the amount of microcracks is given in Eq. (8) (Ref. 6).

$$d*^2 = (\frac{C}{K_o} \cdot \varepsilon_{f_o}^2) \cdot \Delta\varepsilon \cdot F \tag{8}$$

Accumulation factor, F, increases with strain cycling, then takes its satulated value $\sqrt{12l/\rho}$. In such a condition as the authors adopted in the test, F takes values from 10 to 50. Notch radius can be further assumed to be proportional to l, $\Delta\varepsilon$, ε_{f_o}. Morphology of cracks gave a rough relation as $F \simeq \sqrt{12/\Delta\varepsilon\ \varepsilon_{f_o}} \times 10^2$.

Unknown constants χ and B are determined to be 2 and 100 respectively, so as to fit the calculated rate of crack propagation to observed ones shown in Fig. 9. Both local fracture strain ε^*_f and the rate of crack propagation are calculated using Eqs. (4) and (5). Applicability of the analysis to austenite stainless steels (SUS 304, SUS 347) are good as shown in Fig. 10.

4.3 Application of the analysis to thermal fatigue

If the rate of microcrack formation during thermal fatigue is same to that during isothermal fatigue tested at the temperature

equal to the maximum cycle temperature of thermal fatigue, this
analysis easily calculates the rate of thermal fatigue crack propagation, taking the tensile ductility at the temperature where the
tensile strain takes its maximum value; either the minimum cycle
temperature T_1 or the maximum cycle temperature T_2 according to the
phase relation, out-of-phase or in-phase.

It was obtained by microscopic observation of partially fatigued
smooth specimens that the rate of thermal fatigue crack formation is
slightly larger than that of isothermal fatigue tested at T_2.
Furthermore, thermal fatigue crack formation under the in-phase
condition is much more active than out-of-phase condition both in
quantity and quality at higher temperature and larger strain range.
As shown in photo 1, many small cracks (30 μm - 50 μm) are observed
in the specimen tested under the in-phase, 400-600°C condition. The
specimen seems to be very brittle. Crosssectional observation shows
homogeneously distributed voids in the specimen (Photo. 2) (Ref. 10).

Further study on the temperature-phase-dependence of microcrack
formation and the effect of both shape and arrangement of microcracks
will make it possible to analize thermal fatigue crack propagation
quantitatively.

5. Conclusion

Thermal fatigue and isothermal fatigue tests were conducted on
two steels, 0.16% carbon steel and 18Cr-12Ni-Nb austenite steel.
Some of the tests were interrupted at predetermined lapses of time
before the failure for the microscopic observation of cracks. Propagation of crack, which started from notch hole, during thermal and
isothermal fatigue were also examined. The following conclusions
were obtained.

(1) The increase in cycle temperature range as well as temperature level

reduces thermal fatigue life. Combination of the crack initiation dominant at higher temperatures and the crack propagation dominant at lower temperatures will make thermal fatigue strength weaker than the isothermal fatigue strength tested at the temperature equal to the maximum cycle temperature of thermal fatigue T_2.

(2) Thermal fatigue crack propagates at either the same or higher rate, compared with the isothermal fatigue crack tested at T_2.

(3) Under the in-phase condition, thermal fatigue crack can propagate at higher rate, compared with that under out-of-phase condition because of void formation.

(4) Crack propagation rates at elevated temperatures were formulated by the local fracture strain, taking micro-cracks into consideration.

References

1) Taira, S., Fujino, M., and Haji, T., "Testing Machine for Thermal Fatigue with Variable Constraint Ratio", J. Soc. Materials Science, Japan, Vol. 22, 110 (1973)

2) Taira, S., Fujino, M., and Haji, T., "A Method for Life Prediction of Thermal Fatigue by Isothermal Fatigue Testing", Proc. 1973 Symposium on Mechanical Behavior of Materials, 257 (1974)

3) Taira, S., Fujino, M., and Tamai, T., "Distinctive Features of Thermal Fatigue Cracks", Preprints of 22th Annual Meeting of JSM, 99 (1973)

4) Taira, S., Fujino, M., Sakon, T., and Matsuda, T., "Distinctive Features of Fatigue Cracks. Application to Estimating Life for Two-Step Temperature Test", J. Soc. Materials Science, Japan, Vol. 22, 242 (1973)

5) Taira, S., Fujino, M., Sukekawa, M., and Sasaki, R., "Temperature Dependence of Low Cycle Fatigue Strength and Distrinctive Feature of Fatigue Crack of SUS347", J. Soc. Materials Science, Japan, Vol. 23, 208 (1974)

6) Taira, S., Fujino, M., Sakon, T., and Maruyama, S., "Effect of Temperature Levels and Straining Frequency on the Rate of Crack Propagation in a Low Carbon Steel during Low Cycle Fatigue", J. Soc. Materials Science, Japan, to be published

7) Solomon, H.D., "Low Cycle Fatigue Crack Propagation in 1018 Steel", G.E. Technical Information Series, 71-C-327 (1971)

8) Taira, S., Fujino, M., and Maruyama, S., "Effects of Temperature and the Phase between Temperature and Strain on Crack Propagation in a Low Carbon Steel", Preprints of 1974 Symposium on Mechanical Behavior of Materials (1974) Kyoto

9) Weiss, V., "Treates on Fracture", 3, 227 (1971) Academic Press, New York

10) Taira, S., Fujino, M., and Katayama, Y., "Thermal Fatigue Fracture under the In-Phase Condition", Preprints of Annual Meeting of JMS, 45 (1974)

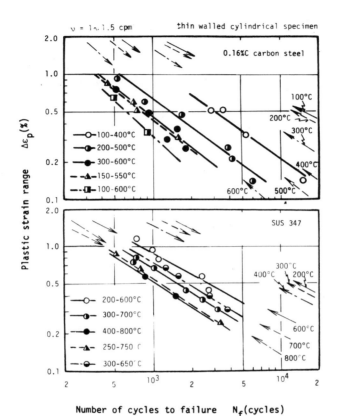

Fig.1   Relation between plastic strain range and the number of cycles
        to failure (low carbon steel).

Fig.2   „   (SUS347).

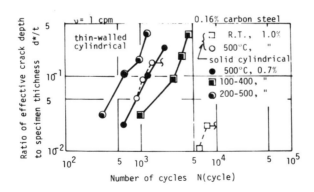

Fig.3   Effect of temperature on crack growth during thermal and isothermal
        fatigue.

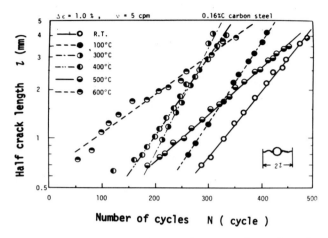

Fig.4　Isothermal fatigue crack length vs. number of cycle plots.

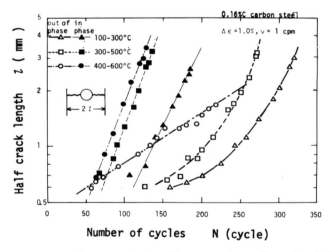

Fig.5　Thermal fatigue crack length vs. number of cycle plots.

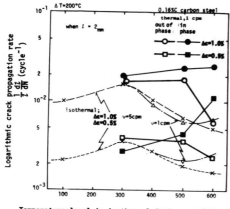

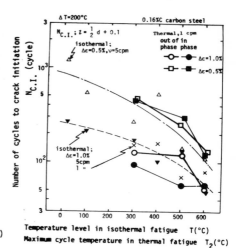

Fig.6 Temperature dependence of the rate of thermal and isothermal fatigue crack propagation.

Fig.7 Temperature dependence of thermal and isothermal fatigue crack initiation period.

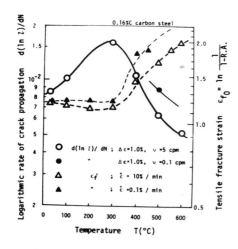

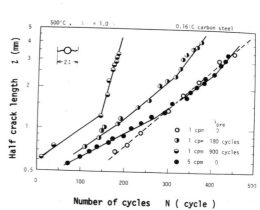

Fig.8 Temperature dependence of logarithmic rate of crack propagation and ductility.

Fig.9 Effect of precycling on crack propagation rate at 500°C.

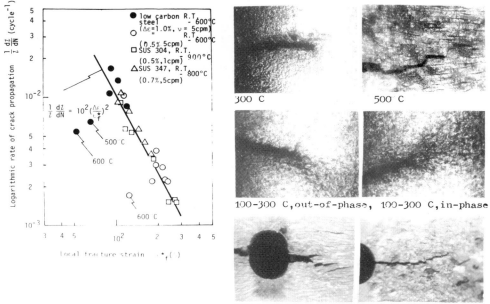

Fig.10 Relation between logarithmic
rate of crack propagation and
the local fracture strain.

300 C   500 C

100-300 C,out-of-phase, 100-300 C,in-phase

400-600 C,out-of-phase, 400-600 C,in-phase

Photo.1 Micro-cracks in the vicinity of
main crack tip ($\Delta\varepsilon=1\%$, $\dot{\nu}=1$cpm).

The graph labels:

Logarithmic rate of crack propagation $\frac{1}{Z}\frac{dZ}{dN}$ (cycle^{-1})

● low carbon R.T.
  steel   600°C
  ($\Delta\varepsilon=1.0\%$, $\nu=5$cpm)
○ "   R.T.
  (0.5% 5cpm)   600°C
□ SUS 304, R.T. 900°C
  (0.5%,1cpm)
△ SUS 347, R.T. 800°C
  (0.7%,5cpm)

$$\frac{1}{Z}\frac{dZ}{dN}=10^2(\frac{\Delta\varepsilon}{\varepsilon_f^*})^2$$

500 C

600 C

600 C

Local fracture strain $\varepsilon_f^*(\ )$

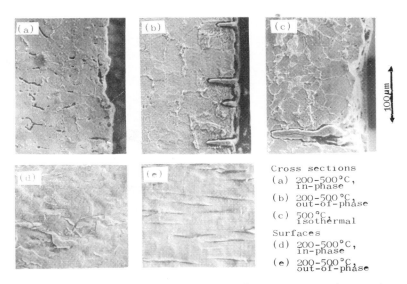

(a)   (b)   (c)

(d)   (e)

100μm

Cross sections
(a) 200-500°C,
    in-phase
(b) 200-500°C,
    out-of-phase
(c) 500°C,
    isothermal
Surfaces
(d) 200-500°C,
    in-phase
(e) 200-500°C,
    out-of-phase

Photo.2 cross sectional views of thermal fatigued specimen ($\Delta\varepsilon=0.7\%$, $\dot{\nu}=1$ cpm).

Fractographic studies on fracture
of metallic materials

by Ryoichi Koterazawa*

It is well recognized that fractographic observation of
fracture surfaces gives some indispensable informations on the
fracture processes in the interior of bulk materials.   This
report presents fractographic data on ·the following subjects on
which relatively few informations are available: (1) quantitative striation analysis of the effects of stress change on fatigue crack propagation, (2) crystallographic mechanism of fatigue crack propagation, (3) fractography of creep-fatigue interaction at elevated temperature.

1. Effects of stress change on fatigue crack propagation

It is known that a maximum stress decrease during fatigue
crack propagation process causes a retardation of crack propagation in the successive stress cycles and a maximum stress increase causes an acceleration in the stress cycle in which maximum stress is increased[1]-[3] It has been suggested that these
effects could be related to the crack closure phenomenon which
was pointed out recently by Elber.[3][4] The aim of this work
was to study these effects quantitatively by means of striation
analysis and to give them some quantitative interpretations.

The test material was an alluminum alloy 2017-T4 and the
specimens were circumferentially notched round bars (notched
section diameter 6 mm, notch depth 0.5 mm and notch root radius

* Dr., Associate Professor, Faculty of Engineering Science,
  Osaka University, Toyonaka, Osaka, JAPAN

0.05 mm) tested under repeated tension loading with an Instron type machine. The fracture surfaces were examined by means of plastic-carbon replication technique with a transmission electron microscope (Akashi Model TRS-50EI) and by a scanning electron microscope (Hitachi-Akashi Model MSM-2).

Figure 1 shows examples of crack propagation curves for the cases where the maximum stress is deceased as illustrated in the figure. Appreciable retardation is observed for some crack length following the point of maximum stress decrease (arrows).

In order to see how the retardation can be interpreted in terms of the crack closure, the effective stress range $\Delta\sigma_{eff}$ was calculated by Elber's equation,[4]

$$\Delta\sigma_{eff} = (0.5 + 0.4R)\Delta\sigma , \tag{1}$$

where $\Delta\sigma$: applied stress range, $R = \sigma_{min}/\sigma_{max}$: stress ratio, and shown in Fig. 1. When the maximum stress is decreased from σ_3 to σ_2, crack closure level will not decrease to the steady state level for σ_2 immediately but stay at the level for σ_3 for some time, so that the effective stress range will decrease from $\Delta\sigma_{eff}^h$ for σ_3 to a small value $\Delta\sigma_{eff}^{h-1}$ instead of steady state value $\Delta\sigma_{eff}^1$ for σ_2, resulting in retardation. The experimental results show that the crack propagation curves following the point of maximum stress decrease lie close to the curves for steady stress (dotted lines) of the same effective value $\Delta\sigma_{eff}^{h-1}$, indicating validity of the above reasoning. The distance within which the prediction is valid is about 100 microns, which is approximately equal to the size of the region of the highest hardness[3] where the largest compressive residual stress, and therefore the largest crack closure, is expected.

Figure 2 shows examples of crack propagation curves where

the maximum stress is increased as indicated in the figure. Remarkable acceleration is observed at the point of maximum stress increase (arrows). In the same way as the case of maximum stress decrease, the effective stress at the point of maximum stress increase (from σ_2 to σ_3 as shown in Fig. 2) is $\Delta\sigma_{eff}^{1-h}$ which is larger than the steady state value of effective stress $\Delta\sigma_{eff}^{h}$ for σ_3. In this case, however, the actual propagation of crack is much larger than the one predicted by $\Delta\sigma_{eff}^{1-h}$. A cause for this descrepancy would be the effect of monotonic loading which appear at the point of maximum stress increase.[6][7] That is, under the steady cyclic stress conditions, the determining factors of crack propagation rate are the range of stress $\Delta\sigma$ and the effective yield stress $2\sigma_y$, σ_y being yield stress under monotonic loading. At the point of the maximum stress increase such as the stroke 0-1-2-3 in Fig. 3, the condition in the stroke 2-3 is the same as the monotonic case, the determining factors being the maximum stress and the effective yield stress σ_y instead of $2\sigma_y$. These effects increase the crack propagation rate over the one predicted by the effective stress range $\Delta\sigma_{eff}^{1-h}$ based on the crack closure effect. Considering crack opening displacement (COD) for these cases, COD for the steady value of $\Delta\sigma_{eff}^{1-h}$ is proportional to $(\Delta K_{13})^2/2\sigma_y E$, whereas COD for the stroke 0-1-2-3 is sum of CODs for the strokes 1-2 and 2-3 which are proportional to $(\Delta K_{12})^2/2\sigma_y E$ and to $(K_3^2/\sigma_y E - K_2^2/\sigma_y E)$, respectively. Therefore, the multiplication factor for COD due to the monotonic loading effect would be

$$r = \frac{(\Delta K_{12})^2/2\sigma_y E + K_3^2/\sigma_y E - K_2^2/\sigma_y E}{(\Delta K_{13})^2/2\sigma_y E} \quad . \tag{2}$$

Since fatigue crack propagation is considered to be directly related to COD, the factor r may be regarded also as the multiplication factor for crack propagation rate. That is, crack propagation rate would be r times as large as the value predicted by ΔC_{eff}^{1-h}.

The above calculation is for the case of small scale yielding, and it is difficult to calculate for the case of large scale yielding. In order to get rough estimate for this case, the equation for COD in infinite plate with transverse crack under tension[8] was employed here, giving

$$r = \frac{2\ln(\sec(\pi\Delta\sigma_{12}/2\cdot2\sigma_y)) + \ln(\sec(\pi\sigma_3/2\sigma_y)) - \ln(\sec(\pi\sigma_2/2\sigma_y))}{2\ln(\sec(\pi\Delta\sigma_{13}/2\cdot2\sigma_y))} \quad \text{------ (3)}$$

Ratios of striation spacing W_{th} calculated by Eq. (3) and experimental spacing W is shown in Fig. 2, and the values of W and W_{th} are plotted against magnitude of maximum stress in Fig. 4. Fairly good agreement is obtained. The yield stress of the heavily deformed material near the crack tip is unknown, and the value of $\sigma_y = 52.5$ kg/mm^2 was used in the calculation. This value is not unreasonable considering that the tensile strength of the notched specimen was 56.4 kg/mm^2.

A fractographic evidence of crack closure was found on the fracture surface in stress change test. An example of fractograph is shown in Photo.1. The stress range was decreased while the maximum stress was maintained constant as shown in Fig. 5. Minimum stress after the stress range decrease was a little below the crack closure level given by Eq. (1) so that the crack will close only a little in this stress cycle (1-2-3). As expected, the space between striations corresponding to this cycle is light indicating little closure, whereas those for other cycles are dark indicating crack closure in these cycles.

2. Crystallographic mechanism of fatigue crack propagation

It has been shown for alluminum alloys that fatigue crack on (001) plane with [110] crack propagation direction and (111),(11$\bar{1}$) operating slip planes is predominant in the range of low crack propagation rate whereas little crystallographic effect is seen in the range of high crack rate.[9][10] This work was carried out to get further informations on this aspect of fatigue crack propagation with special attention being given to the relation between fracture surface topography and crystallographic orientation.

The test material was a 2017-T4 alluminum alloy, fatigue tests being conducted by an Ono-type rotating beam fatigue machine with sharp notched round bar specimens (notched section diameter 12 mm, notch depth 1.5 mm, notch root radius 0.1 mm). Crystallographic orientations were determined by etch pit technique, the etchant being 50 water, 50 HNO_3, 30 HCl, 2 HF and the etching time about 2 seconds. The etchant attacks preferentially (111) planes. The material used in this study was found to have preferred orientations, with about 40 percent of cross section being close to (100) planes and about 60 percent close to (111).

In the range of high crack propagation rate (~ 0.5 micron per cycle), there was moderate crystallographic effect, fracture surfaces close to (100) plane being about 30 percent, those close to (111) plane being about 30 percent and remaining 40 percent having other orientations. On the (100) fracture surfaces, the striations tended to be parallel to [110] direction and striations on the fracture surfaces other than (100) was not as well defined as those on the (100) ones .

In the range of low crack propagation rate (lower than 0.05 micron per cycle), there was a strong orientation dependence, about 70 percent of fracture surfaces being close to (100) plane. Examples of fractographs in this range are shown in Photo.2. Appearance of fracture surfaces are brittle, river- or trough-like markings being predominant. Especially those in the area of (111) texture are noteworthy. As shown in Photo.2 and schematically in Fig. 6, these river- or trough-like markings are made up of narrow facets of (100) orientation, yielding greater percent of (100) fracture surface. Although such fracture surfaces look brittle in low magnifications, the high magnification transmission electron fractograph, Photo.3, shows that there are ductile striations, quasi-striation patterns or slip lines on the fracture surfaces, suggesting that the fracture mode is not cleavage type but ductile in fine scale and the crack propagation mechanism would not be greatly different from that in the range of high crack propagation rate.

3. Fractography of creep-fatigue interaction at elevated temperature

At elevated temperatures, fracture of materials under cyclic stress is influenced by creep as well as fatigue, creep predominating in the case of longer tension hold time.[11] This work was planned in order to get some fractographic informations on the mechanism of the interaction of creep and fatigue in such cases. Experiments were conducted with a 304 stainless steel under square wave repeated tensile stress of various periods. Test specimens were again notched round bars (notched section diameter 6 mm, notch depth 1 mm and notch root radius 0.05 mm) tested with conventional lever type creep testing machines.

Test temperature was 650°C and periods of repeated stress were from 25 seconds to 30 minutes.

As expected, intergranular creep fracture tended to appear under repeated stress of longer periods, where tension hold time was longer. It was also found that the intergranular fracture tended to appear in later stage of crack propagation. That is, the crack generally started as transgranular fatigue crack and changed to intergranular creep fracture at some length of the crack. An example of fractograph is shown in Photo.4. The condition of the transition was found to be

$$2\Delta t \sigma_{net,max}^{n} = C \quad (n = 5.0, \ C = 1.1 \times 10^{8}), \qquad (4)$$

where $2\Delta t$: period of repeated stress (minutes), $\sigma_{net,max}$: maximum value of net section stress σ_{net} (kg/mm^2) (Fig. 7).

There was a power function relation between net section stress and propagation rate of intergranular creep crack under constant stress,

$$(dl/dt)_c - A_c \sigma_{net}^{nc}, \qquad (mm/hr) \qquad (5)$$
$$(A_c = 8.8 \times 10^{-15}, \ n_c = 8.4, \ Fig. \ 8).$$

For the transgranular fatigue crack under repeated stress, propagation rate could be expressed as

$$(dl/dn)_f = A_f \nu^{-1/m} \Delta\sigma_{net}^{nf}, \qquad (mm/cycle) \qquad (6)$$
$$(A_f = 2.3 \times 10^{-8}, \ n_f = 2.9, \ m = 6, \ )$$

where ν is frequency of stress cycle = $1/2\Delta t$ (in cpm). (Fig. 8) If we assume that these two crack propagation processes are possible under repeated stress conditions and that the one with higher rate actually occurs, the transition would occur when propagation rate of creep crack equals that of fatigue crack. The propagation rate of intergranular creep crack in this case is given by Eq. (5) where t is net time under stress, and that of transgranular fatigue crack, Eq. (6), can be expressed in terms of net time as

$$(dl/dt)_f = 2 \times 60 \, \nu A_f \dot{\gamma}^{-1/m}(1 - R)^{n_f} \sigma_{net,max}^{n_f} \qquad (7)$$

where $R = \dfrac{\sigma_{net,min}}{\sigma_{net,max}}$, stress ratio. Equating Eq.s. (5) and (7), we get an equation of the same form as Eq. (4), where

$$n = (m/m-1)(n_c - n_f), \quad C = (120A_f/A_c(1-R)^{n_f})^{m/m-1} \qquad (8)$$

The calculated line is shown in Fig. 7 as dotted line which is fairly close to the experimental points.

Acknowledgments

The author wishes to thank Messrs. D.Shimo, T.Honjo, Y.Iwata and other colleagues at Faculty of Engineering Science, Osaka University for their cooperations in conducting these studies.

References

(1) McMillan, J.C. and Pelloux, R.M.N., ASTM STP-415, 1967,p.205

(2) McMillan, J.C. and Hertzberg, R.W., ASTM STP-436,1968,p.89

(3) Koterazawa, R., Mori, M., Matsui, T., and Shimo, D., Trans. ASME Ser. H, Vol. 95, 1973, p. 202

(4) Elber, W., ASTM STP-486, 1971, p. 230

(5) von Euw, E.F.J., Hertzberg, R.W. and Roberts, R., ASTM STP-513, 1972, p. 230

(6) Rice, J.R., ASTM STP-415, 1967, p. 247

(7) McMillan, J.C. and Pelloux, R.M., Engineering Fracture Mechanics, Vol. 2, 1970, p. 81

(8) Goodier, J.N. and Field, F.A., Fracture of Solids, Interscience, 1963, p. 103

(9) Pelloux, R.M.N., Trans. ASM, Vol. 62, 1969, p. 281

(10) Bowles,C.Q. and Broek,D., Int.J.Fract.Mech.,Vol.8,1972,p.75

(11) Cheng,C.Y. and Diercks,D.R.,Metal. Tr. Vol.4,1973,p.615

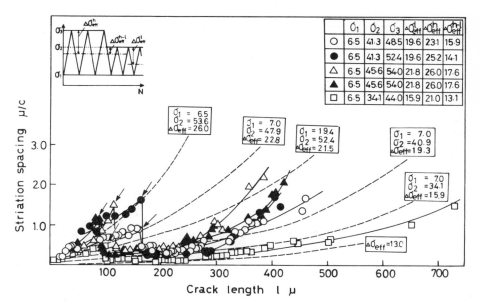

Fig. 1 Crack propagation curves for the case of maximum stress decrease

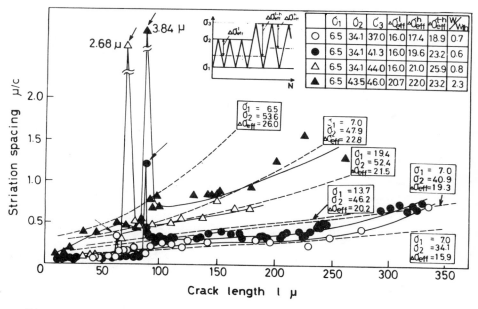

Fig. 2 Crack propagation curves for the case of maximum stress increase

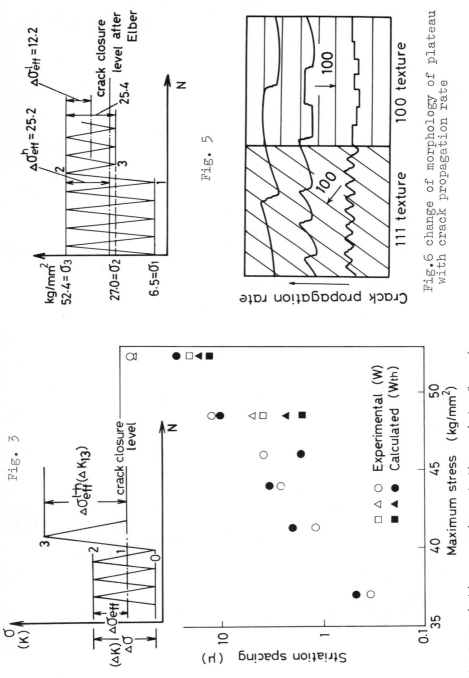

Fig. 3

Fig. 5

Fig.6 change of morphology of plateau
with crack propagation rate

Fig.4 Striation spacing at the point of maximum
stress increase plotted against maximum stress

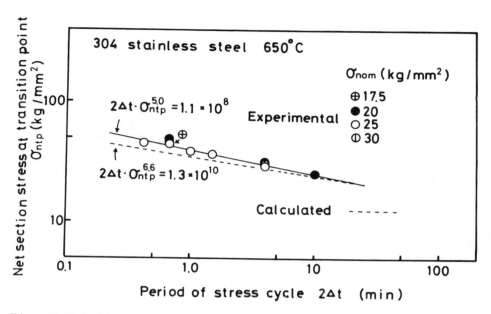

Fig. 7 Relation between net section stress at the transition point and period of stress cycle

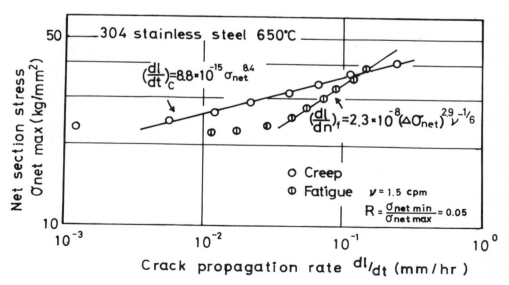

Fig. 8 Relation between net section stress and crack propagation rate

234

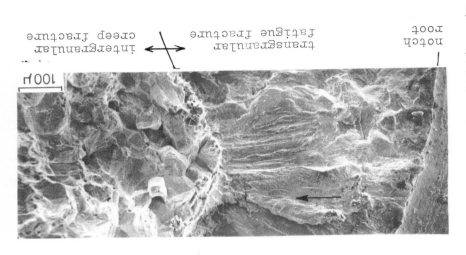

notch
root

transgranular
fatigue fracture

intergranular
creep fracture

Photo. 4 An example of fatigue to creep
transition of fracture

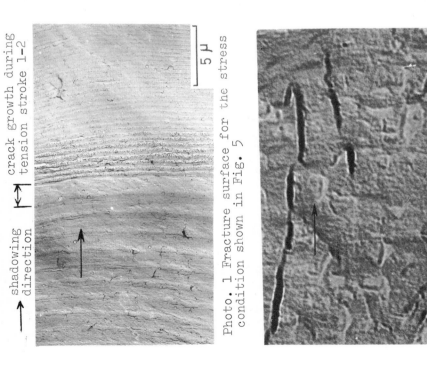

shadowing
direction

crack growth during
tension stroke 1-2

100 μ

5 μ

Photo. 1 Fracture surface for the stress
condition shown in Fig. 5

0.5 μ

Photo. 3 Striations and slip lines in the
range of low crack propagation rate

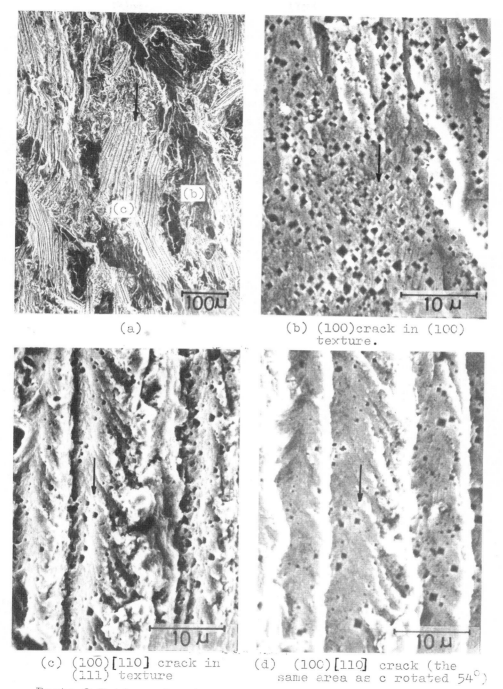

(a)

(b) (100)crack in (100)
texture.

(c) (10$\bar{0}$)[110] crack in
(111) texture

(d) (10C)[110] crack (the
same area as c rotated 54°)

Photo.2 Fatigue fracture surface in the range of low crack
propagation rate.

Evaluation of Thermal Fatigue Strength

for Cast Iron Materials

by   Shigeru Yamamoto* Hidekazu Yamanouchi*

1. Introduction

Cast iron materials are used, in many cases, for the machine component parts whose thermal fatigue strength gives rise to a problem. The cast iron materials, however, are complicated in characteristics and there are not sufficient data about the thermal fatigue strength.

According to literature[1], diesel engine cylinder covers which were exactly the same in shape but different in material, namely, gray cast iron and ferritic ductile cast iron, were manufactured for an actual diesel engine, and were subjected to a test for durability until the initiation of thermal fatigue cracking. Consequently, the cylinder cover of gray cast iron, contrary to expectation, proved more durable. This shows that in order to use cast iron materials for diesel engine cylinder covers, it is necessary to take into consideration not only the results of materials fatigue tests but the total physical properties such as the coefficient of thermal conductivity, the modulus of elasticity, the coefficient of linear thermal expansion, etc.

For that purpose, studies were made of a method of selecting materials against thermal fatigue strength by deriving the thermal fatigue parameter L_{th}. And the low cycle fatigue test with the use of test specimen and the gas heating type thermal fatigue test with the use of test models were conducted on several kinds of cast iron materials, and consequently, it was confirmed that L_{th} could be used for actual materials to evaluate thermal fatigue strength.

---

* Nagasaki Technical Institute of Mitsubishi Heavy Industries, Ltd., Japan

238

2. Thermal fatigue parameter [2) 3)]

The low cycle fatigue strength of materials can be expressed by the following equation (1).

$$Sa = F(N) \tag{1}$$

where, Sa : equivalent elastic stress (kg/mm^2)

N : cycles to failure

$F(N)$: experimental data about low cycle fatigue strength

For qualitative example, the thermal stress σth of a circular plate whose periphery is fixed can be represented by the following equation (2), in the case of the temperature gradient existing in the direction of the thickness.

$$\sigma th = \frac{E\alpha \Delta T}{2(1-\nu)} \tag{2}$$

where, E : modulus of elasticity (kg/mm^2)

α : thermal expansion coefficient

ΔT : temperature gradient in the direction of circular plate thickness

ν : Poisson's ratio

On the assumption that the condition of heat conductivity is generally as shown in Fig. 1, ΔT can be expressed by the following equation.

$$\Delta T = T_1 - T_2 = \frac{Tgm - Tc}{\frac{1}{\alpha g} + \delta + \frac{1}{\alpha c}} \times \delta \tag{3}$$

where, Tgm : mean gas temperature ($^\circ$C)

Tc : cooling water temperature ($^\circ$C)

αg : coefficient of heat transfer on gas side (Kcal/m^{2}h $^\circ$C)

αc : coefficient of heat transfer on cooling water side (Kcal/m^{2}h$^\circ$C)

δ : thickness (m)

λ : thermal conductivity (Kcal/m h °C)

In taking ratio by eq. (1) / eq. (2), it is considered that this value (1)/(2) at any thermal fatigue cracking initiation cycle N is always constant in the same heat transfer condition.

And in the case that the thermal fatigue strength of the circular plate of different quality is estimated, the following equation can be obtained by arranging the term of the constant for simplification of the calculation.

$$\left.\begin{array}{l} T_{gm} - T_c \backsim \dfrac{(1-\nu)(\lambda D + \delta)}{\alpha E \delta} \ F(N) \equiv L_{th} \\[2mm] D = \dfrac{1}{\alpha_g} + \dfrac{1}{\alpha_c} \end{array}\right\} \tag{4}$$

Equation (4) is proportional to $T_{gm} - T_c$ in which a certain material is able to endure thermal fatigue condition at arbitrary numbers of crack initiation cycle. This is to be defined as thermal fatigue parameter L_{th}.

On the other hand, the thermal fatigue parameter L_{th} varies according to the thickness δ. For instance, the diesel engine cylinder cover needs to endure the high cycle fatigue strength due to the repetition of combustion gas pressure. Accordingly, the materials superior in enduring the high cycle fatigue strength can be reduced in thickness δ, then it can be able to endure large L_{th} value. Therefore, on the assumption that the gas pressure is equally distributed to the circular plate whose periphery is fixed, the stress due to the gas pressure P kg/cm^2 can be represented by the following equation (5) qualitatively.

$$\sigma_g = 0.75 \ \frac{Pa^2}{\delta^2} \tag{5}$$

where, σ_g : stress due to gas pressure (kg/mm^2)

       a : diameter of circular plate (mm)

The above equation can be rearranged as follows like in the case of L_{th}.

$$P \backsim \sigma_w \delta^2 \equiv L_g \tag{6}$$

where, σ_w : 10^7 cycle fatigue endurance limit (kg/mm^2)

Like the equation (4), the equation (6) is defined as mechanical fatigue parameter L_g. This means that in the case the material quality σ_w changes, the material thickness can be decided so that the safety factor against the fatigue strength can be made the same by making L_g invariable. Therefore, the true thermal fatigue strength of a material can be evaluated by determining by means of the equation (6) and obtaining L_{th} from the equation (4).

3. Test specimen and low cycle fatigue test

Five different sorts of material were used as shown in Table 1. PF means ductile cast iron with mixed metal structures of pearlite and ferrite. Table 2 and Figure 2 show the tensile test results and thermal conductivity of the test pieces respectively.

The low cycle fatigue test was conducted on the above five different kinds of test material at the temperatures of $400^{\circ}C$, $500^{\circ}C$, and $600^{\circ}C$ (partly). The tests were conducted under the conditions of tension compression and strain control. The test specimen used were 10 mm in gauge diameter and 30 mm in gauge length. The test results are shown in Figs. 3 to 6.

4. Gas heating type thermal fatigue test

Photo 1 shows the apparatus used for the test. This test apparatus is capable of testing two models simultaneously under the same thermal load condition. Figure 7 shows the test models used. Heating was made by means of a gas burner for propane gas and oxygen; and cooling was done by flowing fresh water over the cooling surface continuously.

The temperature distribution of the test specimens during the test was measured by means of a thermocouple; and after that, the temperature distribution and the thermal stress distribution were calculated with the use of the computer program about the axially symmetrical problems through the finite element method. Thermal fatigue test conditions and results are shown in Table 3 and Fig. 8. The color check method was used in ascertaining the thermal fatigue crack; and the cycle obtained at the time when the cracks were propagated to the length of 1 mm from the hole edge at the center of the test models was defined as the failure cycle. The crackings were propagated in the radial direction; Photo 2 shows the typical cracking propagation.

The mark $\diamondsuit$ in Figs. 3 to 6 shows the results of the theoretical estimation of the thermal fatigue life. The stress values were all estimated on the basis of maximum principal stress at the point of crack initiation; and they are mostly in good agreement with experimental data.

Column ⑩ in Table 3 shows the calculation made by the equation (4) of L_{th} which is the result of the gas heating type thermal fatigue test. The calculation results are shown in Fig. 9. As seen from the figure, the thermal fatigue parameter L_{th}, in most cases, can be indicated by one curve irrespective of difference in materials. However, F material and PF material are deviating from this curve; this is, it is considered, because the cracked part of these test points were tested at 700° C or so, and craze cracks occurred. In the case of P material, craze cracks did not occur even when it was tested at 700° C. Consequent upon this test, it was considered that L_{th} could be used in evaluating the thermal fatigue strength of various materials.

5. Evaluation of thermal fatigue strength of materials

Table 4 shows the results of the evaluation made of the thermal fatigue strength of cast iron materials at 400° C by means of L_{th}. This is because

these materials are, in many cases, used at $400°C$ or so. In making calculation, the number of thermal fatigue endurance cycle N was considered as 10^4
cycle (in diesel engines, it corresponds to the numbers of full load - stop
cycles), α_g as 1500, and α_c as 500 Kcal/h m². To know the effect of plate
thickness, results of calculations were shown in columns ⑧ , ⑨ , and ⑩ in
Table 4, in the cases of δ = 10 mm, 20 mm and 50 mm. According to the calculations, there is no great difference in strength between gray cast iron and
ferritic ductile cast iron.

In case that the plate thickness is calculated in consideration of the
fatigue strength due to mechanical stress, the calculation result becomes just
like column ⑪ in Table 4 (as the cylinder cover of the UET 45/80 D type
engine is 28 mm in thickness of SC material, this thickness was considered as
standard.) The calculation results of L_{th} in this state was described in
column ⑫ of Table 4.

With an increase in the diesel engine power rate, gray cast iron, ferritic ductile cast iron and cast steel were used in that order for the cylinder
cover; but pearlitic ductile cast iron has not been used.

But, it is obvious that P material is superior in strength and less in
manufacturing cost as compared with F material. Therefore, P material is
considered to be preferable for cylinder covers.

Incidentally, the author's company is testing for durability the P and F
material cylinder covers of the UET 45/80 D type engine (450 mm in cylinder
bore, 4500 PS) which are installed in a refrigeration ship for testing
purpose.

These methods of evaluating thermal fatigue strength by means of L_{th} is
considered to be applicable to the evaluation of the thermal fatigue strength
of heat resisting steels, etc.

6. Conclusion

(1) In order to evaluate the thermal fatigue strength of cast iron materials, the thermal fatigue parameter L_{th} was defined.

(2) The gas heating type thermal fatigue test was conducted on cast iron materials, and the test results showed that L_{th} could be used in evaluating the thermal fatigue strength.

(3) The thermal fatigue strength of cast iron materials was evaluated by means of L_{th}, and as a result, pearlitic ductile cast iron was found superior in thermal fatigue strength to ferritic ductile cast iron.

Reference

1) A. Dearden; BCIRA, vol. 9, No. 4 (1961) P.540

2) S. Yamamoto, Y. Yamada, K. Honjo; Mitsubishi Tech. Rev. vol. 6, No. 6 (1969)

3) S. Izumi; Proc, 9th CIMAC, (Sweden, 1971), A-21.

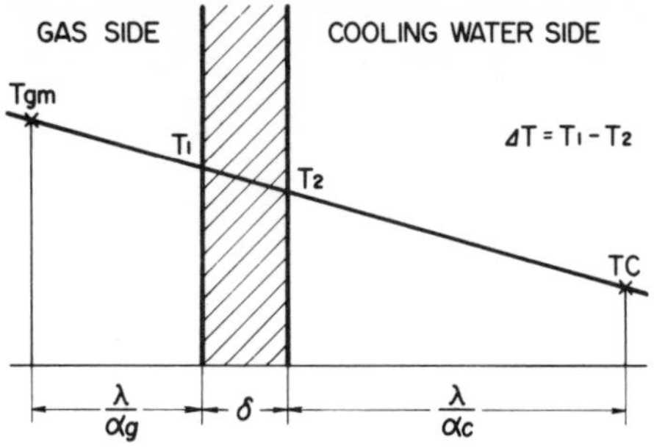

Table 1. Chemical Composition Weight % and Heat Treatment of Tested Materials.

Material	Symbols	C	Si	Mn	P	S	Mg	Heat Treatment
Gray Cast Iron	FC	0.35	1.92	0.67	0.045	0.030	—	540℃x 4ᴴA.C
Pearitic Ductile Cast Iron	P	3.65	1.90	0.68	0.040	0.026	0.068	900℃x 5ᴴA.C 550℃x 4ᴴF.C
Ferritic Ductile Cast Iron	F	3.16	2.62	0.64	0.023	0.010	0.035	900℃x 4ᴴF.C 720℃x10ᴴF.C
As Cast Ductile Cast Iron	PF	3.70	2.32	0.62	0.017	0.012	0.047	550℃x 4ᴴA.C
Cast Steel	SC	0.19	0.36	0.67	0.012	0.010	—	900℃x 5ᴴA.C

Fig. 1. Conditions of Steady Heat Transfer.

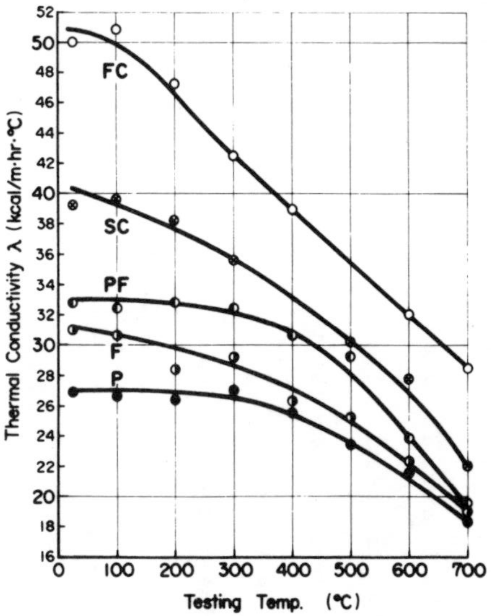

Fig. 2. Results of Thermal Conductivity Test.

Table 2. Results of Tensile Test.

Material	Testing Temp.℃	Yield Stress kg/mm2	Tensile Stress kg/mm2	Elongation %	Contraction %	Young's Modulus kg/mm2
FC	R.T	—	19.5	1.0	—	9500
	300	14.1	19.2	1.0	0.7	8800
	400	12.7	17.1	1.1	0.8	8650
	500	—	15.1	1.8	—	8650
	600	5.8	8.7	3.1	0.9	6650
P	R.T	47.1	82.2	5.0	4.5	16950
	300	36.5	69.3	4.9	5.4	16100
	400	37.2	62.6	4.2	4.5	14800
	500	36.7	46.5	13.2	12.7	14600
	600	14.3	25.2	22.8	21.4	9900
F	R.T	31.3	48.1	21.9	19.5	18100
	300	24.9	42.1	16.2	18.3	16000
	400	23.2	36.5	9.0	10.0	14900
	500	18.7	28.2	29.4	27.2	13100
	600	8.7	15.1	43.2	35.9	10600
PF	R.T	31.7	56.4	12.3	11.3	16200
	300	24.1	44.6	11.1	12.2	15900
	400	22.9	41.6	15.8	14.3	15000
	500	18.3	29.1	22.7	21.8	13000
	600	10.5	17.3	31.4	27.3	10300
SC	R.T	16.9	41.6	38.9	54.1	20200
	300	14.6	38.1	25.8	36.9	19350
	400	14.0	34.6	35.4	49.7	18300
	500	13.2	25.0	28.5	71.0	16300
	600	8.7	15.8	58.7	74.3	13400

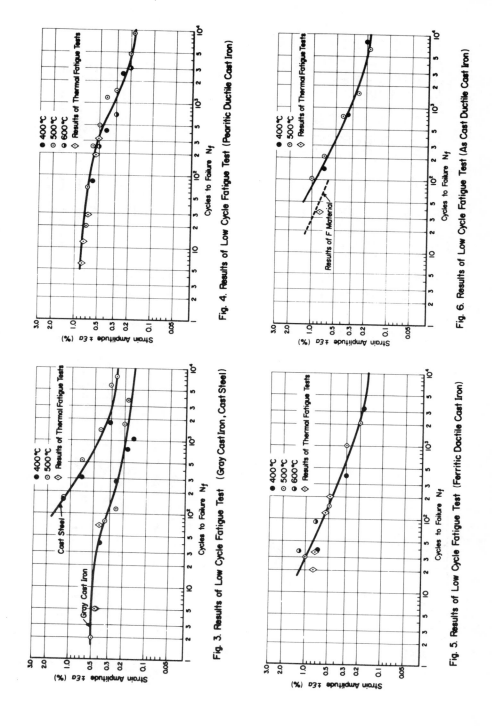

Fig. 3. Results of Low Cycle Fatigue Test (Gray Cast Iron, Cast Steel)

Fig. 4. Results of Low Cycle Fatigue Test (Pearlitic Ductile Cast Iron)

Fig. 5. Results of Low Cycle Fatigue Test (Ferritic Ductile Cast Iron)

Fig. 6. Results of Low Cycle Fatigue Test (As Cast Ductile Cast Iron)

Table 3. Results of Thermal Fatigue Tests and Calculation of Thermal Fatigue Parameter.

Test specimen No.	① LPG-Flow Rate (l/min)	② H (mm)	③ Tmax (°C)	④ Ni Experimental	⑤ ±εαc(%) (calculated)	⑥ Ni estimated	⑦ α (1/°C)	⑧ λ (kcal/mh°C)	⑨ ±εαt(%) (Low Cycle Fatigue)	⑩ Lth
F-3	1300	46	700	20	0.775	56	1.23x10⁵	19.1	1.17	2065
FC-2			468	5	0.450	25	1.20x10⁵	36.4	0.50	1527
P-1	1300	60	700	6	0.795	10	1.23x10⁵	18.5	0.825 1.560 (0.98, onF)	1494 3003 (1790)
PF-3			680	36	0.790	143	1.23x10⁵	20.2		
FC-3			462	5	0.443	25	1.20x10⁵	36.8	0.50	1539
F-4	1300	85	670	35	0.735	63	1.23x10⁵	20.1	0.93	1785
P-2			670	12	0.750	25	1.23x10⁵	19.3	0.795	1484
P-3	1300	110	620	28	0.670	70	1.23x10⁵	20.6	0.74	1442
FC-4			430	70	0.412	42	1.20x10⁵	37.7	0.36	1130
F-2	850	110	540	120	0.550	130	1.23x10⁵	23.9	0.570	1233
P-4			540	192	0.546	240	1.23x10⁵	22.6	0.58	1207
F-5	800	120	520	200	0.485	175	1.23x10⁵	25.2	0.46	1035
P-5			520	320	0.514	320	1.23x10⁵	23.0	0.51	1073

Fig. 9. Results of Thermal Fatigue Test expressed by Thermal Fatigue Parameter.

○—P
●—F
●—PF (⊘ is Calculated from Low Cycle Fatigue data of F)
△—FC

Cycles at Which Thermal Cracks were detected

Thermal Fatigue Parameter Lth

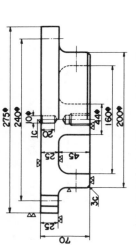

Fig. 7. Model of Thermal Fatigue Test.

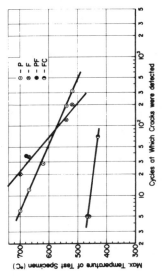

Fig. 8. Results of Thermal Fatigue Tests.

○—P
●—F
●—PF
●—FC

Cycles at Which Cracks were detected

Max. Temperature of Test Specimen (°C)

Table 4. Thermal Fatigue Parameter of Cast iron Materials.

① Materials	② Evaluating Temperature (°C)	③ σₐ·F(N) N=1x10⁴ (kg/mm²)	④ E (kg/mm²)	⑤ ν	⑥ α (°C⁻¹)	⑦ λ (kcal/mh°C)	⑧ Lth δ=10mm	⑨ Lth δ=20mm	⑩ Lth δ=50mm	⑪ δ (mm)	⑫ Lth
FC	400	9.5	8650	0.25	1.20x10⁵	39.0	783	426	307	37.6	258
P		33.7	15300	0.25	1.23x10⁵	25.4	1030	582	433	19.6	598
F		23.8	14900	0.25	1.23x10⁵	27.1	799	448	331	25.7	371
PF		28.5	15000	0.25	1.23x10⁵	30.8	1040	579	425	24.6	502
SC		31.8	18300	0.30	1.14x10⁵	33.1	1050	579	422	28.0	445

Photo. I. Gas Heating Type Thermal Fatigue Tester.

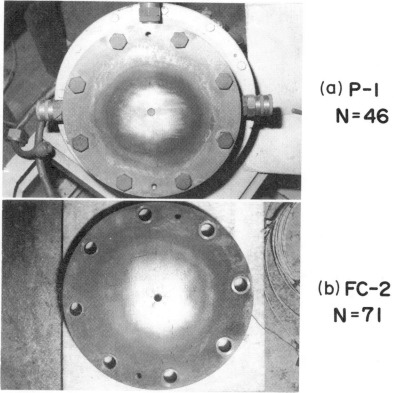

(a) P-I
N=46

(b) FC-2
N=71

Photo. 2. Typical Crack after Thermal Fatigue Test.

SESSION IV

Fatigue

A STUDY ON THE STATISTICAL FLUCTUATION OF FATIGUE CRACK GROWTH RATE
ASSOCIATED WITH THE HETEROGENEITY OF THE MICROSTRUCTURE

Takeshi KUNIO[*], Masao SHIMIZU[**], Kunihiro YAMADA[***] and Yuhji KIMURA[****]

1) INTRODUCTION

Recently, a number of papers have been published on the rate of fatigue
crack propagation from the concept of Fracture Mechanics[1]. Most of them are
emphasized on the rate of crack growth based on the data obtained from relatively
small number of specimens. However, the cracks generated in the specimens of
the same heat do not necessarily show the definite propagation characteristics
under the fatigue loading even at identically the same stress level. Therefore,
a number of curves of crack propagation are drawn in accordance with a number
of the tested specimens[2]. Thus, it is essential to examine the characterization of the probabilistic feature of the fatigue crack propagation.

In general, polycrystalline metallic materials have inevitably the heterogeneity of microstructure such as crystal grains and so on. In addition, it
is well-known that each grain shows the different characteristics of plastic
deformation depending upon the crystal orientation. On the other hand, it is
pointed out that the behavior of crack propagation under fatigue loading should
be associated with the plastic deformation characteristics of the material at
the crack tip[3]-[9]. This leads to a possible consideration that the probabilistic features of fatigue crack growth would be closely related to the heterogeneity of the microstructure in the local region at the crack tip. Hereafter,
such local region will be referred to as the growth process zone.

* Dr., Professor, Dept. of Mech. Engrg., Keio Univ., Yokohama, Japan
** Dr., Associate Professor, Dept. of Mech. Engrg., Keio Univ., Yokohama, Japan
*** Dr., Research Fellow, Dept. of Mech. Engrg., Keio Univ., Yokohama, Japan
**** M.S., Graduate Student, Dept. of Mech. Engrg., Keio Univ., Yokohama, Japan

252

In this paper, the effects of the heterogeneity of the microstructure on the fluctuation of the fatigue crack growth rate will be studied by use of the specimens which consist of low carbon martensitic structure and have the stable crack growth characteristics.

2) STATISTICAL FLUCTUATION IN THE RATE OF FATIGUE CRACK PROPAGATION

The following experiments were conducted, to study the statistical characteristics of the rate of fatigue crack propagation.

Plain carbon steel specimens with 0.17 % C were prepared from the same lot. Dimensions of the specimen are given in Fig. 1. The specimens were fully quenched into martensitic structure by Induction-Hardening with the condition that the time required to reach the quenching temperature $1150^{\circ}C$ is 45 sec.. The average micro-Vickers hardness and mean Prior Austenite Grain size$^{(*)}$ of the microstructure are $Hv_m=490$ and $42\mu m$, respectively. The specimen was bored out into a cylindrical shape before the heat treatment, in order to eliminate the effects of the stress gradient under the rotating bending and the macroscopic residual stress due to the Induction-Hardening. Also, each specimen has an artificial crack starter of circular hole which is 0.2mm in diameter. Rotating bending fatigue tests were conducted at the two stress levels, $\sigma_1=36$ kg/mm^2 and $\sigma_2=50$kg/mm^2, the estimated fatigue lives of which are $N_f=1.5\times10^5$ and 5.0×10^4 respectively. Eighteen and sixteen of specimens were fatigue-tested at the above two stress levels, respectively. Before fatigue testing the surface of specimens were electro-polished for the microscopic observations of the crack propagation. A length of the growing crack was measured by a replication method at regular cycle intervals.

(*) It is pointed out that the Prior Austenite Grain size is closely related to the fatigue strength of the martensitic structure[10].

The rate of fatigue crack propagation were appreciated in terms of Kmax[(*)], and then the statistical characteristics of their fluctuation were pursued in the followings.

Eighteen and sixteen of the crack growth curves were obtained for the stress amplitudes of $\sigma_1 = 36 kg/mm^2$ and $\sigma_2 = 50 kg/mm^2$, respectively. In Fig. 2 and Fig. 3 three kinds of crack growth curves for two stress levels are drawn which correspond to the maximum-, minimum- and median-value of fatigue lives. The rate of fatigue crack propagation was determined by the slope of each crack growth curve at various crack lengths. The statistical characteristics obtained are tabulated in Table 1. The numerical data in Table 1 within a bold line show the characteristics of the tensile mode crack propagation in the region II of the well-known relation between $\frac{dL}{dN}$ and ΔK as illustrated in Fig. 4[(13)]. In this paper, a consideration on the statistical characteristics of crack propagation are restricted to that of the tensile mode crack growth in the region II, where the steady crack propagation appears. The distinction of the tensile mode crack propagation from the shear mode one was appreciated from the existence of the shear lip on the fractured surface.

The rate of tensile mode crack propagation at each Kmax were plotted on the logarithmic-normal probability paper as shown in Fig. 5. It would be recognized from the plots in this figure that the rates of fatigue crack propagation for a given value of Kmax are distributed logarithmic-normally around $(\frac{dL}{dN})$ mean. Also, the average crack growth rate $(\frac{dL}{dN})$ mean was plotted as a function of Kmax

---

(*) Kmax was calculated from the following equation:
$$K_I = k_1 \cdot k_2 \cdot \sigma_{max} \sqrt{\pi L}$$
where L: crack length
          k_1: the correction factor for the effect of a hole[(11)]
          k_2: the correction factor for the effect of cylindrical
             shape[(12)]

in Fig. 6. It is found from this figure that the average rate of tensile mode crack propagation $(\frac{dL}{dN})$mean in the region II can be expressed by the following equation:

$$\left(\frac{dL}{dN}\right)_{mean} = \alpha \, (K_{max})^{\delta} \tag{1}$$

in which the exponent δ was determined as 2.5. Therefore, the ordinary expression of the fatigue crack growth rate in terms of stress intensity factor range ΔK might be interpreted as the equation in which $(\frac{dL}{dN})$mean takes the place of the conventional $\frac{dL}{dN}$.

To investigate the probabilistic feature of the rate of fatigue crack propagation in more details, a coefficient of variation (C.V.) of the fatigue crack growth rate was examined with respect to the various values of Kmax. The relationship between the C.V. of the rate of tensile mode fatigue crack propagation in the region II and the values of Kmax are shown in Fig. 7. This figure shows the tendency that the C.V. decreases with the increase of Kmax value. The increase of Kmax value would correspond to the extension of the growth process zone. Therefore, it seems from the figure 7 that there might exist some factors associated with the fluctuation of the fatigue crack growth rate in this zone, and that the less effect of the factor appears as the zone size becomes larger. This might lead to the considerations that the heterogeneity of microstructure plays an important role in the fluctuation of the fatigue crack growth rate.

3) THE MICROSTRUCTURAL FACTORS ON THE FLUCTUATION OF FATIGUE CRACK GROWTH RATE

3.1 A model for the fluctuation of the rate of the fatigue crack propagation

In this section, a model for the fluctuation of the rate of the fatigue crack propagation will be proposed involving the size of the growth process zone and some measure of the heterogeneity of the microstructure.

Though there might be a lot of unknown parameters to be considered, the special emphases are placed on the effects of the Prior Austenite Grain (PAG) size d_m as a heterogeneity parameter and the plastic zone size r_p given by the continuum mechanics calculation as the size of the growth process zone.

The significance of the contribution of the heterogeneity of microstructure in the growth process zone to the fluctuation of the fatigue crack growth rate will be studied in the followings.

The relative dimension of r_p to d_m is emphasized in the model as illustrated in Fig. 8(a) and (b). In the case (a), r_p has about the same size as d_m. In the case (b), r_p is much greater than d_m. This model indicates that the fluctuation of the microstructural factors in the growth process zone in the case (a) is extremely remarkable, compared with that in the case (b). In other words, the contribution of each grain to the characteristics of the plastic deformation in the zone in the case (a) is more significant than that in the case (b). This model is applicable to the qualitative explanation of the experimental results in Fig. 7, as described in the following section.

3.2 An examination on the validity of the model

In this section, the validity of the above model was examined through the experiments by use of the specimens with different PAG size d_m.

The low carbon steel specimens as same as the previous ones were repeatedly induction-hardened two times with the condition that the time required to reach the quenching temperature 950°C is 4 sec., to obtain the different grain size from the previous specimens. These specimens are referred to as B-series, while the previous ones are referred to as A-series. This heat treatment brings about a fully hardened martensitic structure with the finer PAG size than that of A-series (Photo 1). The properties of both A and B series specimens are tabulated in Table 2. The same experimental procedures as those of A-series were employed for present B-series specimens.

The results for B-series specimens are tabulated in Table 3, where the data for the steady tensile mode crack propagation are surrounded by a bold line. Fig. 9 shows the dependency of the C.V. of the $\frac{dL}{dN}$ on Kmax for B-series specimens in comparison with that for A-series specimens. It is recognized from this figure that the fluctuation of $\frac{dL}{dN}$ depends clearly upon the grain size. A possible explanation for this result can be given through the present model in Fig. 8 as follows. Following this model, the fluctuation of $\frac{dL}{dN}$ should be controlled by the degree of the heterogeneity of the microstructure within the growth process zone. These facts would imply that the C.V. of the fatigue crack growth rate are not governed by the values of r_p and d_m individually but it depends on the relative size of r_p to d_m. There would be considered several parameters to express relative relationship of r_p to d_m. Here, for the sake of the qualitative interpretation of the above results, the parameter r_p/d_m was taken up into consideration. The parameter r_p was tentatively calculated by the following equation in continuum mechanics[14]:

$$r_p = \frac{1}{2\pi} \left(\frac{K_{max}}{\sigma_Y} \right)^2 ,$$

while the heterogeneity parameter presented by d_m is obtained from the grain size measurement. Figure 10 showing the relationship between the C.V. of the $\frac{dL}{dN}$ and the proposed parameter r_p/d_m indicates that the C.V., i.e., the fluctuation of $\frac{dL}{dN}$ uniquely depends upon r_p/d_m and tends to approach a constant value asymptotically after the initial rapid decrease with the increase of r_p/d_m. In other words, the contribution of the microstructural heterogeneity to the deformation characteristics would become less as r_p/d_m increases. Then, the fluctuation of the rate of fatigue crack propagation becomes smaller.

It is therefore concluded that the proposed model would be valid for the explanation of the fluctuation of the rate of the fatigue crack propagation.

4) A MONTE-CARLO SIMULATION OF THE FATIGUE CRACK PROPAGATION ASSOCIATED WITH
THE HETEROGENEITY OF MICROSTRUCTURE

In this chapter, it is examined by the method of Monte-Carlo simulation whether the heterogeneity of the microstructure could be one of the dominant factors for the one-to-one correspondence of each growth curve to each specimen. In present simulation, the heterogeneity of the microstructure was presumably represented by the local variation of resistance to the crack propagation on the assumption that it distributes randomly throughout the structure.

Fig. 11 indicates a Monte-Carlo simulation model in which the heterogeneity of the microstructure on the expected path of crack propagation is expressed by the random arrays of the same sized square crystal grains with three kinds of crystal orientations. The number 1, 2 and 3 on square crystals are assigned to denote the orientation of each grain. The difference in crystal orientation represents the local variation in the resistance to crack propagation. However, in more general treatment, any kind of structural element, which gives an appropriate expression for the variation of resistance, would be employed as the factor representing the heterogeneity of the microstructure. The n kinds of arrangements of crystal grains correspond to different n kinds of paths of crack propagation, in other words, to the number of the specimens in the fatigue testing.

In present simulation, the crack lengths at regular cycle intervals of $\Delta N=10$ were calculated by a computer for n=20 kinds of crack paths. The flow-chart for the computer calculation are shown in Fig. 12. It is assumed that (i) the crack propagation follows the equation (1), (ii) the variation of resistance of material due to the crystal orientation is given by the change of the value α in equation (1) and (iii) there is no effect of grain boundary on the crack propagation. The exponent δ was chosen as 2 and $\alpha_1=0.711\times10^{-8}$ mm^4/kg^2cycle, $\alpha_2=1.42\times10^{-8}mm^4/kg^2cycle$ and $\alpha_3=2.13\times10^{-8}mm^4/kg^2cycle$ were used for the three kinds of grains. The Kmax was calculated by $\sigma\sqrt{\pi L}$ in which L is the current crack length at the number of stress cycles N.

The crack growth curves having maximum and minimum lives for two kinds of grain sizes GS=100μm and 10μm are shown in Fig. 13, where the stress amplitude is σ_a=36kg/mm^2 and the initial crack size is L_o=100μm.

This figure shows that the Monte-Carlo simulation based on the model shown in Fig. 11 brings about a number of curves of fatigue crack propagation, in accordance with the number of the tested specimens. Also, it is recognized that the fluctuation in the process of the fatigue crack propagation depends upon the grain size.

5) CONCLUSIONS

An experimental study has been made to explain the role of the heterogeneity of the microstructure on the fluctuation of fatigue crack growth rate. Also, the computer simulation concerning the fatigue crack propagation was carried out.

The results obtained are summarized as follows:

(1) A new model for the fluctuation of the fatigue crack growth rate was proposed

(2) The coefficient of variation of the fatigue crack growth rate depends upon r_p/d_m in which r_p and d_m are the parameters representing the size of growth process zone and the heterogeneity of the microstructure, respectively.

(3) A Monte-Carlo simulation brings about a number of curves of fatigue crack propagation in accordance with the number of tested specimens. Also, it is resulted that the fluctuation in the process of the fatigue crack propagation depends upon the grain size.

The authors are indebted to the Daido Steel Works Ltd. for the preparation of material. The authors are also extremely grateful to Professor K. Tone, Keio University, for his helpful suggestions in conducting the Monte-Carlo simulation and Mr. T. Makabe for the programming of computer calculation.

REFERENCES

1. Paris, P. C. and Erdogan, F., Trans. ASME, Ser.D, 85-4(1963), 528.

2. Tanaka, S. et al., J. Japanese Soc. Strength and Fracture Materials, 8-2(1973), 56.

3. Liu, H. W., Trans. ASME, Ser.D, 85-3(1963), 116.

4. Erdogan, F. and Robert, R., Proc. 1st Int. Conf. Fracture, Sendai, Japan, Vol.1(1965), 341.

5. Yokobori, T. et al., J. Japanese Soc. Strength and Fracture Materials, 5-4(1971), 106.

6. Taira, S. and Hayashi, K., Trans. Japan Soc. Mech. Engrs., 33-245(1967), 1.

7. Ogura, T. and Karashima, S., J. Japan Inst. Metals, 34-7(1970), 746.

8. Taira, S. and Tanaka, K., J. Soc. Materials Science, Japan, 18-190(1969), 620.

9. Hahn, G. T. et al., Met. Trans., 3(1972), 1189.

10. Kunio, T. et al., Proc. 3rd Int. Conf. Fracture, München, III-234, (1973).

11. Paris, P. C. and Sih, G. C., Fracture Toughness Testing and Its Applications, ASTM STP 381, (1965), 30, ASTM.

12. Erdogan, F. and Ratwani, M., Int. J. Fracture Mechanics, 6-4(1970), 379.

13. Liu, H. W. and Iino, N., Proc. 2nd Int. Conf. Fracture, Brighton, (1969), 812.

14. McClintock, F. A. and Irwin, G. R., Fracture Toughness Testing and Its Applications, ASTM STP 381, (1965), 84, ASTM.

Crack length: L(μm)	100	150	250	350	450	600	750	1000	1250	1500	1750	2000	2500
σa=36 $K_{max.}$: kg/mm$^{3/2}$	29.3	32.0	37.5	42.4	46.9	53.2	58.7	67.7	74.3	82.9	91.2	106	116
$\left(\frac{dL}{dN}\right)$mean :10^{-5}mm/cycle	0.538	0.770	1.22	1.70	2.23	3.13	4.15	6.11	8.10	10.8	13.3	17.6	27.8
Standard deviation :10^{-5}mm/cycle	0.218	0.258	0.393	0.522	0.657	0.867	1.07	1.62	2.05	2.67	3.18	5.51	11.3
Coefficient of variation: %	40.5	33.4	32.1	30.7	29.3	27.7	25.8	26.5	25.3	24.8	23.9	31.4	40.6
σa=50 $K_{max.}$: kg/mm$^{3/2}$	40.7	42.4	52.1	58.9	65.2	73.9	81.4	94.0	103	116	126	138	161
$\left(\frac{dL}{dN}\right)$mean :10^{-5}mm/cycle	1.51	2.06	3.98	4.10	5.31	7.76	9.57	13.2	18.7	26.7	32.8	46.2	71.0
Standard deviation :10^{-5}mm/cycle	0.656	0.688	1.03	1.41	1.66	2.13	2.17	2.97	5.09	9.09	10.9	26.8	35.3
Coefficient of variation: %	43.4	33.5	33.4	34.4	31.3	27.5	22.6	22.5	27.1	34.1	33.2	58.0	49.7

Table 1 The characterization of the probabilistic feature of fatigue crack growth rate: A-series

Specimen	Chemical composition (%)	Induction-hardening condition	Prior Austenite grain size number (ASTM)	Average Prior Austenite grain size (mm)	Average Vickers hardness (100gr)	Yield stress: 0.2% proof stress (kg/mm2)
A	C :0.17 Si:0.29 Mn:0.42 S :0.010 P :0.009	1150°C 45sec.	6.4	0.042	490	100
B		950°C 4sec. repeated twice	9.6	0.014	518	112

Table 2 Specimen specifications

Crack length: L(μm)	100	150	250	350	450	600	750	1000	1250	1500	1750	2000	2500
σa=36 $K_{max.}$: kg/mm$^{3/2}$	29.3	32.0	37.5	42.4	46.9	53.2	58.7	67.7	74.3	82.9	91.2	106	116
$\left(\frac{dL}{dN}\right)$mean :10^{-5}mm/cycle	0.500	0.662	0.828	1.13	1.80	2.14	3.01	4.55	6.14	8.54	11.1	13.8	18.7
Standard deviation :10^{-5}mm/cycle	0.222	0.247	0.237	0.284	0.476	0.526	0.491	0.638	0.893	1.34	1.59	1.91	3.06
Coefficient of variation: %	44.5	37.4	28.6	25.1	26.4	24.6	16.3	14.0	14.6	15.7	14.3	13.9	16.4
σa=50 $K_{max.}$: kg/mm$^{3/2}$	40.7	42.4	52.1	58.9	65.2	73.9	81.4	94.0	103	116	126	138	161
$\left(\frac{dL}{dN}\right)$mean :10^{-5}mm/cycle	1.46	1.92	3.02	4.34	5.75	8.20	10.9	15.7	19.9	25.9	31.1	36.9	49.4
Standard deviation :10^{-5}mm/cycle	0.573	0.651	0.709	0.621	0.599	0.755	0.925	1.26	1.99	2.40	3.15	4.36	8.50
Coefficient of variation: %	39.2	34.0	23.5	14.3	10.4	9.20	8.52	8.01	10.0	9.26	10.1	11.8	17.2

Table 3 The characterization of the probabilistic feature of fatigue crack growth rate: B-series

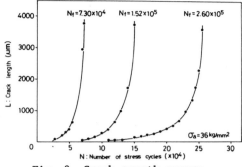

Fig. 2　Crack growth curves

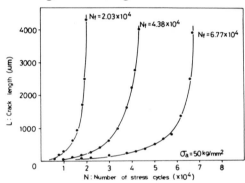

Fig. 3　Crack growth curves

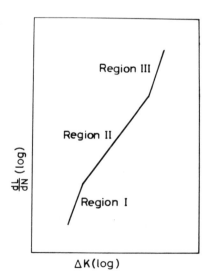

Fig. 4　Crack growth rate vs.
stress intensity factor range

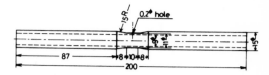

Fig. 1　Specimen geometry (mm)

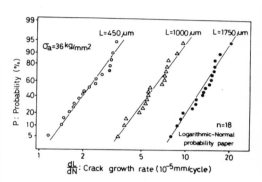

Fig. 5　The plots of the date of
the fluctuation of dL/dN
on the logarithmic-normal
probability paper

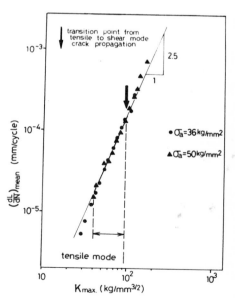

Fig. 6　The average crack growth
rate (dL/dN)mean vs. Kmax

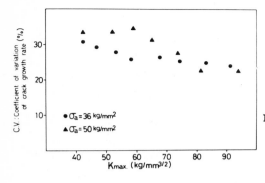

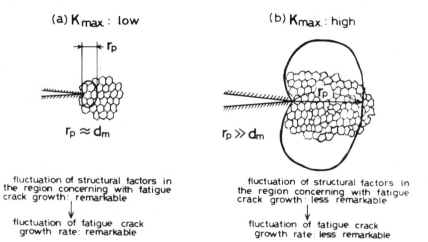

Fig. 7 Dependency of the coefficient of variation of crack growth rate on Kmax

(a) K_{max}: low

(b) K_{max}: high

r_p

$r_p \approx d_m$

$r_p \gg d_m$

fluctuation of structural factors in the region concerning with fatigue crack growth: remarkable

↓

fluctuation of fatigue crack growth rate: remarkable

fluctuation of structural factors in the region concerning with fatigue crack growth: less remarkable

↓

fluctuation of fatigue crack growth rate: less remarkable

Fig. 8 A model for the fluctuation of fatigue crack growth rate

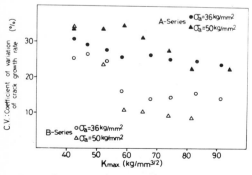

Fig. 9 Dependency of the coefficient of variation of crack growth rate on Kmax

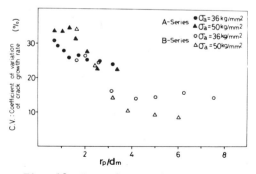

Fig. 10 Dependency of the coefficient of variation of crack growth rate on r_p/d_m

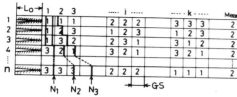

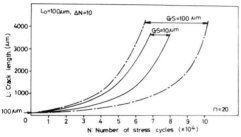

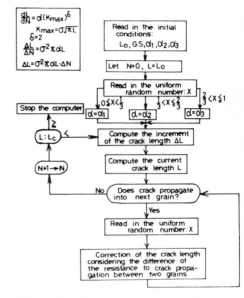

↑ : Positions of fatigue crack front in each crack path when cyclic stress are repeated N_1, N_2 and N_3 times

GS : Grain size

Fig. 11   A model for the Monte-Carlo simulation

Fig. 13   Crack growth curves calculated by the Monte-Carlo simulation

Fig. 12   Flow chart of the Monte-Carlo simulation

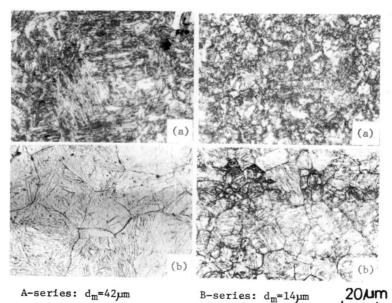

A-series: $d_m = 42\,\mu m$ B-series: $d_m = 14\,\mu m$ **20μm**

Photo 1   Surface microstructure:
(a) martensitic structure
(b) delineated Prior Austenite Grain

Effect of Large Scale Yielding on Fatigue Crack Propagation

A. J. McEvily,* D. Beukelmann,* and K. Tanaka**

1. INTRODUCTION

The correlation between the rate of fatigue crack growth and the stress-intensity factor seems to be well established in the linear elastic range where the size of the plastic zone is small relative to other critical dimensions. However, for large scale yielding in the elastic-plastic region as well as for fully plastic behavior, crack growth relationships have as yet not been as well established. It is the purpose of the present paper to explore possible analytical expressions for fatigue crack growth for such cases involving large-scale plasticity effects. The approach is based upon the use of an assumed relationship between the crack opening displacement (COD) as given by Bilby, Cottrell and Swinden (BCS) and the rate of fatigue crack growth (1).

2. LARGE-SCALE YIELDING IN THE ELASTIC-PLASTIC RANGE

The BCS expression for the COD is as follows:

$$\text{COD} = \frac{8}{\pi} \frac{\bar{\sigma} a}{E} \, \ell n \, \sec \frac{\pi \sigma}{2 \bar{\sigma}} \tag{1}$$

where $\bar{\sigma}$ is a characteristic flow stress, E is Young's modulus, a is the semi-crack length for a central slit, and σ is the applied stress. When σ is small with respect to $\bar{\sigma}$ this expression reduces to

$$\text{COD} = \frac{\sigma^2 \pi a}{\sigma_y E} = \frac{K^2}{\sigma_y E}, \tag{2}$$

where K is the stress-intensity factor. To relate this expression to the rate of fatigue crack growth in the linear-elastic range certain modifications have been introduced to incorporate the effects of a threshold stress intensity, K_{TH}, as well as the influence of fracture toughness, K_c, on the introduction of static modes of fracture into the crack advance process. The overall equation

*Professor and Graduate Assistant, respectively, Metallurgy Department, University of Connecticut, Storrs, Conn. 06268. **Department of Mechanical Engineering II, Kyoto University, Kyoto, Japan.

becomes

$$\frac{\Delta a}{\Delta N} = \frac{A}{\sigma_y E} (\Delta K^2 - \Delta K_{TH}^2)(1 + \frac{\Delta K}{K_c - K_{max}}) \tag{3}$$

where A is a material constant, σ_y is the monotonic yield strength, ΔK_{TH} is a function of mean stress, and K_{max} is given by $\frac{\Delta K}{1-R}$, where R is the ratio of the minimum to maximum stress in a loading cycle (2). In materials of high toughness, mean stress effects become relatively unimportant except for an influence on the threshold level. Eq. 3 has been found to be in good agreement for high strength aluminum alloys and steels over a range of toughness values (2).

In considering large-scale plasticity effects, it is preferable to use a form of Eq. 1, rather than Eq. 3, since σ is no longer small with respect to σ_y. In the light of Rice's analysis for the effective yield stress under cyclic conditions it seems reasonable to approximate $\bar{\sigma}$ in Eq. 1 by taking it to be twice the monotonic yield strength (3). Additional modifications are needed in order to compare with available experimental evidence. Firstly, the BCS solution is for an infinite plane body in which net section and gross section stresses are equal to net section stresses. We assume that in a sheet of finite width the net section stresses are more important than are the gross-section stresses, since yielding on the net section will place an upper limit on Eq. 1. Secondly, a finite width correction for the COD expression must be introduced to compare with experiment, and we take this correction to be of the form sec $\pi a/w$, where w is the specimen width. In the analysis we will neglect mean stress and static mode effects on crack growth. The resultant expression for the rate of crack growth becomes:

$$\frac{\Delta a}{\Delta N} = \frac{A'}{\sigma_y' E} \left[\frac{8}{\pi} \sigma_y'^2 a \left(\frac{w-2a}{w}\right)^2 \left(\ell n \; sec \; \frac{\pi}{2} \frac{\Delta\sigma_{net}}{\sigma_y'}\right)\left(sec \; \frac{\pi a}{w}\right) - \Delta K_{TH}^2 \right] \tag{4}$$

where $A' = 2A$, σ_y' equals $2\sigma_y$, and σ_{net} equals $\sigma_{gross} \frac{w}{w-2a}$. The quantity

$\frac{8}{\pi} \sigma_y'^2 a \left(\frac{w-2a}{w}\right)^2 \left(\ell n \; sec \; \frac{\pi}{2} \frac{\Delta\sigma_{net}}{\sigma_y'}\right)\left(sec \; \frac{\pi a}{w}\right)$ can be considered at an equivalent

$(\Delta K_{eq})^2$ by analogy with Eq. 3. We will compare with results for a 0.5 wt.% carbon steel in which rates of crack growth were analyzed in terms of the usual stress intensity factor modified by a finite width correction (4), that is

$$\Delta K_Y^2 = (\Delta \sigma_g)^2 \; \pi a \; \sec \frac{\pi a}{w} \tag{5}$$

Therefore

$$K_{eq}^2 = \Delta K_Y^2 \left[\frac{8}{\pi^2} \left(\frac{\sigma_y'}{\Delta \sigma_g} \right)^2 \left(\frac{w-2a}{w} \right)^2 \; \ell n \; \sec \frac{\pi}{2} \frac{\Delta \sigma_{net}}{\sigma_y'} \right] \tag{6}$$

The extent of crack growth per cycle can then be expressed as:

$$\frac{\Delta a}{\Delta N} = \frac{A'}{\sigma_y' E} \left[\Delta K_Y^2 \frac{8}{\pi^2} \left(\frac{\sigma_y'}{\Delta \sigma_g} \right)^2 \left(\frac{w-2a}{w} \right)^2 \; \ell n \; \sec \left(\frac{\pi}{2} \frac{\Delta \sigma_{net}}{\sigma_y'} \right) - \Delta K_{TH}^2 \right] \tag{7}$$

In Ref. (5), an empirical relationship between A and σ_y/E was observed for tests in inert environments. We will assume that a relationship of a similar type holds in the present case, that is

$$A' = C \; \sigma_y'/E \tag{8}$$

where C is a constant independent of yield stress for a given alloy, for we will next compare the predictions of Eq. (7) with the interesting experimental results obtained by Yokobori, Kawada, and Hata for a 0.5 carbon steel in which yield stress was varied by varying the grain size. Yokobori et al. analyzed their results by means of the Paris–Erdogan equation (6).

$$\frac{\Delta a}{\Delta N} = A_o (\Delta K_Y)^\delta \tag{9}$$

Table 1 summarizes the test conditions and lists the δ and $\log A_o$ values they observed. In their tests, in certain cases as the crack extended the value of $\Delta \sigma_{net}$ approached twice the monotonic yield stress and an effect on the shape of the curve of crack growth plotted against ΔK_Y is to be expected on the basis of Eq. 4. Figures 1-5 provide a comparison between experiment and Eqs. 4 and 9. (A constant value of ΔK_{TH} equal to 22 Kpmm$^{-3/2}$ was used in conjunction with Eq. 4. The value of C in Eq. 8 was taken to be equal to 6. In materials of low yield strength such as alloy F, the S-shape of the rate data plot is evident. As the yield strength increases this effect is less pronounced. It is noteworthy that Eq. 4 yielded reasonably good agreement with the experimental data in all cases. It would be difficult to account for observed deviations from alloy to alloy as there does not appear to be any systematic trend involved. It is clear that the S-shape results from threshold as well as yield-associated effects, neither of which is accounted for by Eq. 9, which of course plots as a straight line in the figures. In materials of low toughness, static modes of separation also contribute to the S-shape of the curves, but such a contribution is assumed not to be a factor in the present case.

A main purpose of the above analysis is to demonstrate the feasibility of the COD approach to the case of large scale yielding. It is considered that this aim has been achieved. However, certain seemingly arbitrary aspects of the analysis are worthy of further study to put the approach on firmer grounds. Further work should include consideration of the cyclic stress-strain behavior as it affects the value of $\bar{\sigma}$, and also further consideration of the role of net section as compared to gross section stresses in influencing crack growth behavior.

4.   FULLY PLASTIC BEHAVIOR

We have just seen how large-scale yielding can introduce complexities into the analysis of crack growth in the elastic-plastic range. In the fully plastic region the analyses become even more arbitrary because of the inherent difficulties in establishing well-founded analytical expressions for crack growth in this range. The general approach is to modify and reinterpret analyses obtained for elastic-plastic behavior, as exemplified by the work of Tomkins (7). We will employ a similar approach, but will use the COD equation and apply it with modification to the fully plastic range.

Eq. 1 provides the basis for this analysis. As does Tomkins, we take $\bar{\sigma}$ to be equal to the ultimate tensile strength, σ_u. We also employ a cyclic stress-strain relationship

$$\sigma_a = K' \left(\frac{\Delta\epsilon_p}{2} \right)^{n'} \tag{10}$$

where σ_a is the stress amplitude, $\Delta\epsilon_p$ is the plastic strain range, and K' and n' are constants. σ_u is then given by $K(n')^{n'}$. It is also necessary to re-place E by an equivalent value, and we take E to be equal to the secant modulus, E_{sec} (2). The resultant expression for the rate of crack growth in terms of strain becomes:

$$\frac{\Delta a}{\Delta N} = B \left[\frac{16}{\pi} (n')^{n'} \left(\frac{\Delta\epsilon_p}{2} \right)^{1-n'} a \; \ln \sec \frac{\pi}{2} \left(\frac{\Delta\epsilon_p}{2n'} \right)^{n'} \right] \tag{11}$$

We next compare equation 11 with experiment. In the low cycle fatigue range Stage II fatigue crack growth occupies much of the total lifetime (8). If it is assumed that the growth of cracks, in fact, takes up the entire life-time, then an integrated form of Eq. 11 can be used to compare with low cycle fatigue lifetime results. The integrated equation is:

$$\ln \frac{a_f}{a_i} = \left[\frac{16}{\pi} (n')^{n'} \left(\frac{\Delta\epsilon_p}{2} \right)^{1-n'} \ln \sec \frac{\pi}{2} \left(\frac{\Delta\epsilon_p}{2n'} \right)^{n'} \right] N \tag{12}$$

Figure 6 is a plot of Eq. 12 for two typical values of the strain hardening exponent, n'. It appears from this plot that the slopes are higher than usually observed in low cycle fatigue tests, and we therefore inquire further into this matter. Three factors can serve to lessen the observed slopes over these in-dicated to Fig. 6. These are:

1. Static modes of separation are more operative the higher the strain level.

2. The final crack length, a_f, increases with decreasing strain amplitude.

3. Crack initiation and Stage I growth becoming more important components of the total life the lower the strain amplitude.

We can illustrate the effect of the second of these if we assume $\Delta\epsilon_p \sqrt{a_f}$ is the constant. This relationship indicates that the crack length at failure increases with decreasing strain, a factor which is generally neglected in this

type of low-cycle fatigue analysis. In Fig. 7, the lifetime curves for two
ratios of a_f/a_i are shown for a constant value of n'. If a_i remains constant
but a_f varies with the above $\Delta\varepsilon_p \sqrt{a_p}$ relationship then the predicted lifetime in
the mid-range of the plot would vary as indicated by the dashed line (a ten-
fold reduction in strainrange results in a 100-fold increase in a_f). This
leads to a reduction in slope from about 0.8 to about 0.5, and this latter
value is more in accord with experimental observation. This analysis therefore
raises some questions about certain approaches to lifetime prediction based
upon crack propagation considerations.

Results of crack growth rate determinations in low cycle fatigue can be
used to check the rate Eq. 11 more directly (9). Fig. 8 is a plot of the rate
of growth determined from experiment as a function of crack length for two
strainranges, ε_R. From Eq. 11 we can also calculate the slopes of these lines
for each of these strainranges and compare with experiment. The values are as
follows:

ε_R	Calculated Slope	Observed
.02	0.03	0.0064
.05	0.104	0.0315

It is seen that the calculated slopes are higher than the observed by an average
factor of approximately 4. Therefore the value of B in Eq. 11 can be set at
1/4, so that

$$\frac{\Delta a}{\Delta N} = \frac{4}{\pi} (n')^{n'} \left(\frac{\Delta\varepsilon_p}{2}\right)^{1-n'} a \ln \sec \frac{\pi}{2} \left(\frac{\Delta\varepsilon_p}{2n}\right)^{n'} \tag{13}$$

Further work in this general area would be useful to establish the validity of
this approach.

4. ACKNOWLEDGMENT

The support of this work under Air Force Grant No. AFOSR-74-2703 is gratefully acknowledged. One of the authors (A.J.M.) would also like to express
his thanks to the Japan Society for the Promotion of Science and to

Professor S. Taira for their support while a Visiting Professor in the Department of Mechanical Engineering II at the Kyoto University where this study was initiated.

5. REFERENCES

(1a) B. A. Bilby, A. H. Cotrell, and D. H. Swinden, "The Spread of Plastic Yield from a Notch, Proc. Royal Soc., A272, (1963), p. 304.

(1b) B. A. Bilby, and K. H. Swinden, "Representation of Plasticity at Notches by Linear Dislocation Arrays," Proc. Royal Soc., A285, (1965), p. 22.

(2) A. J. McEvily, "Phenomenological and Microstructural Aspects of Fatigue," Proc. of the Third Int. Conf. on the Strength of Metals and Alloys, 2, Cambridge, England, (1973), The Institute of Metals and The Iron and Steel Institute, London, p. 204.

(3) J. R. Rice, "Mechanics of Crack Tip Deformation and Extension by Fatigue," ASTM, STP No. 415, (1967), p. 247.

(4) T. Yokobori, I. Kawada, and Hideo Hata, "The Effects of Ferrite Grain Size on the Stage II Fatigue Crack Propagation in Plain Low Carbon Steel," Strength and Fracture of Materials, 9, No. 2, Tohoku Univ., (1973).

(5) R. J. Donahue, H. M. Clark, P. Atanmo, R. Kumble, and A. J. McEvily, "Crack Opening Displacement and the Rate of Fatigue Crack Growth," Int. J. of Fracture Mech., 8, (1972), p. 209.

(6) P. C. Paris, and F. Erdogan, "A Critical Analysis of Crack Propagation Laws," J. Basic Eng. Trans. ASME, 85, (1963), p. 528.

(7) B. Tomkins, "Fatigue Crack Propagation - An Analysis," Phil. Mag., 18, (1968), p. 1041.

(8) C. Laird, and G. Smith, "Crack Propagation in High Stress," Phil. Mag., Ser. 7, (1962), p. 847.

(9) R. C. Boettner, C. Laird, and A. J. McEvily, "Crack Nucleation and Growth in High Strain-Low Cycle Fatigue," Trans. AIME, 233, (1965), p. 379.

TABLE I

TEST CONDITIONS

Specimen Mark	Mean Diameter of Ferrite(μ)	ASTM GSNO	Stress (K_p/mm^2)	Lower Yield Strength (K_p/mm^2)	Ultimate Tensile Strength (K_p/mm^2)	δ	Log A_o Eq. 9
M 5	12.1	9.6	10.08 ± 6.17	25.9	36.1	2.99	− 9.328
M 14			10.70 ± 4.63				
M 17			10.19 ± 6.17				
M 18			10.55 ± 4.63				
A 2	15.9	9.0	10.80 ± 4.63	21.6	30.9	3.56	−10.402
A 4			9.98 ± 6.17				
B 3	25.0	7.6	10.13 ± 6.17	20.7	30.2	3.47	−10.194
B 4			10.19 ± 6.17				
B 5			10.55 ± 4.63				
B 6			10.39 ± 4.63				
G 3	28.5	7.3	10.13 ± 4.63	19.6	29.3	4.03	−11.103
G 4			10.19 ± 6.17				
G 7			9.52 ± 6.17				
F 4	205.6	1.6	4.63 ± 4.63	10.4	27.1	3.63	−10.380
F 5			4.89 ± 4.63				
F 6			4.32 ± 4.63				

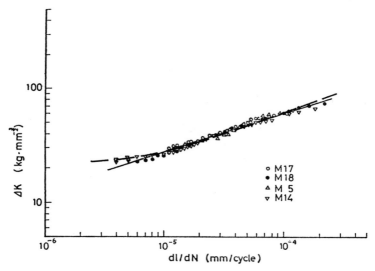

Fig. 1. The rate of fatigue crack growth, dℓ/dN, as a function of the stress intensity range, ΔK, for 0.5 C steel in condition M. Solid line equation (9), dashed line equation (7). Data from Ref. 4. (ℓ=2a)

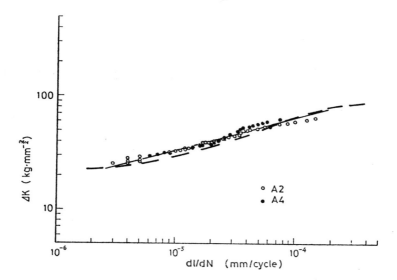

Fig. 2. The rate of fatigue crack growth, dℓ/dN, as a function of the stress intensity range, ΔK, for 0.5 C steel in condition A. Solid line equation (9), dashed line equation (7). Data from Ref. 4.

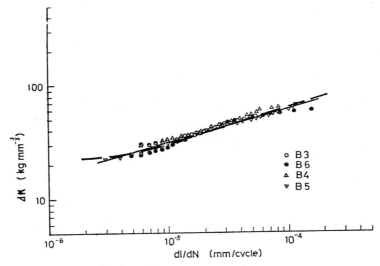

Fig. 3. The rate of fatigue crack growth, dℓ/dN, as a function of the stress
intensity range, ΔK, for 0.5 C steel in condition B. Solid line
equation (9), dashed line equation (7). Data from Ref. 4.

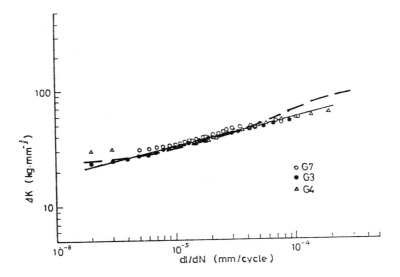

Fig. 4. The rate of fatigue crack growth, dℓ/dN, as a function of the stress
intensity range, ΔK, for 0.5 C steel in condition G. Solid line
equation (9), dashed line equation (7). Data from Ref. 4.

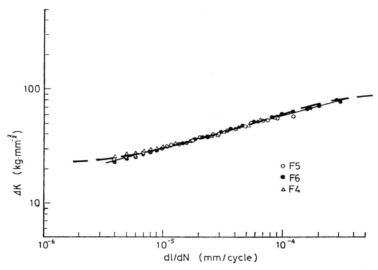

Fig. 5. The rate of fatigue crack growth, dℓ/dN, as a function of the stress intensity range, ΔK, for 0.5 C steel in condition F. Solid line equation (9), dashed line equation (7). Data from Ref. 4.

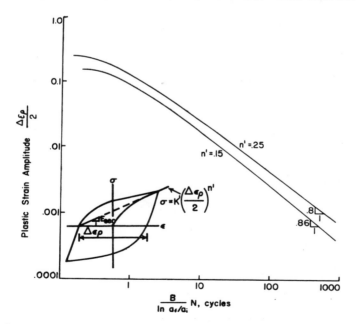

Fig. 6. Life-time predictions based upon equation (12) for two typical strain hardening exponents, n'.

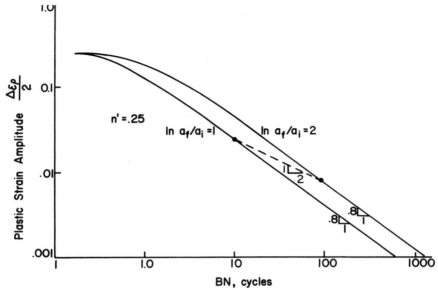

Fig. 7.   Life-time curves for two ratios of a_f/a_i for constant strain
hardening exponent, n'.   The dashed line indicates the predicted
life-time if a_f is determined by the relation $\Delta\varepsilon_p\sqrt{a}$ = constant.

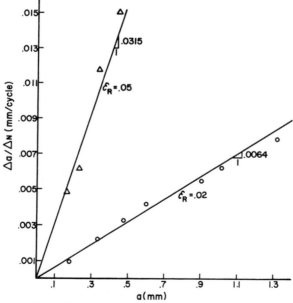

Fig. 8.   The rate of crack growth as a function of crack length for two strain
ranges $\Delta\varepsilon_p$.   Experimental points from Ref. 9.

Refinements of the Crack Models for Subcritical Crack Growth

by

Hideo Kitagawa*, Ippei Susuki*, Ryoji Yuuki* and Osamu Sakazume*

Abstract

   Various refinements of a crack model in fracture mechanics are
now being tried, so that the subcritical crack growth can be examined
in more details. Some of them are discussed, with introduction of
some developed methods of analysis.

   The first refinement is a statistical fracture mechanics
analysis of the fracture process in which randomly distributed
plural cracks interact mutually, instead of the analysis based on a
single isolated crack. A corrosion fatigue process is analized as
an example, and a general basic model or procedure is proposed.

   The second refinement is a crack-morphological fracture
mechanics approach to the crack growth which is accompanied by
branching or folding. Prepared analytical solutions of the stress
intensity factors for such nonlinear shaped cracks are applied.
Various models can be proposed.

1.   Significance of Refinements of Crack Model in Material Science

   When the fracture of solid materials is considered in relation
to the structures of materials, refined approaches are often tried.

   In metallurgical approaches, many various factors are usually
taken into consideration. For instance, the rolls of grain
boundaries, shapes and sizes of grains, and local chemical composition,

* Tokyo University, Institute of Industrial Science, Tokyo

physical properties and orientation of microstructures are discussed.

In mechanical approaches, however, only a very simple crack
model, that is, a single isolated straight line crack of the first
mode is usually applied to the analyses of the fracture.

Very many cases of subcritical crack growth have been well
explained with the results of the elastic analyses of this simple
crack. But when the phenomena of the micro-structure scale are
introduced, the analyses with such a simple crack model may not be
enough, as the observed actual cracks are not so simple. Important
phenomena are possibly missed or misunderstood. Refinements of the
model will be required in this meaning.

2. Various Refinements of the Crack Model in Fracture Mechanics

From another point of view, the refinements of the crack model
are required.

In general, fracture mechanics is composed of two procedures,
those are, (1) an analysis of a crack model for obtaining some
characteristic parameters of the model, and (2) a characterization
of the behaviors of an actual crack, which is assumed to be similar
to the analized model, with the parameters stated above.

Whether the selected model or parameter is suitable to the
given actual crack or not should be determined by the experiments
and the survey of actual fractures in service. If the selection is
determined as improper, a refinement of the crack model will be
required.

The following refinements of the crack model, shown in Table 1,
are thought to be important at present.

Many works have recently been done in relation to i) and ii)

Table 1   Important Refinements of Crack Model in
           Subcritical Crack Growth

refined model	classification of analysis	relating problems (applicable problems)
i) plastically deformed crack	material non-linearity problem	speed effect, temperature effect, bi-axial stress, hysteresis effect (variable loads) non-propagating crack, and creep-fatigue interaction
ii) partially closed crack	indeterminate crack problem (crack closure problem)	residual stress effect, stress ratio effect, hysteresis effect, non-propagating crack, and compression and shear
iii) non-linear shaped crack	geometric non-linearity problem (crack morphology) statistical problem	branching and secondary crack zig-zaging mix-modes effect early stage crack growth intergranular cracking apparent K-independency
iv) plural cracks with or without interaction (randomly distributed cracks)	statistical fracture mechanics	mix-modes effect, random initiation, corrosion fatigue, stress corrosion cracking, merging of preceding cracks, fracture of cast iron, concrete and defective welds
v) crack with non-linear crack front	three dimensional crack problem	internal crack, surface crack, thickness effect, plane bending and local stress intensity
vi) boundary crossing crack and interface crack	dissimilar media crack problem and statistical problem	composite, lamination, plating, welds, intergranular cracking and transgranular cracking

in Table 1. As for vi), only some analyses are found. As to v),
many works are now being carried out. The problems iii) and iv) are
discussed here. The problem vi) will be discussed later.

3. Fracture Due to Interaction or Coalescence of Randomly
    Distributed Cracks (An Example of No.4 Refinement)

3.1 Points of the Problem and Our Proposal

    (1) In the tests of corrosion fatigue, stress corrosion
cracking, some low cycle fatigue, fretting fatigue or some thermal
fatigue, randomly distributed many small cracks can be seen in usual.
Sometimes the cracks are distributed in high density (Fig.1 and 2),
and when the interaction among these cracks and succeeding coalescence
occur, an accelerated fast crack growth or a fast fracture can often
be observed at the final stage. And even then a little bit before
the break-down, we can find only small cracks in many cases.

(2) Only fairly small cracks are found on the surface of machines
and structures, when they are under a satisfactory inspection or
quality control. In these cases, the conditions for significant
growth of a gingle crack are not easily attained, because the stress
intensity factors of these small cracks are often lower than or near
to the threshold, e.g., ΔK_{TH} (shown in Fig.3) or K_{1SCC} , and only a
very slow crack growth or no crack growth can be expected. In such
cases, we can not well explain an extraordinary loss of strength or
life probably due to cracks or flaws, if we do not assume any
interaction among them.

(3) Many fine cracks can often be detected which are located close
to the fracture part of actual broken components in severe service
conditions. These cracks in themselves seem not to have contributed

to the final fracture. But on the fracture surface along its periphery the traces of small cracks, those are, stepwise faceted surfaces can be observed, which are assumed to have contributed to the fracture as origins of a final crack growth or a fast fracture. (4) Deducing from (1), (2) and (3) stated above we can assume the following process: Increase of the distribution density of randomly distributed cracks induces the interaction among the cracks, and is accompanied with extensive increase of stress intensity factors (K) of crack tips and emergence of the second mode of K, K_{II} . These two above seem to accelerate the coalescence or the coalescence-like interaction state of the cracks. But when such a kind of coalescence is achieved, these high K values at the neighbouring crack tips disappear or fall down. If such coalescences can succeed, some of the values of K at the other crack tips of the coalesced or interacted cracks will increase. If the K value at a certain crack tip or tips considerably exceeds the threshold stress intensity factors (ΔK_{TH} , K_{1SCC}) for crack growth or other similar conditions, the crack tip or tips can start to grow with significant growth rate. After then the extraordinary acceleration of crack growth due to the accelerated interaction and coalescence and a final fracture can also be expected, which means the relative lowering of strength of the cracked material. (5) As a suitable example of this kind of fracture process, corrosion fatigue was selected. The crack distribution pattern and the increasing rate of the distribution density of cracks due to corrosion fatigue were taken as input data. Not only the case of corrosion fatigue, the interaction among small cracks and succeeding coalescences and coalescence-like interactions can be observed also in the case of fatigue in dry conditions[1] or stress corrosion cracking[2].

The analysis was carried out by a statistical simulation as a Monte Carlo method combined with linear fracture mechanics.

3.2 Calculation of "Equi-Interaction Factor Diagrams" of Mutually Interacted Distributed Cracks

There have been some papers on the analysis of the strength of the material with distributed cracks. But in most cases the distribution density of the cracks was fixed and the cracks were treated as independent of each other and the interactions among them were not taken into consideration. The present model includes the treatments of the behaviors of the increasing or growing randomly distributed cracks with their mutual interaction or connection. So the determination of the interaction is a meaningful step of the procedure for the present model.

Being based on the Isida's solution[3], we made a computor program which can follow the variation of stress intensity factors (K) of the continuously shifting or growing parallel cracks with any given relative positions (as Fig.4) and crack sizes, by means of a Laurent series expansion and a perturbation technique.

The values of K of modes I and II at four crack tips in Fig.4 were determined. The eight values of non-dimensional stress intensity factors for each combination of the positions of crack-center and the crack sizes

$$f_k^{i,j} \equiv \frac{K_k^{i,j}}{\sigma \sqrt{\pi a_j}} = 2 \sum_{q=0}^{\infty} \alpha_i^q \sum_{p=0}^{q} \left[(p+1)(p+2) a_j^p \left({}_k L_{p,j}^{(q-p)} + {}_k M_{p,j}^{(q-p)} \right) S(p) \right]$$

$$i = \begin{cases} A \text{ (right tip of crack)} \\ B \text{ (left tip of crack)} \end{cases} \qquad j : \text{ name of crack}$$

$$\alpha = \begin{cases} 1 & (\text{if} \quad i = A) \\ -1 & (\text{if} \quad i = B) \end{cases} \qquad k = \begin{cases} I & (\text{for mode I}) \\ II & (\text{for mode II}) \end{cases}$$

$$S(p) = \begin{cases} S_0\left(\dfrac{p}{2}\right) & (p : \text{even}, 0) \\ S_e\left(\dfrac{2p-1}{2}\right) & (p : \text{odd}) \end{cases}$$

where

$$S_0\,(p) = \sum_{m=0}^{p} A_{p-n,\,2n+1}$$
$$S_e\,(p) = \sum_{m=0}^{p} A_{p-n,\,2n+2}$$

$A_{p-n,\,2n+1}$, etc., are the real constant in the conformal mapping functions, and $_k\angle_{p,j}^{(g-p)}$, etc., are the complex factors depending on the free boundary conditions of the cracks.

In Figs.5 and 6 are shown several examples of the results of the computation on the crack-interaction-region within which the value of the function $\int$ is larger than the given value indicated at each curved line. When the left tip of crack 2, B2, is located on the curved line $\int_{1}^{A2} = 1.06$ in Fig.6, the value of K_I at the crack tip A2 is $K_{1}^{A2} = (1.06) \times 6\sqrt{\pi a_2}$. Here we will call the " $\int$ " as "interaction factor". Eight figures for $\int_{I}^{A2}$, $\int_{II}^{A2}$, $\int_{I}^{A1}$, $\int_{II}^{A1}$, etc., can be given for each crack length ratio, a_2/a_1 . Each curved line can be called as a "Equi-interaction factor diagram".

For the higher values of $\int_{I}^{A2}$, etc., than the highest value indicated in these figures, the similar shaped curves continue toward the crack tip with rapidly increasing density of the curves. This means that the value of $\int$ increases rapidly at the vicinity of the crack tip.

The value of $\int$ has a maximum, as the crack shifts along x-direction except for the case of exact colinear cracks.

3.3 Experimental Determination of Input Data and the Characteristics of the Process

It is considered that the distribution function of crack size $[\,\mathcal{F}(a)\,]$ and the increasing rate of the number of cracks (dm/dt or dm/dN) and other input data and the characteristics of the fracture process required to simulate have to be determined by experiments or statistical surveys of accidents. Repeated tension loads were applied to the unnotched plate specimens (70^{mm} x 440^{mm} x 2.3^{mm}) of a high tensile strength steel (HT 50, 50kg/mm^2 grade) in the well water in our university campus, and the cracks were observed (Figs.1, 7 and 8) and measured.

As shown in these photographs, the cracks are essentially parallel each other. And they were very small, and the observed maximum values of their half length ($a$max) were about 0.2^{mm}, and the K values of these cracks were much less than the expected value of ΔK_{THCF} (threshold stress intensity factor range of corrosion fatigue cracks for the given environment).

The distribution density of the cracks was more than 150/100 (the number of cracks/mm^2) at the final stage for fracture, which seems to be high enough for the interaction among them.

The observed process of their interaction or connection were geometrically similar to that of the large through parallel cracks which we observed in the other tests.

The several criteria for the connection can be selected. Here, we took the criterion of "constant interaction factor", considering the observed geometrical similarity stated above. In the present calculation, the interaction or connection region was taken as Fig.9 for f = 1.50 .

The total number of the cracks observed increased linearly with respect to the number of load cycles, N , as shown in Fig.10.

It is the most interesting that the distribution pattern of the crack size (a) did not change as N increases up to Nf, and all

of the data fell upon one log-normal distribution line as shown in
Fig.11.

It seems natural to consider that the cracks originated without
the change of distribution pattern, and they hardly started to grow
or grew very slowly. Nf in Fig.11 is the number of stress cycles
to fracture, and it was given as the mean value of the results of
several specimens.

In addition to the tests stated above, the crack growth of a
single large crack in the same wet conditions was measured.
Origination of the other cracks was prevented by coating. The
results show that the growth rate of a single crack in wet conditions
follows the $\Delta K \sim \frac{da}{dN}$ relation of the same kind as that in dry
conditions, straight and almost parallel, as shown in Fig.12. This
assures that the present growth rate $\frac{da}{dN}$ shows the below K_{1SCC}
behavior and it is a simple function of ΔK .

The other fatigue tests were carried out on the wide plate
specimens with two interacted large through cracks. The results
show that the values of $\frac{dn}{dN}$ at each crack tips follow the same
function of ΔK as a single crack.

3.4 Simulation Model and Calculation Procedures

A model for one trial of the statistical simulation for the
increasing process of the maximum value of K with increase of the
number of given cracks (m) (the number of crack initiation) was
constructed as the followings, taking the test results stated above
into consideration.

① Any crack tip does not start to grow remarkably, if the value of
its K does not exceed a threshold stress intensity factor (ΔK_{TH} ,

ΔK_{THCF} , ΔK_{1SCC}).

② The positions of cracks (x, y) at their initiation are the random variables following a given distribution function, $\Phi(x,y)$. (In the present calculation, it was assumed as uniformly random both in x and y directions)

③ The sizes of cracks (a ; half crack length) at their initiation are the random variable following a given distribution function $\Psi(a)$. (In the present calculation, it was put as a log-normal function obtained from the test results.)

④ The cracks generate one by one at a given increasing rate (obtained from the test results), following ② and ③ stated above.

⑤ When a tip of a crack falls in a given interaction region (Fig.9) at a tip of another crack, these two cracks come rapidly into the state of interaction or connection or into the same conditions as connection. In other words, the values of K at the outer tips of a pair of the two cracks become nearly equal to the values of K at the tips of one equivalent crack which is assumed to generate by the connection.

⑥ Even before the next crack initiates, the interaction region can be enlarged only by the connection stated above, and accordingly the second or the third connection can occur with the other cracks which initiated in the outside of the interaction region and could not come into interaction before this time. This is also computed.

⑦ When this one series of successive connections have finished without any maximum K value which exceeds K_{TH} , or when any connection has not occured, the initiation of the next crack succeeds.

⑧ When the maximum value of K exceeds K_{TH} , a stable crack growth begins, leading rapidly to final fracture. Here, this one trial ends.

A flow chart for this one trial is shown in Fig.13.

The trials for the calculation of the increase of the maximum value of K, Kmax, or the maximum value of crack size, $a$max, with respect to the increase of the number of given cracks, m , are repeated. And the accumulated probability for $K > K_{TH}$, $P\{ K > K_{TH}\}$, is obtained with respect to " m ", and accordingly, to " N ", by means of the given relation between m and N.

3.5 Calculation and its

Results

An example of a part of one trial is shown in Table 2. " m " is the total number of the observed cracks, which was chosen as the number of initiation of the cracks in this case for the reason stated previously. The number of connection and the value of $a$max increased very slowly with increase of the number of cracks, " m ", over the almost whole period of crack initiation or fatigue life. And, however,

Table 2 An example of the trial of the calculation on the increase of maximum crack length and the number of the connection of cracks

numbers of crack initiation m	numbers of crack connection	maximum crack length a (mm)
0	0	0
10	0	
20	0	
30	0	
⋮	⋮	⋮
100	0	0.092
110	0	0.092
120	1	0.107
130	2	0.126
140	2	0.126
150	2	0.126
160	2	0.126
170	2	0.126
180	3	0.126
190	3	0.126
200	4	0.126
210	5	0.126
220	5	0.126
230	5	0.126
240	5	0.126
250	6	0.126
260	7	0.126
270	7	0.126
280	7	0.126
290	7	0.126
300	8	0.126
310	8	0.126
320	8	0.126
330	8	0.126
340	8	0.126
350	8	0.126
360	205	100.205

being of great importance, at a certain value of "m" ($350 < m \leqq 360$, in this case) the number of crack-connections and the maximum crack size, $a$max, increased abruptly and explosively up to about 25 times and 1000 times respectively as shown in Table 2.

Some of such trials are shown in Fig.14 for examples.   One folded line represents one trial.   The increasing process of the maximum crack size with respect to "m" can considerably change from a trial to a trial. Most of the values of "m" which are required for fracture were scattered between 100 and 400, and even at more than 400, there were the cases in which any unstable connection did not occur.

This can well explain many of the actual accidental fractures, and the properties of the extraordinary lowering of the strength in some corrosion fatigue processes.   For an example, in the corrosion fatigue tests with the unnotched specimens of a mild steel or a low alloy steel in some mild corrosive water, an abrupt fall-down of S-N curves can be observed after the duration of tests for more than 40 days.   And again, in the long time stress corrosion cracking tests of a stainless steel in a certain chloride solution, an abrupt breakdown can also be recognized at some stage of the tests when the surface of the specimen has just been covered with many small cracks in high density (Fig.2).

Fig.15 shows the probability (P) that the value of $a$max exceeds a_{TH} , or the value of ΔKmax exceed ΔK_{TH} (or ΔK_{THCF}) before the given value of "m".   The value of P varies with respect to "m" as a normal distribution.

This P can be approximately taken as the probability of the occurence of fracture in our case.   The reliability, R , is 1-P.

3.6   Crack Growth Accelerated by Interactions

After starting to grow, the growth of each connected crack can be also accelerated extensively by the interaction between cracks. As the growth rate follows well ΔK_I in spite of the existence of ΔK_{II}, the crack growth curves ($a \sim N$ relations) can be computed.

The results showed that the crack growth with interaction is much faster than the crack growth with no interaction. This is the reason why the time of fracture in the simulation in the preceding section was chosen as the time when $\Delta Kmax$ exceeds ΔK_{THCF}.

4. Crack Morphology in Subcritical Crack Growth

4.1 Available Criteria for Two-Dimensional Crack Path

Careful observations and fractographic experiences tell us that a two-dimensional crack path in subcritical growth is not always straight and it includes considerable amount of non-linear shaped portions. Typical examples of such geometric non-linearity are branching, zig-zaging and merging.

Available criteria controlling the direction of sub-critical crack growth are follows ;

(1) σ_θ is max. or $\tau_{r\theta} = 0$ at the tip of original straight crack

(2) S_c - criterion ($\frac{\partial S}{\partial \theta} = 0$) ---- ditto

(3) K_I is max. at the presumed bent tip of original straight crack

(4) $K_{II} = 0$ - - - - - - - - - - - - - - ditto

The solutions required to answer the latter two, (3) and (4), have recently been prepared by the authors[4] by an analytical method. Some of the numerical values of the solutions have recently been reported[5]. Some of them will be referred again for the present discussion.

In the case of a single straight crack in the uniaxial stress

field, the selection of the criteria does not give any significant
effect upon the conclusion on the crack path. But for applying them
to general cases further researches will be required. A few
particular cases will be discussed later.

4.2 Available Criteria for the Growth Rate in Two-Dimentional Crack Growth

Available general criteria for the subcritical crack growth
rate which include mixed mode effects have not been found both in
fatigue and SCC.

As for a sharply bent slant crack, the stress intensity factors
(K_{IB}) at the tip B (in Fig.16) are shown in Fig.16[5]. Comparing
this value of K_{IB} with the value of growth rate da/dN obtained
from the fatigue tests carried out by Iida and Kobayashi[6], the
relation between ΔK and $\frac{da}{dN}$ for a non-linear shaped crack might
be just the same as the relation in straight line crack perpendicular
to the load. In this case, however, the crack tip B is tracing the
direction in which $K_{II} = 0$, as shown in Fig.17[5], that is,
$\theta = 53 \sim 57°$. The criterion (4) in the preceding section holds
good in this case.

As a present conclusion, it may be said that da/dN in mixed
mode is a function of ΔK_I and follows the same law as for the
first mode crack, i.e., $da/dN = C (\Delta K_I)^m$

Further examinations will be required.

4.3 Non-linear Shaped Crack Growth Path

When a tip of a growing crack sharply bends or branches by
chance, being affected by the microstructures, the value of K_I at
the tip considerably falls down as shown in Figs.18 and 19. This
possibly induces a local or temporary decrease of growth rate.

Both these states, however, are unstable, and they can not continue up to long distance. In the case of the sharp bent, the bent tip will bend again to the original direction, that is, almost parallel to the direction of the main part of the crack, as shown in Fig.20[5], if the criterion (1) can be applied. In the case of the branching, if any unequality occurs in two branches, the value of K_I at the tip of the shorter branch rapidly decreases[7], and the shorter branch will be left behind, if the criterion for the mixed mode holds in this case.

If these proposed processes are repeated, the crack can macroscopically follow its original direction as a kind of stochastic process.

There always exists a possiblity that a microscopic bent or branching occurs within 20 to 30 degrees of bent or branch angle, if the criterion (1) or (3) holds. It can easily be understood from Figs.18, 19 and 17. The authors call it as "Snaking Crack Growth".

Accordingly, the microscopic bent or branching can be induced mainly by non-continuum mechanical reasons, those are, metallurgical and environmental reasons. Actually, remarkable or successive branchings or zig-zagings can be found in the fatigue or stress corrosion cracking in some specific materials and environments.

The macroscopic growing path in subcritical crack growth is controlled by continuum mechanical effects. As a result, a zig-zaging of crack growth (Fig.21) is governed by the combined mechanical, metallurgical and environmental effects.

5. Summary

Refinements of crack model have been stressed for the successful refined studies of fracture and strength of materials, combined with the structures and properties of the materials.

Various kinds of refinements of crack models have been proposed or re-arranged. Some of them have been discussed.

Corrosion fatigue process has been analized by a randomly distributed mutually interacted cracks' model, that is, a Monte Carlo statistical simulation combined with fracture mechanics. An interesting fracture process due to explosive connection of small cracks has been revealed.

Zig-zaging in subcritical crack growth has been discussed by a non-linear shaped crack model which depends on the analytical solution developed by the authors. A "Crack-Morphological Approach in Fracture Mechanics" has been introduced.

6. Acknowledgements

The authors are gratefully obliged to Prof. M. Shinozuka of Columbia Univ., Prof. M. Isida of Kyushu Univ. and Prof. F. McClintock of MIT for their instructive advices and suggestions and fruitful discussions, and to Mr. T. Ohira, Mr. M. Kuroda, Mr. K. Miyazawa and Mrs. K. Yoshioka for their valuable collaborations.

7. References

(1) S.Ueda and others ; presented at the A.E. committee of High Pressure Research Association of Japan, 1974

(2) H.Kitagawa, T.Ohira ; "Possibility of the application of fracture mechanics to the stress corrosion cracking of stainless steels", to be appeared

(3) M.Isida ; "Method of Laurent series expansion for internal crack problems", Methods of analysis and solutions of crack problems, ed. by G.C.Sih (1973) Noordhoff Intern. Publ. p.56

(4) H.Kitagawa, R.Yuuki : "Stress Intensity Factors of Non-linear Shaped Cracks", to be published

(5) H.Kitagawa, R.Yuuki, T.Ohira : "Crack-Morphological Aspects in Fracture Mechanics", to be published in Engng. Frac. Mech.

(6) S.Iida and A.S.Kobayashi : Trans.ASME Ser.D, Vol.91, No.4 (1969) p.764

(7) P.S.Theocaris : J.Mech.Phys.Solids, Vol.20 (1972) p.265

Fig.2 Fracture due to randomly distributed many small cracks in stress corrosion cracking of a stainless steel 304 ($\bar{\sigma} = 25$ k$_g$/mm²)

Fig.1 Randomly distributed many fine cracks, located near the fracture portion of a corrosion fatigue specimen. (50 kg/mm² grade high tension steel in tap water)

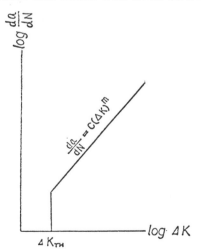

Fig.3 $\frac{da}{dN}$ (fatigue crack growth rate) versus ΔK (stress intensity factor range) diagram, indicating the behavior of crack at a threshold stress intensity factor, ΔK_{TH}

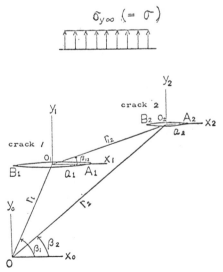

Fig.4 The model of interacted crack used for the calculation of the stress intensity factors

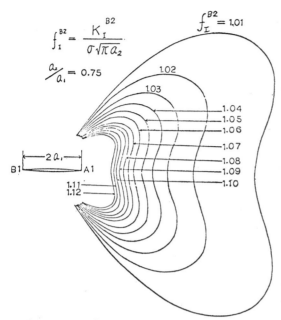

$$f_I^{B2} = \frac{K_I^{B2}}{\sigma\sqrt{\pi a_2}}$$

$$\frac{a_2}{a_1} = 0.75$$

$f_I^{B2} = 1.01$

1.02
1.03
1.04
1.05
1.06
1.07
1.08
1.09
1.10

$2a_1$

B1   A1

1.11
1.12

Fig.5 Equi-interaction factor diagrams for the SIF at the crack tip B2 ($\frac{a_2}{a_1} = 0.75$)

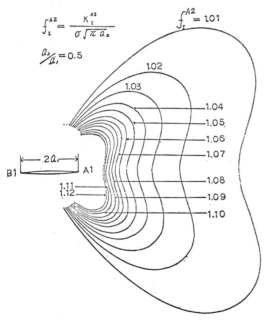

$$f_I^{A2} = \frac{K_I^{A2}}{\sigma\sqrt{\pi a_2}}$$

$$\frac{a_2}{a_1} = 0.5$$

$f_I^{A2} = 1.01$

1.02
1.03
1.04
1.05
1.06
1.07
1.08
1.09
1.10

$2a_1$

B1   A1

1.11
1.12

Fig.6 Equi-interaction factor diagrams for the SIF at the crack tip A2 ($\frac{a_2}{a_1} = 0.5$)

Fig.7 Coalescence of four small parallel cracks

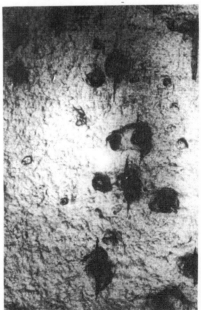

Fig.8 Closely located parallel four cracks are now partially connected

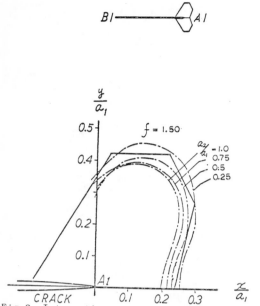

Fig.9 Interaction or coalescence region used for the calculation (The area bounded by the folded solid line)

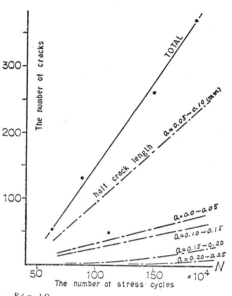

Fig.10 Increase of the number of the cracks with each crack sizes (half length a) and the total number of all cracks

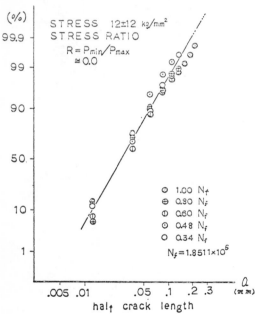

Fig.11 Log-normal distributions of the corrosion fatigue crack lengths at the various stages of stress cycles (Nf : the mean of the number of cycles to fracture)

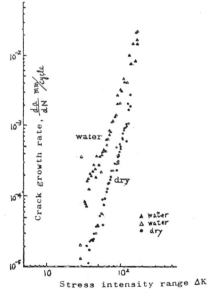

Fig.12 Growth rates of single fatigue crack in high-strength steel (50 k_g/mm^2 grade) in well water and in dry conditions

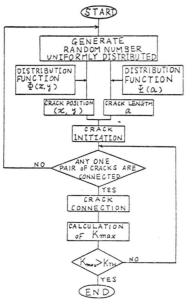

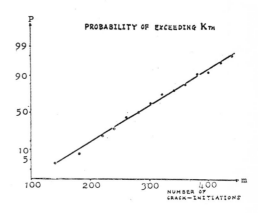

PROBABILITY OF EXCEEDING K_{TH}

Fig.13 Flow chart of the calculation of
the simulation model which gives
the starting of stable growth of
randomly distributed cracks by
their interactions or conections

Fig.15 Normal distribution of the
probability that Kmax (equivalent
to a_{max}) $> K_{TH}$ (equivalent to a_{TH})
against the given number of cracks
(m)

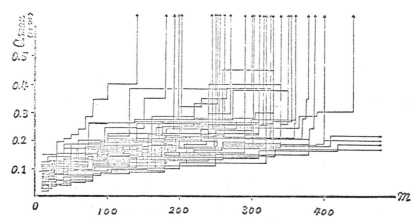

Fig.14 Examples of the trails of the simulation for the increase of
maximum crack length with the increase of the given total
number of cracks

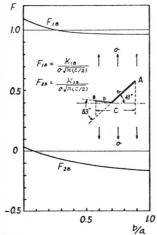

Fig.16 Normalized stress intensity factors of the bent crack for $\theta = 53$, $\varphi_0 = 45°$

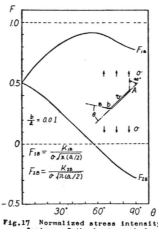

Fig.17 Normalized stress intensity factors of the bent crack for $b/a = 0.01$, $\varphi_0 = 45°$

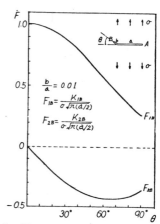

Fig.18 Normalized stress intensity factors of the bent crack for $b/a = 0.01$

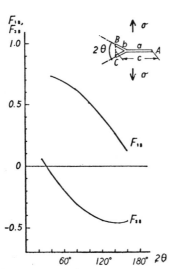

Fig.19 Normalized stress intensity factors of the forked crack for $b/a = 0.1$

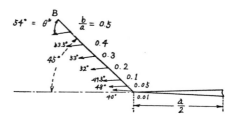

Fig.20 The direction in which the bent crack propagates

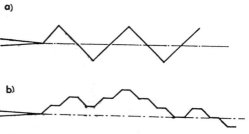

a)

b)

Fig.21 The sketch of the zig-zaging crack

On the Stage I and Stage II Fatigue Crack Propagation Behaviours

by Hajime NAKAZAWA[*] and Hideo KOBAYASHI[**]

1 Introduction

It is generally held that there is no stage I cracking in sharply notched specimens but that stage II cracking is found immediately at the bottom of the notch. On the other hand, evidence of crystallographic fracture surface has been reported in the initial stage of cracking.[1~4] Hitherto, concerning the crack propagation behaviour in notched specimens, the observation of the initial stage of cracking has been insufficient. We however, have now carefully observed this cracking behaviour. All such observations here made on low carbon steels; evidence of stage I cracking, the difference in the mechanisms of stage I and stage II cracking, and other characteristics of stage I cracking are examined minutely by means of metallographical and fractographical methods.[5] In addition, concerning the stage II cracking, the crack propagation behaviour and the fatigue lives to complete fracture of center-slitted low carbon steel specimens were investigated using the techniques of fracture mechanics and fractography.[6][7]

2 Experimental Details

The experiments were performed using a rotating bending fatigue testing machine for the stage I cracking and using a zero-tension fatigue testing machine for the stage II cracking. Two kinds of material were used in these experiments as shown in Table 1. Heat treatments and the tensile properties of the materials are shown in Table 2. Material Y is the same as material X ex-

---

[*] Dr., Professor, Department of Physical Engineering, Faculty of Engineering, Tokyo Institute of Technology, Tokyo

[**] Dr., Associate Professor, Department of Physical Engineering, Faculty of Engineering, Tokyo Institute of Technology, Tokyo

cept that the grain size of material Y was larger than that of material X as the former was annealed at a higher temperature than the latter. The configuration of specimens were shown in Fig. 1. Materials X and Y were used in the rotating bending fatigue test and material Z was used in the zero-tension fatigue test.

3 Results and Discussion

The S-N curves for smoothed and notched specimens of materials X and Y in the rotating bending fatigue test are shown in Fig. 2. As seen from these results, the fatigue limits for smoothed specimens depend on their grain size, but those for notched specimens are independent of their grain size.

Non-propagating cracks were observed on the surfaces of smoothed specimens when 10^7 cycles of the fatigue limit stress was applied. The maximum length of non-propagating cracks did not exceed twice the length of the grain size of the specimens. From these results we may draw the interesting conclusion that the reason for the dependency of fatigue limit on grain size is that the length of non-propagating slip-band cracks at the fatigue limit is effected by grain size. These slip band cracks are no doubt the stage I cracks.

In notched specimens, the fatigue lives to complete fracture and the fatigue limits are decided by the states of the stage II crack propagation. This agrees with the known fact that the stage II crack propagation is not affected by the grain size.[2]

In this work, crack propagation behaviour was examined microscopically at three stress levels, 19.5 kg/mm^2, 14.8 kg/mm^2 and 10.8 kg/mm^2. The crack propagation curves are shown in Fig. 3. It was comfirmed that initial cracks of limited length were observed after limited stress cycles. In the initial stage of crack propagation, the crack propagation curves change stepwise. The crack initiation life N_0 and the delay life in crack propagation N_h of materials X and Y were compared. The results are shown in Table 3. The initiation crack depth

a_0 and the crack depth a_h after delay life N_h are also shown. Although there are a few instances of an increasing rate of depth a_0 and a_h with increased stress levels, depth a_0 and a_h are not materially effected by stress levels. The point we wish to emphasize is that depth a_0 is less than the mean diameter D of the grain of the material. It should be noted that the cracking of depth a_0 is crystallographic cracking, and this fact is supported by the following microscopically observations. Photo. 1 shows a typical example of an micrograph of an initial crack. Many of the initial cracks were inclined about 45 degrees in the direction of the applied stress, and the fracture surface of the initial cracking was fairly smooth. Therefore, it seems safe to assume that these initial cracks in the sharply notched specimens are the cracks of stage I along the slip bands as those in the smoothed specimens. Photo. 2 shows a micrograph of a crack propagated to considerable depth. In this micrograph, it is seen that the initial crack is the one along the slip band and that after this stage I crack has penetrated to a certain depth, stage II crack propagation succeeds stage I at right angles to the direction of the applied stress. It also shows that the zig-zag propagation of cracks occurs independently of grain boundary, but when the crack propagates into a neighbouring grain, there is a change in direction.

We can thus consider that the process of the zig-zag propagation of cracks is performed by crystallographic manner although crack propagation macroscopically seems to take place at right angles to the direction of the applied stress. It has not yet been determined whether the zig-zag propagation of cracks in notched specimens corresponds to the stage I cracking in smoothed specimens or to the stage II cracking in smoothed specimens. The zig-zag propagation of cracks is observed at crack depths of $a = 0.1$ mm for material X and $a = 0.2$ mm for material Y. With the exception of this process, stage II crack propagation behaviour is not effected by grain size, as shown in Photo. 2.

That is to say, since the lives of N_0, N_h and the zig-zag crack propagation process, which are effected by the grain size, are comparatively much shorter than the fatigue lives to complete fracture, the fatigue lives of sharply notched specimens can be considered to be uneffected by the grain size.

The stage II crack propagation behaviour was examined with center-slitted plate specimens under repeated tensile loading. Generally, the crack propagation rates are correlated with the ranges of the stress intensity factor.[8] In this work, the range of the stress intensity factor ΔK used are expressed as follows.

$$\Delta K = \Delta\sigma \, \{ \pi a \, \sec(\pi a/W) \}^{1/2} \qquad\qquad (1)$$

Hence, $\Delta\sigma$ is the range of repeated stress, a is the half length of the crack and W is the width of the specimen.

Some of results are shown in Fig. 4. All the other results were similarly obtained. These results show that crack propagation rates $d(2a)/dN$ over a wide range are formularized as follows.

$$d(2a)/dN = C \, (\Delta K)^m \qquad\qquad (2)$$

In this formura, experimental constants m and C are shown in Table 4. Strictly speaking, it appears that the larger the width of the specimen (from 50 mm to 200 mm), the smaller the value of m, and that the higher the stress ratio (from 0.06 to 0.36), the larger the value of C. But, the thickness of the specimen, slit size, stress wave, tensile/shear fracture mode transition behaviour and reduction of the thickness of the specimen resulting from general yielding do not affect the values of m and C.

Photo. 3 shows a fractograph of initial stage of cracking from the slit. In this fractograph, stage I cracking observed is extremely localized and evidence of crystallographic cracking such as grain boundary facets is also shown. On the fracture surface apart from the slit, striations which show the characteristic behaviour of stage II cracking are observed. They are observed on the surface of the shear mode fracture together with elongated dimples and rub marks. After the reduction of the thickness of the specimen resulting from general yielding has come to an end and the fracture surface has been complete-ly inclined at 45 degrees in the direction of the applied stress, the stria-tions are no longer observed. The characteristic feature of the final fracture surface is the existence of large dimples and rub marks. In general, these re-sults may be effected a little by the levels of applied stress and by the width of the specimen.

The striation spacings s are also formularized as follows.

$$2s = C_s (\Delta K)^{m_s} \tag{3}$$

In this formura, experimental constants m_s and C_s are shown in Table 5. The result reveals the striation spacings to be nearly equal to the macroscopic crack propagation rate d(2a)/dN for a given value of ΔK, as shown in Fig. 5. However, strictly speaking, it appears that the striation spacings exceed the macroscopic rate at low ΔK level, and that lags behind it at high ΔK level. It should be assumed that the former results from existence of the crystallo-graphic fracture, and that the latter results from existence of the dimpled fracture.

The results of a fatigue test are usually expressed by an S/N curve, but in this work, the fatigue lives of the slitted specimens have been expressed by the range of an apparent initial stress intensity factor instead of the range of repeated stress. The range of the apparent initial stress intensity factor are used firstly because the number of unified specimens are too few to deter-mine an S/N curve and secondly because the crack propagation rates over a wide range were expressed using the range of the stress intensity factor as shown in Fig. 4. The range of the apparent initial stress intensity factor ΔK_0 are expressed as follows.

$$\Delta K_0 = \Delta \sigma \{ \pi a_0 \sec(\pi a_0/W) \}^{1/2} \tag{4}$$

Hence, a_0 is the half length of the slit. The relation between ΔK_0 and the fatigue lives N_f obtained is shown in Fig. 6. In Fig. 6, the results obtained Frost, Dugdale and Denton[9][10] are also shown. Thir results were obtained using center-slitted plate specimens of low carbon steel like the ones we used with a constant width and three thicknesses under repeated tensile loading. These results show that the relation between ΔK_0 and N_f can be expressed as follows.

$$N_f = 4.46 \times 10^{-12} (\Delta K_0)^{3.5} \tag{5}$$

By integrating Eq. (2), the predicted fatigue life can be approximately derived. It is shown in Fig. 6 that the predicted fatigue life from Eq. (2) for the cases of widthes of the specimen W = 50 mm and W = 200 mm is consistent with Eq. (5).

From Fig. 6, the threshold value of the range of the stress intensity factor ΔK_{th} for low carbon steel is seen to be about 15 $(kg/mm^2)(mm)^{1/2}$. This value is in agreement with the results obtained by others.[11]

4 Conclusion

From the work described above, we may conclude that:

(1) Stage I crackings are found in both smoothed and sharply notched specimens.

(2) The size of stage I cracking shows apparent dependence on grain size.

(3) The reason that the fatigue limit shows an dependency on grain size is related to the size of non-propagating stage I cracks.

(4) The macroscopic stage II crack propagation rates and the striation spacings correlate with the range of the stress intensity factor, and the tensile/shear fracture mode transition behaviour and reduction of the thickness of the specimen resulting from general yielding do not have a primary effect on this correlation.

(5) Stage II fatigue lives correlate with the range of an apparent initial stress intensity factor, and the predicted lives from the macroscopic stage II crack propagation rates are consistent with the correlation.

References
(1) Forsyth, P.J.E.: Proc. Crack Propagation Sym., 1, 1962, 76.
(2) Birkbeck, G., Inckle, A.E. and Waldron, G.W.J.: J. Mater. Sci., 6 (1971), 319.
(3) Kobayashi, H., Nakazawa, H. and Miyazawa, T.: J. Japan Soc. Materials Science, 10 (1973), 67.
(4) Kobayashi, H., Nakazawa, H. and Komine, A.: Trans. Japan Soc. Mechanical Engineers, 41-341 (1975-1), to be published.
(5) Nakazawa, H., Kobayashi, H. and Morita, A.: Trans. Japan Soc. Mechanical Engineers, 40 (1974), 9.
(6) Kasai, K., Kobayashi, H. and Nakazawa, H.: J. Soc. Materials Science, Japan, 23 (1974), 739.
(7) Nakazawa, H., Kobayashi, H. and Nishimura, A.: Trans. Japan Soc. Mechanical Engineers, 40-338 (1974-10), to be published.
(8) Paris, P.C. and Erdogan, F.: Trans. ASME, Ser. D, 85 (1963), 528.
(9) Frost, N.E. and Dugdale, D.S.: J. Mech. Phys. Solids, 6 (1958), 92.
(10) Frost, N.E. and Denton, K.: J. Mech. Engng Sci., 3 (1961), 295.
(11) Klesnil, M. and Lukas, P.: Engng Fracture Mechanics, 4 (1972), 77.

Table 1   Chemical composition (weight %).

Material	C	Si	Mn	P	S	Cu
X and Y	0.15	0.22	0.43	0.019	0.017	0.04
Z	0.17	0.10	0.71	0.014	0.019	—

Table 2   Heat treatment and tensile properties.

Material	Annealing temperature	Yield strength (kg/mm^2)	Tensile strength (kg/mm^2)	Elongation (%)	Reduction of area (%)
X	900°C x 2hr	25.4	41.5	43.5	64.0
Y	1100°C x 2hr	19.0	40.1	42.2	56.9
Z	950°C x 2hr	20.5	39.0	46.2	51.6

Mean diameter of grain: Material X, 25 μ

Material Y, 96 μ

Table 3   The Effect of grain size on the initial stage of crack propagation in the notched specimens.

	Material X			Material Y		
	σ kg/mm^2			σ kg/mm^2		
	19.5	14.8	10.6	19.5	14.8	10.6
N_0 cycles	2,000	5,000	15,000	3,000	8,000	30,000
a_0 mm	0.020	0.021	0.022	0.031	0.042	0.048
N_h cycles	2,000	4,000	35,000	2,000	7,000	70,000
a_h mm	>0.034	0.055	0.052	0.059	0.077	0.104

Table 4   Summary of results of macroscopic stage II crack propagation rates.

Specimen	W (mm)	T (mm)	$2a_0$ (mm)	σ (kg/mm^2) max.	min.	ΔK_0 (kg/mm$^{3/2}$)	N_f (cycles)	m	C
A1*							27.8×10^4	4.69	2.12×10^{-13}
A2**			14	12.5	0.75	55.3	30.6	4.57	3.52×10^{-13}
A3	200	3.5					39.4	4.07	2.63×10^{-12}
A4			28			78.9	8.0	4.23	1.94×10^{-12}
A5			14	17.5	5.75	55.3	14.4	3.52	7.48×10^{-11}
A6				12.5	4.5	37.6	69.3	4.25	3.24×10^{-12}
B1		3.5	14			55.8	30.5	5.11	2.76×10^{-14}
B2	100			12.5	0.75		69.5	4.45	6.55×10^{-13}
B3		1.75	7			39.1	91.0	5.26	1.56×10^{-14}
B4				17.5	5.75		22.4	4.64	7.69×10^{-13}
C1	50	3.5	7	12.5	0.75	39.4	59.4	5.20	3.85×10^{-14}
C2***							55.0	5.56	7.72×10^{-15}
D1				17.5		78.7	8.9	3.35	6.61×10^{-11}
D2				15.0		67.0	10.0	3.20	2.78×10^{-10}
D3	200	4.0	14	12.5	0.75	55.3	28.5	4.17	1.93×10^{-12}
D4				10.0		43.5	69.1	4.13	2.23×10^{-12}
D5				8.75		37.6	134.3	3.73	1.44×10^{-11}

* Triangle wave   ** Cut-saw wave   *** Fatigue pre-cracked specimen

Table 5   Summary of results of striation spacings.

Specimen	Range of data ΔK (kg/mm$^{2/3}$)	2s (μ)	Number of measurement	m_s	C_s
D1	78.7 - 251.9	0.14 - 8.7	56	3.33	8.67×10^{-11}
D2	67.0 - 196.1	0.22 - 5.7	79	2.96	8.51×10^{-10}
D3	61.0 - 220.9	0.13 - 11.4	97	3.61	3.16×10^{-11}
D4	67.7 - 196.9	0.08 - 4.6	80	3.79	1.17×10^{-11}
D5	65.9 - 196.0	0.08 - 5.3	77	3.41	6.37×10^{-11}

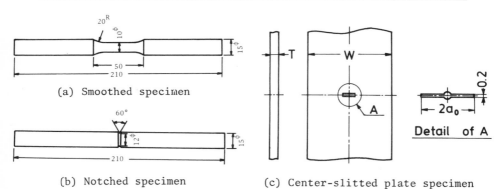

(a) Smoothed specimen

(b) Notched specimen

(c) Center-slitted plate specimen

Detail of A

Fig. 1   Configuration of specimens

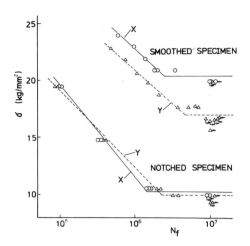

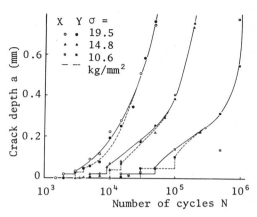

Fig. 3   Crack propagation curves of notched specimens.

Fig. 2   S/N curves for smoothed and notched specimens.

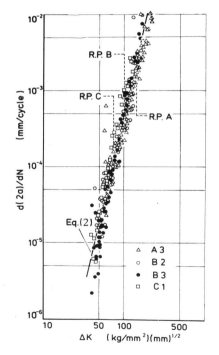

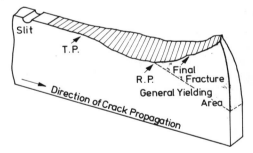

Transition point T.P. represents tensile/shear fracture mode transition.
Reduction point R.P. corresponds to onset of general yielding.

Fig. 4   Influence of range of stress intensity factor ΔK on stage II crack propagation rate d(2a)/dN.

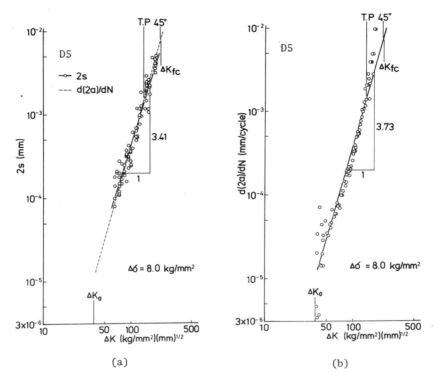

(a)   (b)

Fig. 5   Influence of range of stress intensity factor ΔK on stage II
crack propagation rate d(2a)/dN and striation spacing s for same specimen.

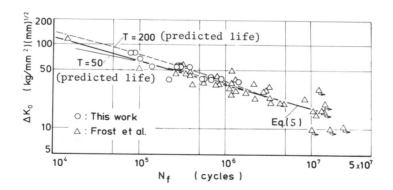

Fig. 6   Relation between range of apparent initial stress intensity
factor ΔK_0 and fatigue lives N_f for various type of specimens.

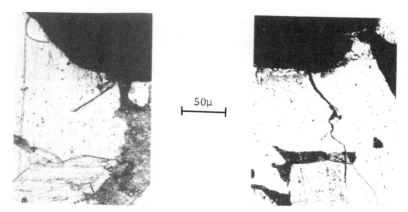

Photo. 1 Typical example of an micrograph of an initial stage I cracks.

Photo. 2 Micrograph of the zig-zag propagation of stage II cracks.

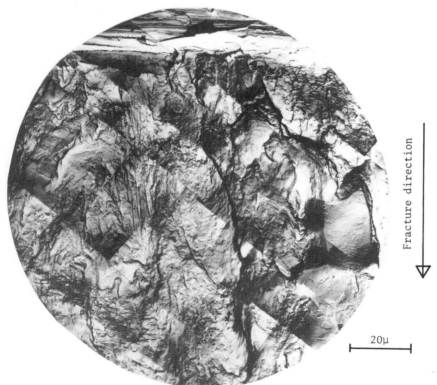

Photo. 3 Fractograph of initial stage of cracking from the slit.

X-ray Diffraction Study of Fatigue and Fracture

by Shuji Taira* and Keisuke Tanaka*

1. Introduction

For an experimental study of the mechanics of fatigue and fracture
in metals, the introduction of the X-ray microbeam diffraction
technique is significant because of the following. First, the high
sensitivity of the method to the microstructure of material gives
useful information on micro-mechanisms of fatigue and fracture when
the observation is carried out on a local area such as the site of
crack initiation, the tip of a fatigue crack and the fracture
surface. Second, the X-ray diffraction technique is powerful as a
tool for experimental stress strain analysis. Third, an analysis
of the cause of failure of structure in service might be possible,
because the different deformation structure in the thin layer of the
fracture surface due to different fracture modes can be detected by
the change in shapes of the diffraction pattern.

The present paper describes the results of the study on the
mechanics of fatigue and fracture in metals using the X-ray microbeam diffraction technique. The problem of nucleation from the tips
of an existing crack and a sharp notch at room and sub-zero temperatures is investigated based on the measured distribution of plastic
strain near the fracture surface. The nucleation of a fatigue crack
is discussed from a viewpoint of microstructure. The stress strain
distribution near the tip of a propagating fatigue crack was measured
and the data is analysed in comparison with the recent results of
fracture mechanics. The materials employed in the present study
were several carbon steels with carbon content ranging from 0.01% to
0.35%. The local and gross features of plastic deformation in the
fatigue and fracture processes, combined with an appropriate fracture
criterion, will presnet a complete picture of the mechanics of
fatigue and fracture.

---

* Dr., Professor and Dr., Assistant, respectively; Department
  of Mechanical Engineering, Kyoto University, Kyoto

2. Measurement of local stress and plastic strain using X-ray microbeam diffraction technique

The X-ray equipment used in the present study was composed of a focused X-ray generator and a specially designed micro-camera [1]. The optical system which was attached to the camera enabled us to take an X-ray photograph of any prescribed region on the specimen surface. The diffraction condition adopted in the measurement of stress was described in a previous paper of the authors [2].

The method of local plastic strain measurement by X-rays is based on the experimental finding for stretched and twisted materials that the equivalent plastic strain is a single-valued function of the total misorientation, measured from the tangential width of diffraction arcs in X-ray films, or the ratio of the half-value breadth of a diffraction ring to its initial value [3]. This method is applied in a preliminary experiment to measure strain distribution in the necked region of a round bar which was stretched monotonically. The distribution was also measured by microscopic observation of grain deformation as performed by Davidenkov and Spiridonova [4]. As seen in Fig. 1, the value measured by X-rays shows a good agreement with those given with the grain deformation measurement. The equivalent plastic strain is seen to be nearly constant in the necked region as assumed in Bridgman's analysis [5]. The constant value nearly equal to the value calculated from the reduction of area.

3. Cleavage crack nucleation at low temperatures

The material used in the experiment was an annealed 0.03%C steel with an average grain size of 40μ. The test specimens had four different shapes, i.e. round bar with a sharp circumferential notch of the radius 0.01 mm (NB), and center-cracked (CCP) and double-edge-cracked plates (DECP) and standard Charpy impact specimen with the notch radius 1 mm. The former three types of specimens were fractured under a monotonically increasing load in a Shimadzu Autograph and the last type specimen was fractured in a Charpy impact testing machine. The temperatures used in the tests were from room temperature to -196°C. The fracture surface of the specimens was examined in a scanning electron microscope and the transition temperature of fracture mode, T_d, from cleavage type to fibrous type was identified as -55°, -99°, -135° and 0°C for the NB, CCP, DECP and Charpy specimens, respectively.

   The X-ray microbeam was irradiated on the middle part of the
specimen thickness close to the initial crack or notch tip on the
fracture surface formed by a cleavage mode. The equivalent plastic
strain was evaluated from the half-value breadth of a diffraction
ring. The plastic strain, ε_p, thus measured is plotted in Fig. 2
against the deviation $\Delta T = T_d - T$ of the testing temperature, T, from
the cleavage-fibrous transition temperature, T_d. It is quite remarkable that the fracture surface strain, ε_f, is a single-valued
function of the temperature deviation ΔT for all four different
types of specimens. The differences in the specimen geometry and
the loading rate, i.e. slow tension or impact bending, can be regarded as the differences in the mechanical properties and the
plastic constraint of deformation in the near-tip strain field,
which might cause the variation of the transition temperature with
the specimen geometry and the loading rate. Therefore, the result
given in Fig. 2 means that the relation between the fracture surface
strain and the temperature deviation is independent of the plastic
constraint and the mechanical properties as far as the same material
is concerned. This might be an attractive finding in the field of
X-ray fractography for determining the condition of fracture of
structures in service from X-ray information on the fracture surface.
   In order to measure the distribution of plastic strain along
the depth from the cleavage fracture surface, the surface layer of
the fracture surface was removed by electro-polishing and the X-ray
beam was irradiated on the new surface. For the experimentally
obtained curves of plastic strain distribution, the distance R from
the fracture surface to the point where the strain value is equal to
Lüders strain, ε_L, of smooth specimens was measured. In Fig. 3, the
ratio of plastic strain ε_p to ε_L is plotted against the $1/(1+n)$th
power of the ratio of R to the distance y from the fracture surface
for the cases of the NB, CCP and DECP specimens, where n is the
work-hardening exponent of the material at a corresponding temperature. In the figure, the strain values measured on the fracture
surface and in its close neighbourhood are not included. It is
worthy to note that plastic strain distribution observed near the
fracture surface is expressed by the following equation similar to
the equation of strain singularity given by Rice-Rosengren [6] and
Hutchinson [7]:

$$\varepsilon_p \quad = \quad \varepsilon_L (R/y)^{1/(1+n)} \tag{1}$$

The results of the plastic strain distribution presented above present significant information on the criterion of the initiation of cleavage fracture at a crack tip. The start of the growth of an initial crack in a cleavage mode might occur when the ustable condition is satisfied in the ligament between main crack and micro-cracks formed just ahead of the tip [8]. If we assume, like McClintock [9], that cleavage fracture takes place when the plastic strain reaches a critical value at a certian structural distance ρ_s from the crack tip and further assume that the plastic strain measured on the fracture surface corresponds to the critical value, the value of ρ_s can be calculated as a specific value of y by sub-stituting ε_f for ε_p in Eq. (1). The values evaluated for the NB, CCP and DECP specimens at temperatures below T_d were found to nearly equal to a constant value of 1μ. The structural size used in the above criterion is thus confirmed to be a material constant as far as the fracture mode of the cleavage type is concerned.

4. Local criteria for fatigue crack nucleation

Unnotched specimens of annealed 0.01%C steel with the grain size 50μ were fatigued under completely reversed cyclic loading. The results of X-ray microbeam observation showed that three micro-structural parameters, i.e., the total misorientation, β, of a grain, the micro-lattice-strain in a grain, $\Delta d/d$, and the subgrain size, t, were nearly constant when fatigue cracks were observed with an optical microscope for the cases of stress amplitudes above the endurance limit [10]. Table 1 presents the plastic strain ε calculated from the total misorientation β, the value of which is considered to indicate the strain value stored in the material according to the progress of the fatigue process. Since most dislocations which are introduced in the material in the early stage are supposed to be concentrated at the region of subboundaries [11], the plastic strain at subboundary, ε_b, can be estimated by dividing the grain size, t_o, times plastic strain, ε, by the number of subboundaries in a grain, $[(t_o/t)-1]$, as

$$w \, \varepsilon_b = \varepsilon \, t \, t_o / (t_o - t) \tag{2}$$

where w is the subboundary width. Table 1 contains the calculated

results. It should be noticed that the product of ε_b and w takes
a value around $1\sim3\times10^{-4}$ mm at the time of crack initiation in
contrast to ε and t, which change their values by an order of
magnitude under the same range of stress level change.

5. Mechanics of plastic deformation near fatigue crack tip
5.1 Experimental strain analysis

The distribution of plastic deformation near.the tip of a growing
fatigue crack under completely reversed loading was examined on the
specimen side surface with an optical microscope and the X-ray microbeam diffraction technique. Dark and dispersed slip bands, which
were characteristic of slip deformation under cyclic stressing, were
detected near the crack tip and the size of slip band zone in the
direction of crack growth was measured. The size of the slip band
zone ahead of the crack tip, ξ, was found to be correlated to the
gross stress amplitude, σ, and the crack length, a, as

$$\xi/a = C_2 \{\sec(\pi\sigma/2C_3\sigma_Y) - 1\} \tag{3}$$

where σ_Y is the lower yield stress in monotonic tension tests, and
C_2 and C_3 are empirical constants depending on material [12, 13].
Figure 3 presents a typical example of the data, where the material
is annealed 0.03%C steel. The values of C_2 and C_3 are 0.060 and
0.92, respectively. The dashed line in the figure is the curve
predicted using the rigid plastic strip model (BCS model) [14]:

$$s_0/a = \sec(\pi\sigma/2\sigma_Y) \tag{4}$$

The value of ξ is much smaller than s_0.

The distribution of strain in the plastic zone was obtained
from detailed X-ray microbeam observation near the fracture surface.
Figure 5 presents the result for annealed 0.03%C steel. The distribution can be expressed by the equation

$$\varepsilon = \varepsilon_0\exp(-\alpha y) \tag{5}$$

where α is a decreasing function of the stress intensity factor and
y is the distance from the fracture surface. According the analysis
based on deformation plasticity, the strain is proportional to the
$[1/(n+1)]$th power of the inverse of the distance, where n is the
work-hardening coefficient of the material. Therefore, for non-
work-hardening material as assumed in BCS analysis, the strain is
proportional to the inverse of the distance. The date presented in

Fig. 5 were replotted in a log-log graph, which showed that the strain distribution has the same form of singularity as predicted, except in the close vicinity of the fracture surface [15]. The following two reasons are at least responsible for this deviation of strain distribution in the close vicinity, i.e., the work-hardening and the tip-blunting. Both factors reduce the concentration of strain at the close vicinity of the crack tip [15].

In Fig. 5, three parameters which are obstained from strain distribution are plotted against the slip band zone size, i.e., the strain on the fracture surface, ε_f, the size of substructure-developed region, η, which is defined as the distance at the intersection of the straight line of Eq. (5) with $\varepsilon=0.002$, and the integration, Ξ, of the strain distribution curve from 0 to ∞. As is seen in the figure, both values of η and Ξ are proportional to the slip band zone size. The strain on the fracture surface increases with the increase of the slip band zone size. If we regard the integration of strain as the opening displacement at the crack tip derived from the BCS model analysis, this proportional relation is predicted for small scale yielding case.

5.2 Experimental stress analysis

The stress value sensed in a local area of 135μ diameter near the crack tip was measured with the oscillation method of X-ray diffraction [1, 2]. The open circles in Fig. 7(a) shows the distribution of the axial residual stress measured near the crack tip in a 0.31%C steel specimen after the interruption of the fatigue test at a certain number of stress cycle. The residual stress at the crack tip is of the compression type and gradually changes to tension at a certain distance from the crack tip. As a comparison, the figure also includes the residual stress distribution measured in a cracked specimen which was loaded up to the same maximum stress and then unloaded to zero stress. The cracked specimen used for the latter experiment was prepared in the following manner. The specimen was first cracked under cyclic loading and then annealed at 600°C for 90 min for residual stress relief formed in pre-fatigueing. The stress at the point 100μ from the crack tip in the specimen with a growing crack was measured with X-rays under the application of one cycle of loading and reverse-loading and the result is presented in Fig. 7(b).

The distribution shown by the dashed line in Fig. 7(a) is calculated based on BCS model assuming that the compression cycle did not affect the residual stress distribution. The shapes of the residual stress distributions ahead of the tips of both stationary and growing cracks agree with the shapes of analytically derived curve. The calculated distance is nearly equal to the measured distance in the case of a stationary crack, while it is larger than the distance measured in the case of a growing fatigue crack. The variation of the stress value at the point 100μ ahead of the crack tip during one cycle was calculated using BSC model under the assumption that the crack starts to close behind the crack tip in compression cycle [15]. The analytical result is compared in Fig. 7(b) with the experimentally measured value. As seen in the figure the observed variation of stress is approximately in accord with the result of analysis.

6. <u>Local fracture criteria and fatigue crack propagation law</u>
The relations of the propagation rate against the stress intensity factor, the slip band zone size and the strain on the fracture surface were examined under various experimental conditions of stress level, loading mode and specimen geometry and materials. The results are summarized in Table 2. The propagation behavior of a fatigue crack under cyclic tension-compression is taken as the standard of comparison. The range of the propagation rate observed is between 3×10^{-7} and 3×10^{-4} mm/cycle. Main conclusions are summarized as follows:

(1) High value of the applied stress causes a deviation of the data from $da/dN = CK^m$ obtained in the rate range of 3×10^{-7} to 3×10^{-4} mm/cycle. On the other hand, the relation $da/dN = C'\xi^{m'}$ was confirmed to be applicable for the whole range of the propagation rate examined. The combination of the above relation and Eq. (3) yields a new propagation low [12].

(2) The propagation rate under plane bending was found to be much smaller than that under tension-compression when compared at the same propagation rate. This was explained by the difference in strain distribution near the crack tip in two cases. A small effect on da/dN vs. ξ relation was ascribed to the effect of the non-singular term on the slip band formation [11].

(3) The propagation behavior of a fatigue crack in the vicinity at the notch root under reversed loading is complex, which can not be explained from the variation of the K value. While the relation between the rate and the slip band zone size was found to be similar to that obtained in the case of tension-compression, but with a smaller C' value [16].

(4) The rate of crack growth increases with decreasing R value, while the mean stress does not affect the relaiton between da/dN vs. ξ [15].

(5) For the date of various carbon steels with various carbon contents and heat-treatments, the value of m' in the relation $da/dN =C'\xi^{m'}$ changes from about 2 to 1 as the carbon content of the material decreases and the hardness of the material increases [12, 1.

(6) The rate of crack propagation was found to be a single-valued function of the strain on the fracture surface for various experimental conditions [12, 13] (See Fig. 8).

(7) The condition of the non-propagation of a fatigue crack can be expressed in term of the critical value of the slip band zone size for fatigue under various loading modes and specimen geometries. The mechanical equation for the non-propagation of fatigue crack was obtained by substituting the critical ξ value in Eq. (3) [13]. The critical value of ξ is a function of microstructure of the material mainly controlled by the grain size of the material. At the critical condition for the non-propagation of a fatigue crack in 0.03%C steel, the subgrain size, t, and the size, n, of a substructure developed region were 3.6μ and 2.2μ, respectively. The latter is smaller than the former which seems to suggest that plastic deformation is reversible at the crack tip.

The fracture criteria used in the theories proposed so far are classified into two categories: (i) non-accumulation criterion, in which the rate of crack growth is controlled by a certain mechanical parameter, such as strain, displacement or energy, in the material at the crack tip; (ii) damage accumulation criterion, where the controlling factor for crack growth is the amount of strain, displacement or energy which is accumulated in the material until it comes to the crack tip. In BCS model analysis, Weertman used a fracture criterion that fracture takes place at the crack tip when the summation of cyclic component of opening displacement becomes a critical value [17, 18]. On the other hand, Lardner assumed that the growth rate of a fatigue crack is determined by the cyclic component of the opening displacement at the crack tip [19].

The present experimental data shown in Fig. 6 indicates that the integrated value of strain, E, is proportional to ξ, and $\dot{E}$ is regarded as the cyclic component of the opening displacement in the terminology of BCS model analusis. Therefore, the exponent m' in the relation $da/dN = C' \xi^{m'}$ indicates the degree of accumulation of damage in the sense that m' equals one for non-accumulation and two for perfect accumulation. The experimental findings showed that m' changes from one to two as the material changes from hard to soft. The difference in m' with material was explained by the difference of distribution of damage in the direction of crack growth on the basis of microscopic observation [15].

The physical meanings of the variation of plastic strain on the fracture surface strain with the propagation rate can be explained from damage accumulation model as follows. Figure 8 shows the plastic strain ϵ_f and the subgrain size t_f on the fracture surface which was observed by the X-ray microbeam method. The fracture surface strain, ϵ_f, at subboundary was calculated by using Eq. (2) for the case of annealed 0.03%C steel. The variation of ϵ_{bf} with the propagation rate is presented in Fig. 9. It is found that the product of plastic strain at subboundary ϵ_{bf} and the subboundary width w is nearly constant when the rate is smaller than 3×10^{-5} mm/cycle. The constant value of $w \epsilon_{bf}$ is 1.5×10^{-4} mm, which is approximately equal to the value at fatigue crack nucleation in unnotched specimens given in Table 1. This suggests that the fracture criterion for crack initiation expressed in terms of the plastic strain at subboundary is applicable as the criterion for crack propagation. In other words, a fatigue crack cannot propagate until the plastic strain at subboundary reaches a critical value at the tip for the cases of low rates of crack growth. The threshold condition for crack growth indicates the limit of the occurrence of this growth mechanism.

The electron microscopic observation of fatigue fracture surfaces done with the same specimens of 0.03%C steel showed that the topography of the fracture surface which was made at lower rates were not striations, but some other patterns which were considered to correspond to the above-mentioned damage accumulation micro-mechanism

[20].   At higher rates of $5 \times 10^{-5} \sim 2 \times 10^{-3}$mm/cycle, striations and cleavage fracture mechanisms were confirmed through electron fractography for low-carbon steels.

Further studies are required to derive the law of fatigue crack propagation combining the mechanics of plastic deformation at the crack tip and local fracture criterion. The fatigue crack propagati law to be established should work for explaining complex engineering problems of fatigue fracture, such as the nucleation and the growth of a fatigue crack from notches, the growth of a crack under variabl stresses, high strain cycling and multiaxial stresses.

References

[1]   Taira, S. edited, *X-ray studies on mechanical behavior of materials* (1974) The Society of Materials Science, Japan.

[2]   Taira, S., Tanaka, K., and Shimada, T., J. Soci. Mat. Sci. Jap., 21, 1141 (19

[3]   Taira, S., Ryu, J.G., and Tanaka, K., J. Soci. Mat. Sci. Jap., 23, 58 (1974).

[4]   Davidenkov, N.N., and Sprinova, N.I., Proc. ASTM, 46, 1147 (1946).

[5]   Bridgman, P.W., *Studies in large plastic flow and fracture* (1952) McGraw-Hill

[6]   Rice, J.R., and Rosengren, G.E., J. Mech. Phy. Solids, 16, 1 (1968).

[7]   Hutchinson, J.W., J. Mech. Phy. Solids, 16, 13 (1968).

[8]   Wells, A.A., Proc. Roy. Soci. London A285, 34 (1965).

[9]   McClintock, F.A., *Fracture* edited by H. Liebowitz, 3, 47 (1971).

[10] Taira, S., Tanaka, K., Shimada, T., and Kato, Y., Proc. 16th Jap. Cong. Mat. Res., 174 (1973).

[11] Taira, S., and Tanaka, K., Eng. Frac. Mech., 4, 925 (1972).

[12] Taira, S., and Tanaka, K., Proc. Int. Conf. Mech. Beh. Mat., Kyoto, 2, 48 (1972).

[13] Taira, S., and Tanaka, K., 3rd Int. Cong. Frac., Munich, V-61 (1973).

[14] Rice, J.R., ASTM Spec. Tech. Pub. No. 415, 247 (1967).

[15] Taira, S., and Tanaka, K., Proc. 1974 Symp. Mech. Beh. Mat., Kyoto (1974).

[16] Taira, S., Tanaka, K., and Ryu, J.G., Proc. 16th Jap. Cong. Mat. Res., 36 (1973).

[17] Weertman, J., Int. J. Frac. Mech., 2, 460 (1966).

[18] Weertman, J., Int. J. Frac. Mech., 5, 13 (1969).

[19] Lardner, R.W., Phil. Mag., 17, 71 (1968).

[20] Taira, S., and Tanaka, K., Proc. 15th Jap. Cong. Mat. Res., 43 (1972).

Table 1. Product of plastic strain in subboundary and subboundary width at crack initiation in unnotched specimen. (Annealed 0.01%C steel).

Stress amplitude, σ_a kg/mm^2	Cycle number at fracture, N_f	Plastic strain, ε	Subgrain size, t	Plastic displacement at subboundary $\phi_b = w\varepsilon_b$, mm
16.0	2.0×10^6	0.012	17	3.1×10^{-4}
20.0	4.4×10^5	0.014	8.0	1.3×10^{-4}
22.0	3.0×10^4	0.040	3.5	1.5×10^{-4}
25.0	5.2×10^4	0.152	1.9	2.9×10^{-4}

Table 2 Effects of experimental conditions on the propagation law in carbon steels (The propagation rate range observed is 3×10^{-7} to 3×10^{-4} mm/cycle).

-STANDARD- Tension-compression, fully reversed, without notch effect		Loading mode		Notch Vicinity	Mean stress effect R<0	Material effect (0.001<C%< 0.55) carbon content heat-treatment
$3 \times 10^{-7} \sim 3 \times 10^{-5}$ mm/cycle	Scale of yield zone	In-plane bending	Plane bending			
$da/dN = CK^m$	YES become non-linear in log-log plot due to deviation from SSY	YES small effect	YES much smaller C	YES complex effect	YES increasing rate with decreasing R	YES decreasing m with increasing carbon content, hardeness
$da/dN = C'\xi^{m'}$ ($3 \times 10^{-7} \sim 3 \times 10^{-4}$ mm/c)	NO	YES small effect of non-singular term	YES large effect due to non-singular term	YES A similar equation can be obtained, but with a smaller C' value	NO	YES
$da/dN = C''(\varepsilon_f)^{m''}$	YES Become non-linear in log-log plot	NO	NO	NO	(NO)	NO
Non-propagation condition given by the critical ξ value		NO	NO	NO	NO	YES microstructure decreasing critical value with decreasing grain size

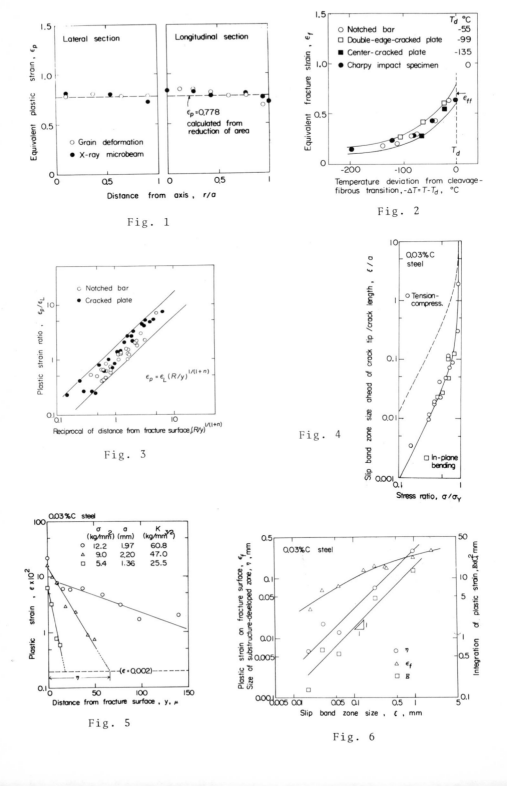

Fig. 1

Fig. 2

Fig. 3

Fig. 4

Fig. 5

Fig. 6

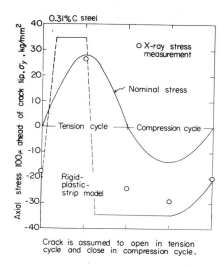

Crack is assumed to open in tension
cycle and close in compression cycle.

Fig. 7(a)

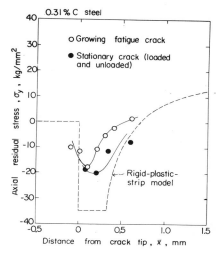

Fig. 7(b)

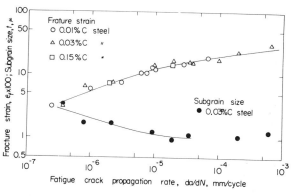

Fig. 8

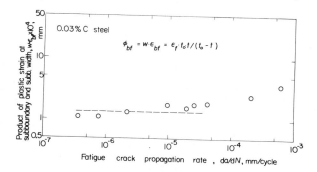

Fig. 9

CRACK TIP DEFORMATION AND FATIGUE CRACK PROPAGATION

by

H. W. Liu* and Albert S. Kuo**

1.   INTRODUCTION

Fatigue crack growth is the result of the interplay between the cyclic stresses and cyclic deformation ahead of a crack tip. In this study, the accumulated strain as well as the cyclic strain range near a crack tip were measured, and the measurements were correlated with fatigue crack growth rate. In addition, crack opening displacements were measured under a static load. Crack growth can be in shear or normal separation mode. Based upon the basic concepts of slip plane decohesion and crack tip blunting, an unzipping model was illustrated. In addition, the effects of cyclic yield strength and cyclic ductility on crack growth resistance of a material were also investigated.

2.   CRACK TIP STRAIN

Near a crack tip, there exists a region of the characteristic crack tip elastic stress and strain fields having the singularity of 1/2. When the crack tip plastic zone is very small in comparison with the region of the characteristic crack tip stress and strain fields, the plastic deformation inside the zone does not greatly disturb the stresses and strains outside of the zone so that the stresses and strains outside the zone are essentially those given by the singular terms of the elastic solution. This is known as the case of small scale yielding.

The essential feature of small scale yielding can be easily seen, if one considers the case of a semi-infinite crack in an infinite plate. The stress

*Professor of Materials Science, **Research Assistant,
 Syracuse University, Syracuse, N.Y. 13210

field in the region at a finite distance away from the crack tip is dominated by the singular terms of the elastic solution with the sigularity of $r^{-1/2}$. The stress relaxation within a small plastic zone will not significantly change the stresses and strains in the massive elastic region which surrounds the small plastic zone. If the conditions of small scale yielding and plane strain prevail, the crack tip regions can be scaled by their respective cyclic plastic zone sizes, $r_{p(c)}$'s, such that the stresses and strains experienced by the materials are the same at the geometrically similar points in these regions. When one measures fatigue crack growth rate, a crack increment, Δa, and its corresponding increment of load cycles, ΔN, are measured. If the crack increment is scaled by $r_{p(c)}$, then the material within the increment $\Delta a/r_{p(c)}$ must have experienced the same stress and strain cycles regardless of the size of $r_{p(c)}$. If the "average" deformation and fracture properties over the crack increments can be considered homogeneous, the increment of load cycles, ΔN, necessary to propagate the crack through the increment, must be the same. Hence, da/dN is proportional to $r_{p(c)}$ which in turn is proportional to ΔK^2. If the conditions of small scale yielding and plane strain prevail, and if the "average" deformation and fracture properties of the material can be considered homogeneous, the elasto-plastic analysis leads to the conclusion that

$$\frac{da}{dN} = f(R)\Delta K^2 \quad , \tag{1}$$

where $f(R)$ is a function of the stress ratio R. Any deviation from the deduced crack growth relation must be the result of the fact that one or more of these conditions, i.e. material homogeneity, small scale yielding and plane strain condition, are not satisfied. A more detailed analysis is given in Reference (1).

Figure 1 shows the typical crack growth data for Al2024-T351 aluminum alloy. In the low ΔK region, region I, the slope of the curve is more than four. As ΔK decreases, da/dN rapidly decreases. The data in region I often indicate the existence of a limiting ΔK value for a non-propagating crack[2]. The limiting ΔK value is commonly known as the threshold stress intensity factor range, ΔK_{th}, for crack growth. In the intermediate ΔK region, region II, the slope of the line is close to or slightly higher than two. The slope of the line in the high ΔK region, region III, is often equal to or higher than four. Often a smooth curve can correlate with the data more closely than can the three line segments shown in Figure 1.

Factors which can account for the deviation in region III are material inhomogeneity, crack tip necking, and extensive crack tip plastic deformation so that the condition of small scale yielding does no longer exist. One possible reason to account for the deviation in region I is material inhomogeneity particularly in plastic deformation process. Other possible causes for the deviation in region I, are corrosion at the crack tip or some other thermally activated process, which has not been taken into consideration in this analysis. If region I and region III overlap each other, region II may disappear entirely. The transition from region I to region II often occurs at a crack growth rate of several micro-inches per cycle. The crack growth rate at the transition point is close to the size of the undeformed packet of a glide lamella in an aluminum single crystal[3], see Figure 2.

This study deals primarily with the effects of crack tip deformation on the crack growth rate in region II and region III. Kang and Liu[4] have measured the cyclic deformation near a crack tip in region II and region III of crack growth. They observed that as a crack propagates toward a point ahead of the crack tip both the cyclic strain range as well as the maximum strain at the point increase. The positive mean stress induces cyclic creep, and the accumulated cyclic creep strain increases the maximum strain at the point. Kang and Liu measured both the total accumulated strain at K_{max}, ε_{max}, and the cyclic strain range, $\Delta\varepsilon$, ahead of a crack tip. $\Delta\varepsilon$ is the strain range as K increases from K_{min} to K_{max}. The strain at K_{min}, ε_{min}, is the difference between ε_{max} and $\Delta\varepsilon$.

Figure 3a and 3b show the typical measurements of $\Delta\varepsilon$ and ε_{max}. The slopes of all the lines for $\Delta\varepsilon$ in Figure 3a are -1/2. The slope of the lines for ε_{max} in Figure 3b decreases as K_{max} increases. The cyclic plastic zone, $r_{p(c)}$, is defined as the region in which $\Delta\varepsilon$ exceeds the cyclic yield strain range of the material. Figure 4 shows the plot of both da/dN and $r_{p(c)}$ against ΔK. The data indicate that da/dN is proportional to $r_{p(c)}$ in both region II and region III. The transitions of both da/dN and $r_{p(c)}$ from region II to region III take place at ΔK approximately equal to 23 ksi $\sqrt{in}$. Below this ΔK-value, both da/dN and $r_{p(c)}$ are proportional to ΔK^2. Above this value, both are proportional to ΔK^4.

In Figures 3a and 3b, the strains are plotted against the distance from crack tip, r. If the horizontal axis is scaled by $r_{p(c)}$, and if the conditions of small scale yielding and plane strain prevail, all of the ε_{max} and $\Delta\varepsilon$ data should fall onto two lines: one for ε_{max} and the other for $\Delta\varepsilon$. Since all of the $\Delta\varepsilon$ data have a slope of -1/2, all of the data automatically fall on the same line, even in the region of high ΔK, when crack tip necking increases $\Delta\varepsilon$. Figure 5 shows that the ε_{max} data at the lower ΔK values lie in a very narrow band. At ΔK = 22.8 ksi $\sqrt{in}$, the ε_{max} data are considerably higher. This increase in ε_{max} is caused by crack tip necking, which increases strain concentration at a crack tip and thus causes faster crack growth rate. The ΔK value of 22.8 ksi $\sqrt{in}$ corresponds to the transition point of da/dN from region II to region III.

The data in Figure 4 and 5, as well as the data from thicker (0.25")
specimens and the data at a higher stress ratio (1/3)[5] indicate that if
da/dN is proportional to ΔK^2, the crack tip region can be scaled by $r_{p(c)}$
such that both ε_{max} and $\Delta\varepsilon$ are the same at geometrically similar point. On
the other hand if the crack tip region cannot be scaled by $r_{p(c)}$, the expo-
nent of ΔK in Equation (1) is higher than two[4]. One of the causes accounting for the deviation from Equation (1) is crack tip necking in thin specimens at high levels of ΔK. In order to avoid the geometric effect of a crack
tip necking on crack growth, one should use thick specimens to evaluate the
crack growth resistance of a material.

3.   SHEAR AND NORMAL SEPARATION MODES OF CRACK GROWTH

Crack growth can be either in a shear separation or a normal separation
mode or in a combination of both. Based upon the crack tip blunting model
of fatigue crack propagation together with the crack tip opening displacement, COD_t, given by the Dugdale model, the crack propagation rate can be
written as[6,7]

$$\frac{da}{dN} = \frac{1}{2} COD_t = (1 - \nu^2) \frac{\Delta K^2}{2E\sigma_{Y(c)}} \tag{2}$$

Both da/dN and COD_t are inversely proportional to $\sigma_{Y(c)}$. $\sigma_{Y(c)}$ is the cyclic
yield strength range. The same inverse relationship should persist for any
modified crack growth model based on the concept of crack tip blunting such
as the model of slip plane decohesion[8]. In the case of a shear separation
mode, Equation (2) indicates that cyclic yield strength plays a major role.
Crack growth rate increases as $\sigma_{Y(c)}$ decreases and vice versa.

Based upon the basic concept of crack tip blunting and slip plane decohesion[8,9,10,11], an unzipping model of crack growth is illustrated in
Figure 6[1]. In the case of plane strain, slip planes are inclined 45° as

shown in Figure 6a. The slip process is not homogeneous. Extensive plastic deformation is concentrated in slip bands separated by lightly deformed "slabs". When the slip plane α is activated under a tensile loading, the crack tip moves forward to slip plane b, (Figure 6b). After slip takes place on plane b, the crack assumes the configuration of Figure 6c. The crack opens like a zipper. As the unzipping process proceeds, slip decohesion takes place successively on the alternate planes β, c, γ, and d. The crack at K_{max} is schematically illustrated in Figure 6e. The crack upon unloading is shown in Figure 6f. Numerous fractographic studies have substantiated crack tip blunting as the basic mechanism of striation formation and fatigue crack growth in ductile materials.

The concept of damage accumulation for fatigue crack growth was first introduced by McClintock[12], and it was subsequently modified by Lehr and Liu[13] and Liu and Iino[14]. In the analyses by Liu and Iino, the Manson-Coffin's law of strain controlled fatigue and the Miner's law of damage accumulation were used. Their analyses give the relation

$$\frac{da}{dN} = C[\frac{\Delta K}{\sigma_{Y(c)}}]^2 [\frac{\varepsilon_{Y(c)}}{M}]^2 ,$$ (3)

where $\varepsilon_{Y(c)}$ and M are respectively cyclic yield strain range and the strain range at a fatigue life of one cycle for a smooth specimen. M is a measure of the ductility of a material. In this case, da/dN is proportional to $\sigma_{Y(c)}^{-2}$ and M^{-2}. In their analyses the effect of the triaxial state of stress at a crack tip in a thick specimen is not taken into consideration. The value of M would be lowered by the triaxial state of tensile stress at a crack tip. The reduced value of M increases the crack growth rate. Equation (3) indicates that da/dN is inversely related to the ductility of a material.

Recently Ritchie and Knott[15] have studied the effect of temper embrittlement on fatigue crack growth. They found that the room temperature tensile properties of the steel were not affected by the embrittlement process. But the crack growth rate was increased ten fold. According to the blunting model, this increase in growth rate can be justified only by a reduction of $\sigma_{Y(c)}$ by a factor of ten. Certainly, such a reduction is not likely. On the other hand, this increase in crack growth rate agrees with the basic concept of "exhaustion of ductility" of the damage accumulation model. As the ductility of the steel is reduced by the temper embrittlement process, da/dN increases.

Kang and Liu[16] have studied the effect of prestress on subsequent crack growth. Each specimen was prestressed at a certain stress range for a given number of cycles. After the prestress cycling, a slot was machined into the specimen and the crack growth rates were measured. When the prestresses were below the static yield strength, the effects were negligible. When the prestress levels were above the static yield strength and caused gross yielding, crack growth rate increased considerably. The prestress cycles will not change much the cyclic yield strength. Therefore, the increased growth rate cannot be explained in terms of crack blunting model, but it can be understood in terms of the damage accumulation model.

Based upon the observed increase in crack growth rate caused by both prestress cycling and temper embrittlement and the fractographic studies on striation formation and crack growth in ductile materials, one can see both the damage accumulation and the crack tip blunting mechanisms are valid models of crack growth.

Figures 7a and 7b are schematic diagrams of the two types of crack growth. For the ductile shear separation mode, the crack growth, δa_d, is equal to one half of the crack tip opening displacement. For the brittle

normal separation mode, the crack growth, δa_b, is much larger than the crack
tip opening displacement as schematically shown in Figure 7b. Both types of
crack growth may be active in separate regions of a crack, or both may be
active at the same region of a crack. Conceptually, the total crack increment per cycle, δa consists of two components: the brittle and the ductile
component, δa_b and δa_d. The ductile component of a crack increment grows by
slip plane decohesion or the crack tip blunting mechanism, and it is controlled by the cyclic yield strength of a material. The brittle component
grows by the normal separation mode, and it is controlled by the ductility of
a material. δa_d is more dominant in ductile materials and δa_b is more
dominant in brittle materials.

4.   CRACK OPENING DISPLACEMENT MEASUREMENTS

Crack opening displacements in Al2024-0, Al2024-T3, and Al2024-T351
aluminum alloys were measured. The tensile yield strengths of these three
alloys are 7.8, 45, and 57 ksi respectively. Centrally slotted specimens of
0.016 inches and 0.25 inches thick were tested. The width of the slot is
0.009 inches. The measurements were made under static tensile loads. The
pertinent test conditions are summarized in Table I and the results are shown
in Figure 8. Specimens 4 and 6 were slotted and subsequently fatigue cracked.
All the other specimens were only slotted. Both specimens 3 and 4 were thin
and were loaded to high K-values, and the values of the quantity $(K/\sigma_Y)^2/t$
were 17.5 and 18.8 for these two specimens. Specimen 3 was slotted
and specimen 4 was fatigue cracked. The measurements of both specimens 3 and
4 agree well with the Dugdale calculations. Both specimens 6 and 7 were
thick and loaded to a lower level of K-values. The values of the quantities
$(K/\sigma_Y)^2/t$ of these two specimens were close, 0.52 and 0.46. Specimen 6 was
cracked, and 7 slotted. The measurements of both specimens 6 and 7 agree
well with the elastic calculations. The data clearly indicate that the slot

and the fatigue crack gave essentially the same results in the region of the measurements. It is conceivable that the difference could be significant at a distance closer to the crack tip.

The value of $(K/\sigma_Y)^2/t$ is a measure of the deviation from the plane strain condition, and it is also an index of the relative sizes of plastic zone and specimen thickness, t. When the value of the quantity $(K/\sigma_Y)^2/t$ is large, as in the case of thin specimens, No. 1, 3 and 4, the COD measurements agree well with the Dugdale model calculations. When the value of the quantity is low, as in the case of thick specimens, No. 2, 5, 6 and 7, the measurements agree with the elastic crack calculations. For specimen 5, the quantity, 0.20, is very low. The Dugdale and the elastic plane stress calculations deviate only slightly from each other in the region of the measurements. The measurements of specimens 5 and 7 agree well with the elastic plane stress calculations, while those of specimens 2 and 6 agree well with the elastic plane strain calculation. However, the difference between plane stress and plane strain calculations is only 10%. The results clearly indicate that the Dugdale model is applicable only to thin specimens at high level of stress intensity factor. Therefore the Dugdale model is not suitable for a quantitative calculation of fatigue crack growth rate in a thick specimen in the plane strain condition.

5.   SUMMARY

1. Crack tip necking increases local strain and crack growth rate. Therefore, one should use thick specimens to evaluate the resistance to crack growth of a material in order to avoid the geometric effects of crack tip necking of thin specimens.

2. Fatigue crack growth can be in shear separation or normal separation mode, or in a combination of both. The shear separation mode is dominant in a ductile material, and it is controlled by the cyclic yield strength of a

material. The normal separation mode is dominant in a brittle material, and it is controlled by the ductility of a material.

3. Crack opening displacement measurements in thin specimens loaded to high K levels agree with Dugdale calculations. COD in thick specimens at low K levels agree with elastic crack calculations. Therefore, the crack tip opening displacement calculated with Dugdale model is not suitable for a quantitative calculation of fatigue crack growth rate and fracture toughness of a thick specimen in the plane strain condition.

REFERENCES

(1) H. W. Liu, "Analysis of Fatigue Crack Propagation," NASA Contractor Report, NASA CR-2032, (May 1972) 40 pp.

(2) H. W. Liu, Discussion of a paper by P. C. Paris, "The Fracture Mechanics Approach to Fatigue," Fatigue, Proc. Tenth Sagamore Army Matls. Res. Conf., Syracuse University Press, Syracuse, N.Y. (1964) pp. 127-132.

(3) R. D. Heidenreich and W. Shockley, Report on Strength of Solids, London Physical Society, (1948) pp. 57-75.

(4) T. S. Kang and H. W. Liu, "Fatigue Crack Propagation and Cyclic Deformation at a Crack Tip," International Journal of Fracture Mechanics, Vol. 10, No. 2, (June 1974) pp. 201-222.

(5) T. S. Kang, "Fatigue Crack Tip Deformation and Fatigue Crack Propagation," Ph.D. Dissertation, Syracuse University, Syracuse, N. Y., (Feb. 1972) 118 pp.

(6) J. R. Rice, "Mechanics of Crack Tip Deformation and Crack Extension by Fatigue," Fatigue Crack Propagation, ASTM STP 415, (1967) p. 247-311.

(7) F. A. McClintock, Discussion of a paper by C. Laird, "The Influence of Metallurgical Structure on the Mechanism of Fatigue Crack Propagation," Fatigue Crack Propagation, ASTM STP 415, (1967) pp. 169-180.

(8) R. M. N. Pelloux, "Mechanisms of Formation of Ductile Striations," ASM Transaction Quarterly, Vol. 62, (1969) pp. 281-285.

(9) C. Laird, "The Influence of Metallurgical Structure on the Mechanisms of Fatigue Crack Propagation," Fatigue Crack Propagation, ASTM STP 415, Am. Soc. for Testing and Materials, (1967) pp. 131-168.

(10) F. A. McClintock, "Local Criteria for Ductile Fracture," International Journal of Fracture Mechanics, Vol. 4, No. 2, (June 1968) pp. 101-130.

(11) D. Brook, "Some Considerations on Slow Crack Growth," International
      Journal of Fracture Mechanics, Vol. 4, No. 1, (March 1968) pp. 19-34.

(12) F. A. McClintock, "On the Plasticity of the Growth of Fatigue Crack,"
      Fracture of Solids, Met. Soc. AIME Conference,Interscience, New York,
      (1963) pp. 65-102.

(13) K. R. Lehr and H. W. Liu, "Fatigue Crack Propagation and Strain Cycling
      Properties," International Journal of Fracture Mechanics, Vol. 5, No. 1,
      (March 1969) pp. 45-55.

(14) H. W. Liu and N. Iino, "A Mechanical Model for Fatigue Crack Propaga-
      tion," Proceedings of the Second International Conference on Fracture,
      Brighton, England, Champman Hall, (April 1969) pp. 812-823.

(15) R. D. Ritchie and J. P. Knott, "Brittle Cracking Processes During Fa-
      tigue Crack Propagation," Third International Congress on Fracture,
      Munchen, Germany, (1973) V-434/A.

(16) T. S. Kang and H. W. Liu, "The Effect of Prestress Cycles on Fatigue
      Crack Growth - An Analysis of Crack Growth Mechanisms," Engineering
      Fracture Mechanics, Vol. 6, No. 4, (December 1974) pp. 631-638.

TABLE I. EFFECT OF SPECIMEN THICKNESS ON DUGDALE MODEL

Specimen No.	Materials	Strain Hardening Exponent	Yield Strength σ_Y	Thickness t, (IN)	Applied Stress (KSI)	Stress Intensity Factor, K (KSI√IN)	$\dfrac{(K_I/\sigma_Y)^2}{t}$	Measured Plastic Zone Size (IN)	Agree with Dugdale Model
1	2024-0	0.307	7.8	0.016	3.9	4.13	17.5	0.112	Yes
2	2024-0	0.307	7.8	0.25	3.9	4.13	1.12	--	No
3	2024-T3	0.12	45	0.016	22.5	23.8	17.5	0.086	Yes
4	2024-T3	0.12	45	0.015	24	25.7	18.8	0.137	Yes
5	2024-T351	0.097	56	0.25	10	12.1	0.20	0.047	*
6	2024-T351	--	52	0.25	--	18.7	0.52	--	No
7	2024-T351	0.097	56	0.25	18	19.1	0.46	0.018	No

*In this case, Dugdale model and elastic calculation are coincided with each other.

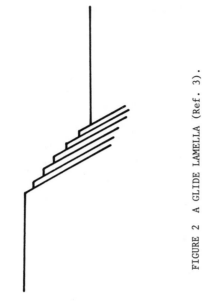

FIGURE 2   A GLIDE LAMELLA (Ref. 3).

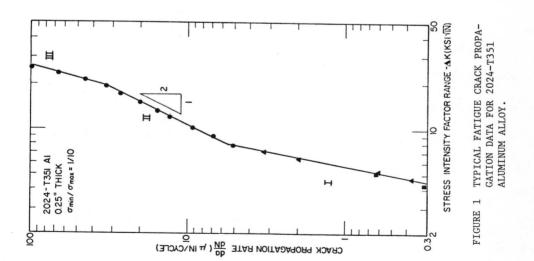

FIGURE 1   TYPICAL FATIGUE CRACK PROPA-
GATION DATA FOR 2024-T351
ALUMINUM ALLOY.

338

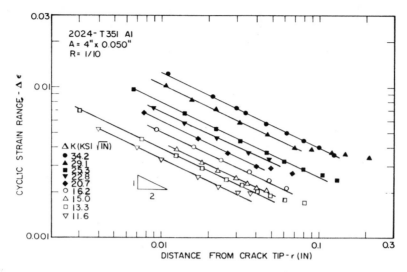

FIGURE 3a CYCLIC STRAIN RANGE AHEAD OF A CRACK TIP.

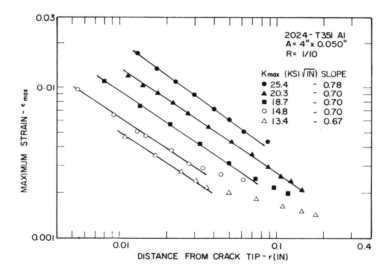

FIGURE 3b MAXIMUM STRAIN AHEAD OF A CRACK TIP.

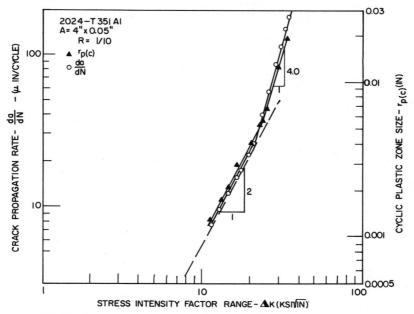

FIGURE 4   STRESS INTENSITY FACTOR RANGE VERSUS CYCLIC
PLASTIC ZONE SIZE AND CRACK PROPAGATION RATE
(2024-T351): CYCLIC YIELD STRENGTH RANGE
134 KSI.

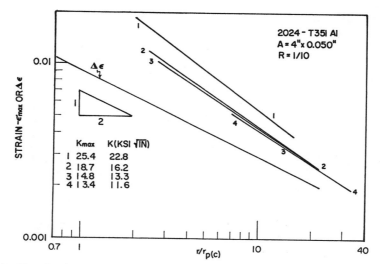

FIGURE 5   STRAIN VERSUS $r/r_{p(c)}$ FOR 0.050 INCHES THICK SPECI-
MENS AT R = 1/10.

340

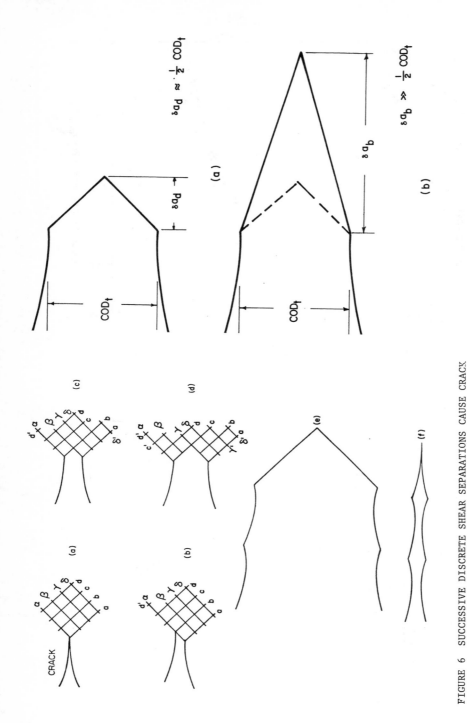

FIGURE 7   SCHEMATIC DIAGRAMS OF TWO TYPES OF
CRACK GROWTH: (a) SHEAR SEPARATION
MODE, (b) NORMAL SEPARATION MODE.

FIGURE 6   SUCCESSIVE DISCRETE SHEAR SEPARATIONS CAUSE CRACK
TIP OPENING. (a), (b), (c), (d) SHEAR SEPARATION
MOVEMENTS ALONG SLIP LINES DURING THE LOADING
CYCLE. (e) CRACK OPENING AT MAXIMUM LOAD. (f)
CRACK PROFILE UPON UNLOADING.

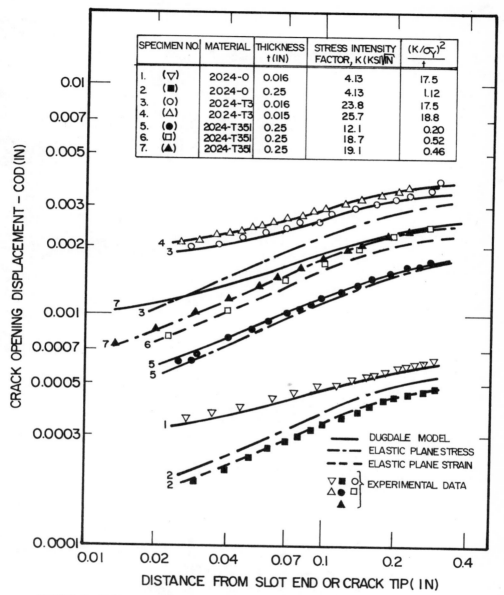

FIGURE 8 CRACK OPENING DISPLACEMENT VERSUS THE DISTANCE FROM A SLOT END OR CRACK TIP FOR A12024-0, A12024-T3 and A12024-T351 ALUMINUM ALLOYS UNDER STATIC LOAD.

Fatigue Crack Growth in a Structural Member
Subjected to Combined Tensile and Bending Stress

Nobu Iino*

1. Introduction

Fatigue characteristics of structural elments must be obtained
in order to assess safety of structures subjected to repeated loading. Estimation of total life for a known geometry and load can
be done to a certain degree of accuracy. Therefore design stress
can be determined using stress-life relation for a given material.
However, in actual welded structures, misalignment, angular distortion and other stress raisers around weldments cause stress
concentrations, thus yielding uncertainty in life prediction.
In many cases, initiation of a flaw is not vital to structures
but often flaws could grow to a critical size without being detected
and result in a catastrophic failure. The fail-safe design concept
may be adopted in structures such as containers when a flaw of
a reasonable size may be allowed. To apply this concept to
pressurized vessels, detection of flaw is done by detecting the
leak from the container, then the size of the flaw at the time it
is detected after its penetration becomes an important factor in
evaluating safety of the container.

Experimental investigation on fatigue was done on 5083-0
aluminum plates for various modes of applied stresses, the results
were then analysed and their applicability in leak before failure
criterion is discussed.

---

*Research staff, Ishikawajima-Harima Heavy Industries
Co. Ltd., Tokyo, Japan

2. Experimental Investigation

2-1. Materials and testing

Fatigue crack growth rates were obtained for material 5083-0 aluminum. Using specimen shown in Fig. 1(a) (referred to as AT specimen hereafter). Surface notched specimen (referred to as CP specimen hereafter) is shown in Fig. 1(b).

Tables 1 and 2 show the chemical composition and the mechanical properties of the material.

To obtain the shape of crack front configuration before penetration, load amplitude in CP specimen was decreased to half that of test condition at certain intervals.

2-2. Experimental results

AT specimens were used to obtain relation between the crack growth rate da/dN and the stress intensity factor range ΔK, which is plotted in Fig. 2 for two temperatures. ΔK is defined as,

$$\Delta K = \Delta \sigma \sqrt{a} \, f'(\frac{a}{W}) \tag{1}$$

where
$\Delta \sigma$ stress range
a crack half length
W half width of the plate
$f(\frac{a}{W})$ correction for finite plate width
     $= 1.77 + 0.227(\frac{a}{W}) - 0.51(\frac{a}{W})^2 + 2.7(\frac{a}{W})^3$
N number of load cycles

Fig. 2 shows that 5083-0 aluminum has marked effect of the mean stress on crack growth rates both at room temperature and at cryogenic temperature. The solid and the dashed lines in the figure are calculated curves based on the empirical equations.

A typical fractured surface of a CP specimen is shown in Fig. 3, with markings showing crack front configurations.

3. Analysis of Fatigue Crack Growth Rates

3-1. Effect of mean stress on crack growth rate

There are many quantitative expressions for fatigue crack growth rates.[2,3] A majority of them are in the form,

$$\frac{da}{dN} = f(K) \tag{2}$$

and more specifically Paris[2] proposed an m-th power law

$$\frac{da}{dN} = C'(\Delta K)^m \tag{3}$$

Another expression with K_{max} in term is[3]

$$\frac{da}{dN} = C(\frac{K_{max}}{\Delta K})(\Delta K)^m \tag{4}$$

Toyosada et al.[4] expresses the crack growth rates in terms of crack opening displacemnt. Equation (4) is shown to express crack growth rates for high strength aluminum.[5,6]

For the materials tested, it is readily seen from the results in Figs. 2(a) and (b) that Eqn. 3 is not appropriate and must include the effect of mean stress. When the following relation is applied to Eqn. 4,

$$\frac{K_{max}}{\Delta K} = \frac{1}{1 - R} \tag{5}$$

$$R = \frac{\sigma_{min}}{\sigma_{max}}$$

a good agreement is seen between the equation and the experimental results. Deviation of experimental data from the curve in the higher ΔK values seems to coincide with the general yielding of specimens, where Eqn. 4 is not applicable.

3-2. Fatigue crack growth by combined tensile and bending stresses

If stress intensity factor is known at each location of the surface flaw, the fatigue crack growth rate may be obtained for a semi-elliptical surface flaw subjected to tensile and

bending stress, provided that the constants in Eqn. 4 do not
change greatly according to shape and location.

Application of solution by Kobayashi et al.[7] is utilised
to express the intensity factor for the surface crack shown in
Fig. 4.

$$K = \sqrt{\frac{\pi b}{a}} \left(a^2 \cos^2\beta + b^2 \sin^2\beta \right)^{\frac{1}{4}} \frac{1}{E(k)}$$

$$\times \left[m_1 \left\{ \left(1 + \frac{2b}{t} \right) \frac{Mt}{2I} + \sigma_n \right\} - m_2 \frac{2b}{t} \frac{Mt}{2I} \left\{ 1 - \frac{k^2 \cdot E(k) \cos\beta}{(1 + k^2) E(k) - k^2 k(k)} \right\} \right]$$

$$(6)$$

$$k^2 = (a^2 - b^2)/a^2$$

$$k'^2 = b^2/a^2$$

$$k(k) = \int_0^{\frac{\pi}{2}} \frac{d\beta}{\sqrt{1 - k^2 \sin^2\beta}}$$

$$E(k) = \int_0^{\frac{\pi}{2}} \sqrt{\frac{1 - k^2 \sin^2\beta}{1 - k^2 \sin^2\beta}} \, d\beta$$

where σ_n tensile stress

            M   bending moment

            I   moment of inertia

            t   thickness of the plate

         a,b   radius of major and minor axis of ellipse

                         angle from the minor axis

      m_1, m_2 coefficients for free surface effect for
                         tension and bending

3-3.   Analysis of fatigue crack growth from a surface flaw

If fatigue crack is semi-elliptical, application of Eqn.6
to Eqn. 4 will yield the relationship between the size of the
crack and the number of loading cycles for 5083-0 aluminum plate
specimens when integration is done at an appropriate interval ΔN.

The comparison of the experiment and calculated crack front
configurations for CP specimens are shown in Fig. 5. The number
of cycles corresponding to crack fronts are also shown in this
figure. The analytical results indicate that crack configuration
is in good agreement with the experiment.

Use of Eqn. 6 in analysis of crack growth beyond plate penetration is not valid. But assuming semi-infinite thickness yielded results shown by dashed lines.

4. Application of the Analysis to Safety Evaluation of Structural Elements

Fatigue behavior of semi-elliptical surface crack is taken into consideration to evaluate safety of structural elements on the basis of leak before failure.

4-1. The effect of initial surface crack length

Fig. 6 shows the relations between initial surface crack length and the crack length at penetration. The figure also shows the relation between the initial surface crack length to number of load cycles required for penetration. For a given stress condition, the longer the initial surface crack length the smaller the number of load cycle to penetration. In thin plate structures, initial crack length is important in evaluating the safety.

4-2. Fatigue crack growth in structural members having distortion caused during fabrication

Distortion such as misalignment and angular distortion cannot be avoided in fabricating welded structures. Even thin plates in such a case, will experience bending stress so that the initiation of a fatigue crack from a surface flaw, if any, will be accelerated as well as affecting the fatigue crack growth rate and crack dimensions at penetration. When a plate having angular distortion w, is stressed in axial direction the strains on both surfaces of the plate is given by[8]

$$\left.\begin{array}{c}\varepsilon_F \\ \varepsilon_B\end{array}\right\} = \varepsilon_m \left\{ 1 \pm 6(1-\nu^2)\frac{w}{t}\ \frac{\tanh m}{m} \right\} \tag{7}$$

where

$$m^2 = 12(1-\nu^2)\frac{\sigma_m}{E}\left(\frac{\ell}{t}\right)^2$$

$\varepsilon_m, \varepsilon_F, \varepsilon_B$	mean strain, strain on front surface and strain on back surface respectively
2ℓ	span (both ends supported)
t	plate thickness
w	angular distortion
E, ν	Young's modulus and Poisson's ratio

Nominal surface stresses for CP specimen are given by

$$\left.\begin{array}{c}\varepsilon_F \\ \varepsilon_B\end{array}\right\} = \varepsilon_m \left\{ 1 \pm 6(1-\nu^2)\frac{\delta}{t}\ \frac{D_1 m_1 \tanh m_2 \ell_2}{D_1 m_1 \tanh m_2 \ell_2 + D_2 m_2 \tanh m_1 \ell_1} \right\} \tag{8}$$

where

$$m_i^2 = \frac{12(1-\nu^2)\sigma_m}{E\,t^2} \qquad (i = 1, 2)$$

δ	misalignment between centers of the plate at gripping and test sections in thickness direction
D_1, D_2	bending stiffness
ℓ_1, ℓ_2	half span length

suffix 1 and 2 refer to the test section and the gripping section of the plate

Fatigue crack growth behaviors are calculated for various angular distortion and various stress conditions. Calculated results showing the relation between the amount of angular distortion and the surface crack length at penetration are shown in Figs. 7(a) and (b).

When the stress condition on the front surface is kept constant, Fig. 7(a) indicates that the larger the angular distortion the longer is the growth in the plate width direction and also the number of cycles to penetration is larger.

For a constant applied load condition, Fig. 7(b) indicates that

that the larger the amount of angular distortion the smaller is
the number of cycles required for crack penetration, but the
surface crack length at penetration a_t is larger.

Since the fracture toughness of the material dealt here
is sufficiently high, the leak before failure condition is
satisfied for the loading conditions considered. But in low
toughness materials unstable fracture can occur before the crack
penetrates the plate to be detected. It must be noted that the
thicker the plate the dimension of the crack at penetration
becomes larger, and consequently more toughness is required as
compared to thinner material if leak before failure condition
should be satisfied.

5. Conclusion

Fatigue crack growth rates for 5083-0 aluminum were obtained
at room temperature and at liquid nitrogen temperature. Surface
notched specimens subjected to tensile and bending stresses
were fatigue tested. And following results were obtained.

(1) The fatigue crack growth rates can be expressed in terms of
stress intensity factor taking into account of the mean stress,

$$\frac{da}{dN} = C \frac{K_{max}}{\Delta K} (\Delta K)^m \qquad (4)$$

(2) Fatigue testing of plate specimens having a semi-elliptical
surface flaw by combined axial and bending stresses grow as
semi-elliptical cracks. And the larger the bending stress component becomes the longer the crack becomes in the plate width
direction.

(3) Fatigue crack growth characteristics of a semi-elliptical
crack can be obtained by integrating Eqn. 4 using the stress

intensity factor obtained by Kobayashi et al.[7]

(4) Safety evaluation of containers can be done based on the leak before failure concept applying the results obtained. And problems to be noted in this evaluation are pointed out and discussed.

Acknowledgement

Thanks are due to Dr. K. Sakai of IHI for support and discussions. The author is also indebted to Mr. Uemura, Mr. Sakano, and Mr. Kobayashi of IHI for the help rendered in carrying out the experiments and advice.

References

(1) ASTM STP 410.

(2) P. C. Paris, et al., "A Critical Analysis of Crack Propagation Laws," Trans. ASME Ser. D., 85(1963), 528.

(3) L. Tall, "Initial Findings from a Study in Low Cycle Fatigue of Welded ASTM A 514 Steel," Fatigue of Welded Structures Conf., (1970).

(4) M. Toyosada, "Fatigue Crack Propagation in Tension-tension Loading Cycles" (in Japanese), Jour. Soc. Nav. Arch., 133 (1973), 209.

(5) M. Matoba, et al., "A Study on the Fatigue Strength of a Linear Type Elbow Plate in Tankers" (in Japanese), Jour. Soc. Nav. Arch., 132.

(6) Uchimoto, et al., "Study on Fatigue Crack Propagation in Aluminum Alloy Sheet," MTR, vol. 7, No. 5.

(7) R. C. Shah, et al., "Stress Intensity Factor for an Elliptical Crack Approaching the Surface of a Plate in Bending," ASTM STP 513.

(8) Y. Akita, et al., "Effect of Angular Distortion in Welded Joints on Brittle Fracture Initiation," IIW, X-569-70 (1970).

Table 1 Chemical composition of 5083-0 aluminum plates

Thickness, mm	Chemical composition (%)							
	Cu	Si	Fe	Mn	Mg	Zn	Cr	Ti
25	0.03	0.13	0.17	0.69	4.61	0.01	0.13	0.01
35	0.03	0.16	0.22	0.68	4.53	0.01	0.13	0.01

Table 2 Mechanical properties

Thickness, mm	Temperature, °C	0.2% offset yield strength 0.2, kg/mm²	Tensile strength T, kg/mm²	Elongation, %
25	20	15.8	33.5	30.0
	-196	18.3	44.0	32.0
35	20	15.5	33.0	27.0

Table 3 Test conditions for fatigue specimens

Specimen No.	Stress condition kg/mm²	Stress ratio min/max	Test temp., °C
AT-1	9.0±4.0	0.38	20
AT-2	4.0±4.0	0	"
AT-3	0±4.0	-1.0	"
AT-4	9.0±4.0	0.38	-196
AT-5	6.0±2.7	0.38	"
AT-6	4.0±4.0	0	"

	Front surface	Back surface	(Front surface)	
CP-1	9.0±2.0	9.0±2.0	0.64	20
CP-2	8.1±1.8	5.1±1.2	0.63	"
CP-3	9.9±2.6	-1.0±0.1	0.58	"

352

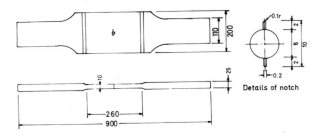

Fig.1-(a)   Specimen with a through thickness notch (AT specimen)

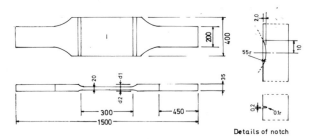

Fig.1-(b) Specimen with a surface notch having a dislocated axis in
thickness direction (CP specimen)

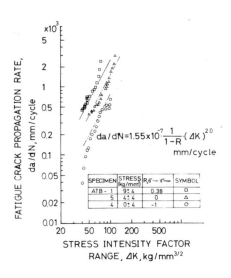

$$da/dN = 1.55 \times 10^{-7} \frac{1}{1-R} (\Delta K)^{2.0}$$
mm/cycle

SPECIMEN	STRESS (kg/mm)	R, $\sigma_{min}/\sigma_{max}$	SYMBOL
ATB - 1	9 ± 4	0.38	□
5	4 ± 4	0	△
4	0 ± 4	-1	○

Fig. 2-(a) da/dN VS. ΔK FOR BASE METAL
(5083-0) AT 20°C

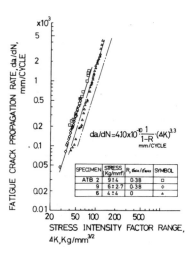

$$da/dN = 4.10 \times 10^{-10} \frac{1}{1-R} (\Delta K)^{3.3}$$
mm/CYCLE

SPECIMEN	STRESS (Kg/mm)	R, $\sigma_{min}/\sigma_{max}$	SYMBOL
ATB 2	9 ± 4	0.38	□
9	6 ± 2.7	0.38	◇
6	4 ± 4	0	△

FIG-2(b) da/dN VS. ΔK FOR BASE METAL (5083-0) AT -196 °C

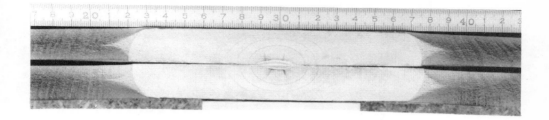

Fig. 3: Fatigue surface of a semi-elliptical surface crack showing crack front configuration

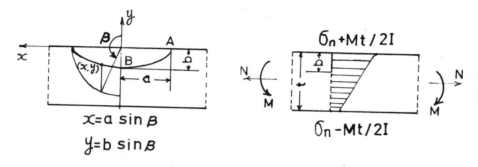

$$x = a \sin \beta$$
$$y = b \sin \beta$$

$$\sigma_n + Mt/2I$$

$$\sigma_n - Mt/2I$$

Fig. 4: Schematic illustration of surface crack and stress distortion under combined tensile and bending stress

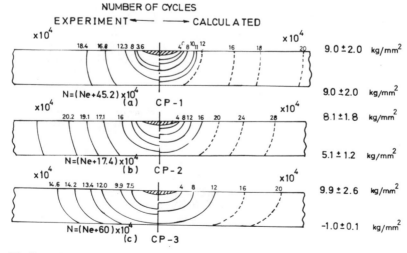

NUMBER OF CYCLES

EXPERIMENT ⟵ ⟶ CALCULATED

$N = (N_e + 45.2) \times 10^4$ (a)   CP-1

9.0 ± 2.0 kg/mm²

9.0 ± 2.0 kg/mm²

8.1 ± 1.8 kg/mm²

$N = (N_e + 17.4) \times 10^4$ (b)   CP-2

5.1 ± 1.2 kg/mm²

9.9 ± 2.6 kg/mm²

$N = (N_e + 60) \times 10^4$ (c)   CP-3

-1.0 ± 0.1 kg/mm²

Fig. 5-a,b,c Comparison of experimental and calculated fatigue crack growth from a semi-elliptical surface crack

354

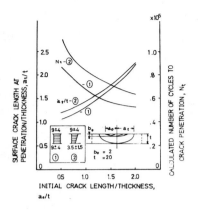

FIG-**6** GEOMETRY OF CRACK FRONT AND NUMBER OF
CYCLES REQUIRED FOR CRACK PENETRATION FOR
DIFFERENT CRACKS.

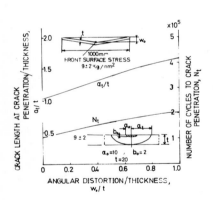

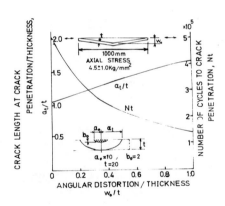

(a) In case 9 ± 2 kg/mm^2 is kept on front surface

(b) In case axial stress is kept
at 4.5 ± 10 kg/mm^2

Fig. 7: Calculated results of fatigue crack propagation and
crack configuration in a member with angular distortion

Some consideration on the fatigue behavior of
Weldable High Tensile Strength steel

by Ishiguro, T., Hanzawa, M., Ishii, N., Yokota, H., Sekiguchi, S.

1.   Synopsis

Some effective way of fatigue design ∧ and use are proposed for the welded structure of High

tensile strength steels (HT), using high cycle and low cycle fatigue characteristics of HT-80 steel.

These method demonstrate the critical dependence of fatigue strength on the stress

concentration at the toe of weld and provide a high fatigue design stress for High tensile strength

steel than usual mild steel.

It is point out that threshold stress intensty for fatigue crack propagation is important to

estimate the maximum allowable defect size based on crack propagation law.

1.   Introduction

With the increasing demand and the expanding application of weldable high tensile strength

steels for the large structures, in the recent years, their fatigue strength in welded joints has become a

very important subjects of design problems.

In many countries, including U. S. A., England, Japan etc., fatigue standard did not give any

merits for design stress of HT-80 steel compared with mild steel.

In the large bridges which consist of the combination of the box section members, the welded

Products Research and Developement Labs.
Nippon steel Corp. KANAGAWA.

parts of stiffner on those members are most important from the point of fatigue design.

There are considerable volume of works in the literature on the fatigue strength of the load carring or non load carring cruciform joints. The non load carring cruciform joint seems to be equivalent to the box member with stiffner. The load carring cruciform joint can either fail from the weld toe or from weld root. In the case of failure from the root, Harrison [1],[2] has annalysed on a crack propagation basis using the concepts of fracture mechanics. From this analysis we have not found any merits of HT-80 steel than mild steel, but using the another material constant of equation to crack propagation for HT-80 steel which have been obtained from many experiment, we can found some merits.

On the load carring cruciform joints, Tajima [4] reviewed many published data of HT-80 steels as shown in Fig. 1. This figure shows almost same fatigue strength of HT-80 steel compared with the design standards for mild steel mentioned in Fig. 1. Therefore, it is a very important problem to fatigue design. The object of this paper is to paint out the some effective way of fatigue design and use for weldable high tensile strength steel, using its fatigue charactenstics.

The most contributing factor to such low fatigue strength in welded joints on HT-80 steel seems to be a stress concentration produced at the toe of welds due to the unfavorable form of welds. Furtheremore these low fatigue strength is caused by a combined effect of such stress concentration and high notchsensitivity of those steels. Therefore it is logical conclusion that if we succeed in reducing such stress concentration by smoothing out the weld profile at the toe of weld by some method. For the purpose of this improving fatigue strength of high tensile strength steel, we have reported that TIG dressing method and additional welding method are highly effective.

2. Fatigue behavior of the high cycle fatigue range

Table 1 and Table 2 shows a chemical composition and its mechanical properties of weldable structural steels used in generally in Japan. Fatigue strength of those steel with mill scale and palished surface are ploted to Fig. 2. Fatigue strength is proportional to the tensile strength of steels with polished surface, but mill sealed specimen do not shows a linear relationship.

For the fatigue testing of the notched specimen, Ono's rotating bending fatigue testing machine with 10kg-m capacity and with loading frequency of 2880 c.p.m. was used. Fig. 3 shows the results obtained in notched fatigue test of mild steel and HT-80 steel. The notch fatigue limits σ'_w, fatigue strength reduction factor K_F to failure life N_f and notch sensitivity factor $q = {K_F - 1}/{K_t - 1}$ have been reported graphically against stress concentration factor K_t. As shown in this figure, when K_t is less than about 2.3, there is relationship of $K_t \doteqdot K_F$ in mild steel as well as HT-80 steel. However, when K_t is bigger than 2.3, there appear great difference in K_F, this difference is more conspicuous at HT-80 steel. Ishibashi's result[5] in Fig. 4 shows schematically this different behavior of both steels.

Iida and Kho[6] have added experimental support to the $K_F - K_t$ relationship to the above mentiond results, but fatigue strength reduction factor to crack initiation life $K_f - K_t$ relationship is linear to mild steel. If this relationship is applied to HT-80 steel too, from point of Nc basis, we can suppose that HT-80 have much merit to mild steel.

3. Low cycle fatigue behavior

The results of the strain-controlled low cycle fatigue tests given to two steels, mild steel and HT-80, are graphically shown in Fig. 6 and Fig. 7. The abscissas indicate the number of cycles to crack initiation Nc and number of cycle to failure Nf. The ordinates indicate $e_{tr.}$, e_{pr} and e_{er}, respectively. Mild steel for which the plastic strain range e_{pr} and life obey Manson-Coffin's relationship over the whole life region, and HT-80 steel for which the e_{pr} and life do not follow Manson-Coffiin's linearity over the whole life region due to substantial lowering of the plastic strain.[7]

The stress amplitude that exhibits the such kink phenomena in plastic strain range v. s. life is closely related to the behavior of the yield strength of HT-80. Cross Point of monotonic and cyclic stress-strain curve which determined from tip of stabilized hysterisis loop in HT-80 steel is correspond to the kink point of e_p—N curve. Except for above mentiond kink region of e_p—N curve, the low cycle fatigue behavior of mild steel and HT-80 steel is nearly same. Estimated points obtained Manson's four-point method are also shown in these figures.

On the application of HT-80 to the welded structure we must avoid the strain range correspond to the kink region of e_p—N curve.

4.   Application of the low cycle test result of the Angular distorted specimens to spiral casing for Hydraulic pump-Turbine.

Fig. 8 shows the geometry of the specimen, the strain distribution at the both sides of specimen were measured with most attention to the toe of weld at the concaved side. The maximum strain at the toe of weld reaches to about 3 to 5 times of the nominal strain. Relation between maximum strain at the concave side and fatigue life are plotted in Fig. 9.[9] This figure shows that comparatively good correlation in noticed e_{max}—N_f relationship.

On the otherhand the relation between nominal stress range and fatigue life of same specimen as shown in Fig. 10 are scattered due to the combined effect of axial stress and bending stress.

It may be useful to estimate the fatigue life of welded joint with maximum strain at the point of stress concentration as reported by Crew and Hardrath [10] with cyclic plastic strain at the notch root.

5.   Improvement of fatigue life of welded joint

The low fatigue strength of high tensile strength steel joints is thought to be caused by a combined effect of stress concentration and high notch sensitivity for fatigue. Therefore if we could reduce such stress concentration by smoothing out the weld profile at the toe of weld bead, we can improve the fatigue strength of such joint.

For the purpose of improving fatigue strength in non load carring fillet welded joint which is most important for box girder of large bridge, we sought the method that will be economical as well as stable. The TIG dressing method and additional bead method are the most useful method which we have found.[11] Fig. 11 shows the test result by TIG dressing method. From this example, it is concluded that by TIG dressing we can markedly improve their fatigue strength to an extent comparable with the effect of dressing by grinding.

6.   Threshold stress intensity (Kth) of HT-80

Harrison[12] and Kitagawa[13] have reviewed literature on a variety of specimens with Kth, but it is not enough for design because of the Kth is effected on the many factors such as loading sequence, stress ratio, kind of stress, size of specimen, kind of material, atmosphere etc.

Harrison concluded that for all material with exception of pure aluminum, crack will not propagate if $\Delta K/E < 1 \times 10^{-4}$ $\sqrt{in}$ (=10.5 kg·mm$^{3/2}$). If an estimation of the size of non propagating crack according to Harrison's Kth Value for bridge which subject to mainly low load cycling will made, it is so small and not economical.

For example our result of Kth with HT-80 steel as shown in Table 3 are higher than Harrison's

Value, but this value obtained by pulsating tension cycling condition in air.

It is evident that a great deal more work is required to obtain a better understanding of the behavior of Kth.

7.   Fatigue Crack propagation of load carring fillet weld

In the case of failure from the weld root of load carring fillet weld. Harrison have concerned with the propagation of a preexisting crack using the concep t of fracture mechanics. In this analysis he assumed that the exporment of ΔK in the paris's crack propagation equation[3] is 4 for mild steel and high tensile strength steel up to HT-80, it will be seen that the results of all tests fall into a scatter band. However if we assume that expornet of ΔK equals 3.2 for HT-80 steel from our experimental results, HT-80 steel shows some different behavior shown in Fig. 12. Therefore from this example it is obvious that the more work is necessary to obtain the constant of crack propagation of steels at various condition.

Conclusion

1)   In high cycle fatigue, when stress concentration factor, K_t is higher than about 2.3, there are so large in fatigue strength reduction factor for failure, K_F with HT-80 steel compared with mild steel.

2)   Mild steel for which plastic strain range ϵ_{pr} and life obey Manson-Coffin's relationship over the whole life range, but HT-80 steel did not follow Manson-Coffin's linearity over the whole life range due to lowering of plastic strain called " Kink phenomenon ".

3)   On the application of HT-80 steel to the welded structure, we must avoid to use strain range

correspond to the " Kink " region of e_p-N curve.

4)   It may be useful to estimate the fatigue life of welded joint with maximum strain at the region

of stress concentration.

5)   TIG dressing method markedly improve the fatigue strength of welded joint of HT-80 steel to an

extent comparable with the effect of dressing by grinding.

6)   Further experimental work should be done to better understanding the threshold stress

intensity and constant of fatigue crack propagation equation.

References

1)   Harrison, J. D., The analysis of fatigue test results for butt welds with lack of penetration

defects using a fracture mechanics approach'. BWRA Report E/13/67)

2)   Harrison, J. D., 'An analysis of the fatigue behaviour of cruciform joints'. Matel Const. and

British Welding J. 1969, 7, 333-335

3)   Paris, P. C., 'The fracture mechanics approach to fatigue'. Syracuse Univ. Press, 1964, 107-127

4)   Tajima, J., ' On Steel for the long suspension bridge between Honsyu to Sikoku'. JSSC Meeting

1973, 8.

5)   Ishibashi, T., 'Fatigue of Metal and its prevension', Yokendo press, 1960.

6)   Iida, K. and KHO, Y., 'A method for calculation of Fatigue strength reduction Factor, Proc. Int.

Conf. Mech. Beh. Mat., Kyoto, 1974, 1-3(4)

7)   IIotta, T., Ishiguro, T., Ishii, N. and Sekiguchi, S., ' Investigation on estimation of Low cycle

fatigue strength of steels', J. Soc. of Naval Architects of Japan, 1968, 124, 341-

8) Manson, S. S., 'Fatigue a complex Subiect some simple approximations', Exp. Mech., 1965, 5, 193

9) Kanazawa, S., Ishiguro, T., Arii, M., Sugiyama, S., ' Fatigue strength of HT-80 spiral casing for Hydraulic Pump-turbine', IIW Doc. XIII-728-74, 1974.

10) Crew, J. H. and Hardrath, H. F., A study of cyclic plastic stresses at a notch root'', Exp. Mech. 1966. 7, 313-320

11) Kanazawa, S., Ishiguro, T., Hanzawa, M., Yokota, H., 'The improvement of Fatigue strength in welded High tensile strength steels', IIW Doc. XIII-735-74.

12) Harrison, J.D., An analysis for data on nonpropagating fatigue cracks on a fracture mechanics basis', Met. Const. Br. Weld. J. 1970, 3, 93-97

13) Ishiguro, T., Mizui, M., Yokota, H., to be publish

14) Kitagawa H. Meeting of JSME 1974, 7. Threshold condition for Fatigue Crack growth.

Table 1 Chemical Composition of Steels

Kind of Steels	Chemical Composition (%)										
	C	Si	Mn	P	S	Cu	Ni	Mo	Nb	Cr	V
SM—4 1	0.17	0.43	1.03	0.013	0.013						
SM—5 0 Y	0.16	0.04	1.09	0.010	0.013				0.026		0.054
COR-TEN50	0.13	0.30	0.99	0.018	0.013	0.30				0.55	0.042
HT—6 0	0.15	0.26	1.27	0.012	0.010						0.016
COR-TEN60	0.14	0.20	1.10	0.015	0.007	0.30				0.54	0.035
HT—1 0 0	0.10	0.26	1.29	0.006	0.006	0.11	0.79	0.74	0.035	0.62	0.059
HT—8 0	0.12	0.31	1.46	0.010	0.002			0.59	0.049		

Table 2 Mechanical properties of Steels

Kind	T.S.	Y.S.	Y.R.	E.L.	JIS No.
	kg／mm	kg／mm	%	%	
SM—4 1	4 6.7	3 1.0	6 6.5	4 2.8	4
SM—5 0 Y	5 3.1	3 8.2	7 2.0	3 0.2	1
COR--T EN5 0	5 2.9	3 5.8	6 7.8	2 8.5	1
HT—6 0	6 1.8	5 3.4	8 6.4	3 2.0	5
COR--T EN6 0	6 9.4	6 1.9	8 9.2	2 1.1	4
HT—1 0 0	1 0 0.1	9 4.7	9 3.7	2 6.0	5
HT—8 0	8 2.4	7 9.0	9 6.0	3 9.0	5

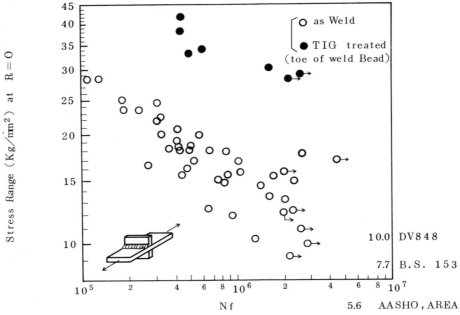

Fig. 1   S—N Curve the fillet welded joint of HT—80 steel.

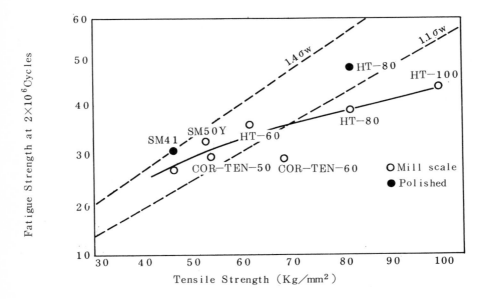

Fig. 2   Corelation of Tensile Strength and Fatigue Strength of Steels with
Mill scale
(Pulsating tension, 500 c.p.m.)

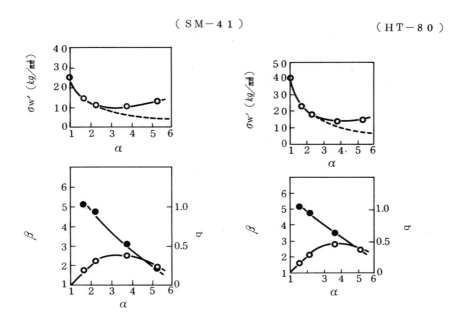

(SM−4 1)　　　　　(HT−8 0)

Fig. 3　Corrilation of Stress Concentration factor(α),notched fatigue
Strength(σw′), fatigue Strength reduction factor(β), fatigue
notch sensitivity(q),Rotating Bending fatigue test

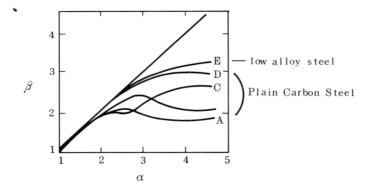

Fig. 4　Schematic Correlation of Stress Concentration factor and fatigue
Strength reduction factor

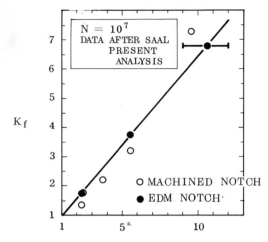

Fig. 5 K_f at 1 x 10^7 Cycles and Predicted Curves[6]

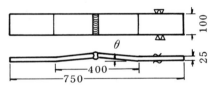

(a) Angular Distortion

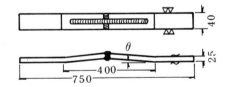

(b) Cross Weld and Angular Distortion

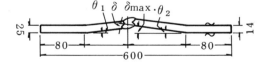

(c) Angular Distortion,Miss Alinement and
    Uneven Plate Thickness

Fig. 8   Fatigue Test Piece of the Welded Joints

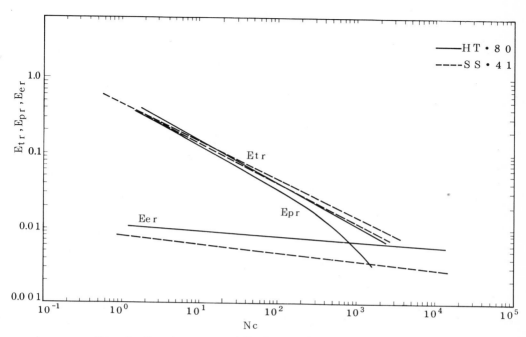

Fig. 6　Strain range and Number of Cycles to Crack initiation

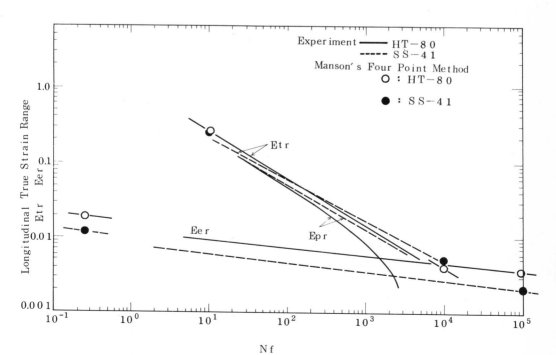

Fig. 7　Strain range and Number of Cycles to failure

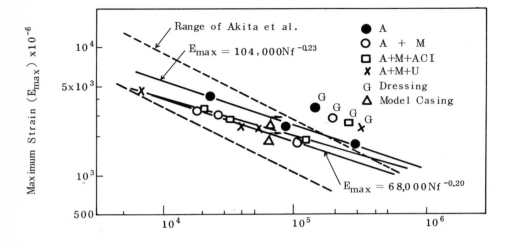

Fig. 9   Maximum Strain to Fatigue Life

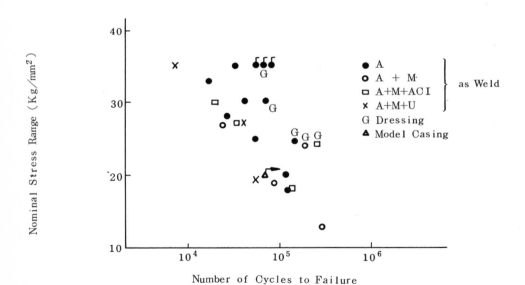

Fig. 10   Nominal Stress Range of Fatigue Life of Test Specimen and Model Casing

A; Angular Distortion, M; Mhss Alinement

ACI; Additive Bead on Weld Toe of Concave Side

U; Uneven Plate Thickness

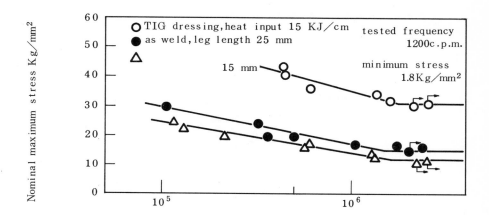

Number of cycles to failure

Fig. 11 Results of fatigue test for fillet welded joints with as
welded condition using electrode L−70 and TIG dressing.

Table 3 Kth of Mild steel HT−60 and HT−80 steel
at pulsating tension test

	Crack length (mm)	Gross stress (Kg/mm^2)	S.I.F. K (Kg/mm$^{3/2}$)	S.I.F. K	at prop.or non prop. (Kg/mm$^{3/2}$)
				prop.	non prop.
S M 4 1	3.9 8	1 6.1	5 8.0	3 1.2	2 9.5
	1 2.0 6	7.9	5 9.2	3 5.2	3 1.5
H T − 6 0	3.9 9	1 7.0	6 1.1	3 6.7	3 4.8
	1 2.1 3	8.1	6 1.4	3 7.6	3 3.7
H T − 8 0	5.1 1	1 6.9	6 7.8	3 7.6	3 5.4
	1 2.1 8	8.1	6 1.4	3 7.6	3 3.6

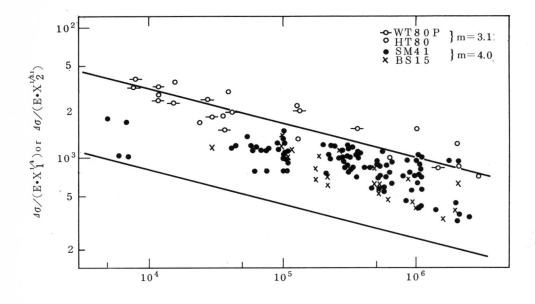

Number of Cycles to Failure (Nf)

Fig. 12   Modified plots of Harrison's calcaration with HT-80 Steel

SESSION V

Non-metals and Composites

On Fracture Behavior of Composite Materials

G. P. Sendeckyj*

1. INTRODUCTION

Even though the fracture behavior of advanced fiber reinforced composite materials has been investigated for a number of years now, theoretically significant work has started appearing only recently. Experimental data, in conjunction with sound data reduction procedures based on exact analyses that take into account material anisotropy and specimen geometry (1)-(6), have documented the crack (or notch) size effect in composites (7)-(10). This effect is manifested in the apparent critical stress intensity factor, K_Q, being a monotonically increasing function of the crack length tending asymptotically to the critical stress intensity factor, K_C, as the crack length increases. For fiber reinforced composite materials, $K_Q \simeq K_C$ when the crack length is greater than one inch; while for high strength brittle metals, $K_Q \simeq K_C$ for crack lengths greater than 0.001 inch. Hence, the crack size effect is hidden in brittle metals by the large crack lengths used in generating the fracture toughness data.

Attempts to explain and quantify the crack size effect in composites have led to the development of a number of two parameter fracture theories (7), (9), (11)-(15). These theories fall naturally into two groups, namely, theories attempting to explain the crack size effect and those attempting to predict the direction of crack propagation. If properly interpreted the theories in the later group can also be used to explain the crack size effect.

A brief review of the more prominent two parameter fracture theories is presented in Section 2. A new fracture theory for fiber dominated composite materials (i. e., composite materials in which fibers must be broken for

* Structural Integrity Branch, Structures Division
  Air Force Flight Dynamics Laboratory
  Wright-Patterson Air Force Base, Ohio 45433

fracture to occur) is presented in Section 3. Theoretical predictions of this fracture theory are in excellent agreement with experimental data. Finally, implications of this theory are discussed in Section 4.

2. REVIEW OF TWO PARAMETER FRACTURE THEORIES

2.1 Intense Energy Region Theory (7)

The first of the two parameter fracture theories was proposed in 1971 by Waddoups et al. (7) in an attempt to explain the absolute hole size effect observed in composite materials. The theory is based on the assumption that an _intense energy region_ exists at the tips of the crack. The intense energy region is assumed to be similar in nature to the plastic zone existing at crack tips in ductile materials and, hence, the critical stress intensity factor is given by

$$K_C = \sigma_C \, [\pi(a + a_o)]^{1/2} \tag{1}$$

where a, a_o, and σ_C are the crack half length, length of the intense energy region, and the fracture strength of the laminate containing a crack of length $2a$, respectively. For an unnotched laminate, $a = 0$ and Eq. 1 gives

$$K_C = \sigma_\infty \, (\pi a_o)^{1/2} \tag{2}$$

where σ_∞ is the ultimate strength of the unnotched laminate. Having experimentally measured values of σ_C, σ_∞, and a, one can solve Eqs. 1 and 2 for a_o and K_C. If the intense energy region correction were not made, we would have

$$K_Q = \sigma_C \, (\pi a)^{1/2} \tag{3}$$

from which it follows that the apparent critical stress intensity factor is given by

$$K_Q = [a/(a + a_o)]^{1/2} \, K_C \tag{4}$$

It should be noted that $K_Q \to K_C$ as $a \to \infty$ and $K_Q \to 0$ as $a \to 0$. This implies that K_Q, defined by Eq. 4, cannot be a material parameter and, hence, fracture data for composite materials cannot be reduced using Eq. 3.

2.2 Whitney and Nuismer Point Stress Fracture Theory (9), (11)

Even though the theory proposed by Waddoups et al. (7) is readily applicable to through-the-thickness cracks in composites, unrealistic assumptions have to be made when using the theory to explain the size effect observed for circular notches. In an attempt to remove the artificialities of the intense energy region theory, Whitney and Nuismer, (9) and (11), proposed that the laminate fractures when the stress, normal to the crack, at some distance a* ahead of the crack reaches the ultimate strength, σ_∞, of the laminate. An explanation of the crack size effect follows from an evaluation of the "exact" stress distribution ahead of the crack. For a crack of length 2a, the normal stress ahead of the crack is given by

$$\sigma_y = K_Q [1 - \xi^2]^{-1/2} (\pi a)^{-1/2} \tag{5}$$

where $\xi = a/(a + \rho)$ and ρ is the coordinate, parallel to the crack, with origin at the crack tip. Upon setting $\sigma_y = \sigma_\infty$ and $\rho = a*$ in Eq. 5, we get

$$K_Q = \sigma_\infty (\pi a)^{1/2} [1 - \xi_o^2]^{1/2} \tag{6}$$

where $\xi_o = a/(a + a*)$. Defining the critical stress intensity factor, K_C, as the limit of K_Q as $a \to \infty$, that is,

$$K_C = \sigma_\infty (2\pi a*)^{1/2} \tag{7}$$

we get

$$K_Q = (a/2a*)^{1/2} [1 - \xi_o^2]^{1/2} K_C \tag{8}$$

As can be seen from Eq. 8, $K_Q \to 0$ as $a \to 0$.

2.3 Whitney and Nuismer Average Stress Fracture Theory (9), (11)

In addition to the point stress fracture criterion, Whitney and Nuismer, (9) and (11), suggested that the fracture strength could also be predicted by

assuming that the average stress over a small region ahead of the crack
attains a critical value equal to the ultimate strength of the laminate.
Upon averaging the stress, defined by Eq. 5, over a distance a_o^* ahead of the
crack and setting the average value equal to σ_∞, we get

$$K_Q = \sigma_\infty (\pi \zeta a_o^*)^{1/2} \tag{9}$$

where $\zeta = a/(2a + a_o^*)$. Upon defining K_C as the limit of K_Q as $a \to \infty$, that is,

$$K_C = \sigma_\infty (\pi a_o^*/2)^{1/2} \tag{10}$$

we get

$$K_Q = [2a/(2a + a_o^*)]^{1/2} K_C \tag{11}$$

If a_o^* is set equal to $2a_o$, we see that the Whitney and Nuismer average stress
fracture criterion is operationally identical to the intense energy region
theory.

2.4 Wu's Strength Tensor Fracture Theory (13)

The three fracture theories, discussed above, assume that the crack
propagates in a self-similar manner. This assumption is seldom applicable
for cracks in laminates under general loading conditions. Hence, fracture
theories that can predict the direction of crack propagation are of considerable interest. One such theory is Wu's generalization of the maximum normal
stress fracture theory for brittle, isotropic solids.

Wu's strength tensor fracture theory is based on the hypothesis that
fracture occurs when the "state of stress" at a point, A, at a distance r*
from the crack tip reaches a critical value. Moreover, the crack will
propagate in a straight line towards point A. The critical stress state is
defined through Wu's stress tensor strength criterion. This hypothesis can
be stated mathematically as follows:

(1) Crack initiation takes place in the direction defined by the
maximum value of the strength function, $F(\sigma_{ij})$, evaluated on a circle of

radius r* about the crack tip, that is,

$$\frac{\partial F}{\partial \theta} = 0, \quad \theta = \theta_o \tag{12}$$

(2) Crack extension occurs when the strength function, $F(\sigma_{ij})$, reaches a critical value, that is,

$$F_C = F(\sigma_{ij}, r^*, \theta_o) = F_{ijk\ell}\sigma_{ij}\sigma_{k\ell} + F_{ij}\sigma_{ij} = 1 \tag{13}$$

where F_{ij} and $F_{ijk\ell}$ are material constants and r^* and θ_o are measured from the crack tip.

It should be noted that this criterion can be considered to be an extension of the Whitney and Nuismer point stress theory. Hence, it can be used to explain the crack size effect.

2.5 Sih's S_C-Theory (14), (15)

Another theory that predicts the direction of crack propagation is Sih's S_C-theory. As used in references (14) and (15), the basic hypotheses of fracture in this theory are:

(1) Crack initiation takes place in the direction defined by the stationary value of the strain energy density factor, that is,

$$\frac{\partial S}{\partial \theta} = 0, \quad \theta = \theta_o \tag{14}$$

(2) Crack extension occurs when the strain energy density factor reaches a critical value, that is,

$$S_C = S(K_1, K_2) \quad \text{for} \quad \theta = \theta_o \tag{15}$$

For a crack, S is given by

$$S = A_{11}K_1^2 + 2A_{12}K_1K_2 + A_{22}K_2^2 \tag{16}$$

where K_1 and K_2 are the stress intensity factors for mode I and II crack extension. The A_{ij} are functions of the angle θ and elastic moduli. In the version of the S_C-theory presented in references (14) and (15), the strain

energy density function is evaluated at a point sufficiently close to the
crack tip that the crack tip stress singularity completely determines S and,
hence, the theory does not predict a crack size effect. Use of the exact
stress distribution in evaluating S would predict the size effect. For self-
similar crack extension, the S_c-theory is equivalent to the K_c-theory.

3. FRACTURE TOUGHNESS OF FIBER DOMINATED LAMINATES

3.1 Theoretical Development

The above mentioned fracture theories require extensive experimental
data to determine all the required constants. At the very least, the
fracture strength of a laminate containing cracks of two different lengths
as well as the static strength of the unnotched laminate are needed for a
reasonable estimate of the parameters in the intense energy region, point
stress, and average stress fracture theories. The strength tensor and S_c
theories require even more data.

It has been noted that the fracture toughness of a fiber dominated
laminate is about one half the tensile strength. This suggests that the
fracture toughness is related to the tensile strength of the laminate and,
hence, it can be expressed as a function of the strength and elastic moduli
of the laminate. With this in mind, let us consider a crack of length, 2a,
in an anisotropic laminate (shown in Fig. 1).

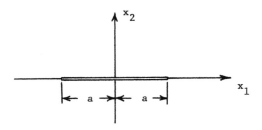

FIGURE 1. Crack in anisotropic solid.

For self-similar crack extension, the energy release rate, G, is related
to the critical stress intensity factor, K_C, by

$$G = K_C^2 (a_{11}a_{22}/2)^{1/2} [(a_{22}/a_{11})^{1/2} + (2a_{12} + a_{66})/2a_{11}]^{1/2} \qquad (17)$$

where the a_{ij} are the elastic compliances (16). The a_{ij} are defined in terms
of the elastic moduli as

$$a_{11} = 1/E_T, \qquad a_{22} = 1/E_L$$

$$(18)$$

$$a_{66} = 1/G_{LT}, \qquad a_{12} = - \nu_{LT}/E_L$$

where E_L, E_T, G_{LT}, and ν_{LT} are the longitudinal Young's modulus, transverse
Young's modulus, longitudinal shear modulus, and major Poisson's ratio,
respectively. If the laminate is sufficiently thick that a state of plane
strain exists, the a_{ij} in Eq. 17 must be replaced by the reduced elastic
compliances, b_{ij}, defined by

$$b_{ij} = a_{ij} - a_{i3}a_{j3}/a_{33} \qquad (19)$$

The energy release rate G can be estimated as the energy released upon
fracturing the fibers that are normal to the crack, that is,

$$G = 0.5(c_o \sigma_{o\infty})^2/E_L = 0.5(c_o \sigma_{o\infty})^2 a_{22} \qquad (20)$$

where c_o is the fraction of plies with fibers oriented perpendicular to the
crack and $\sigma_{o\infty}$ is the longitudinal strength of a 0°-ply. It should be noted
that this estimate of the energy release rate takes into account the
extensive matrix crazing and delamination observed at crack tips in polymeric
matrix composites (17), (18). Moreover, this estimate is valid for crack
extension that is self-similar on the large scale and yet non-self-similar on
the small scale since the energy released upon fracturing the fibers normal
to the crack is independent of where the fibers fracture. If the matrix
crazing is not extensive, then $c_o \sigma_{o\infty}$ has to be replaced by σ_∞ where σ_∞ is the
ultimate strength of the laminate.

Upon substituting Eq. 20 into Eq. 17 and solving for K_C, we get

$$K_C = c_o \sigma_{o\infty} \{(a_{22}/2a_{11})^{1/2} [(a_{22}/a_{11})^{1/2} + (2a_{12} + a_{66})/2a_{11}]^{-1/2}\}^{1/2} \qquad (21)$$

Equation 21 is the expression for the critical stress intensity factor for a laminate exhibiting extensive matrix crazing at the crack tips. For the case of no (or very limited) matrix crazing, Eq. 21 must be replaced by

$$K_C = \sigma_{\infty} \{(a_{22}/2a_{11})^{1/2} [(a_{22}/a_{11})^{1/2} + (2a_{12} + a_{66})/2a_{11}]^{-1/2}\}^{1/2} \qquad (22)$$

Since plane stress assumptions were made in deriving Eqs. 21 and 22, they predict the plane stress fracture toughness of a fiber dominated laminate. Plane strain values can be obtained by replacing the a_{ij} by b_{ij} in Eqs. 21 and 22.

Finally, the crack size effect can be incorporated by using the intense energy region, point stress, or average stress fracture theories. For the intense energy region and average stress theories,

$$a_o = (K_C/\sigma_{\infty})^2/\pi \qquad (23)$$

with K_Q defined by Eq. 4. For the point stress theory,

$$a^* = (K_C/\sigma_{\infty})^2/2\pi \qquad (24)$$

with K_Q defined by Eq. 8.

Let us now compare the theoretical predictions based on Eqs. 21 and 22 with the limited experimental data documenting the crack size effect.

3.2 Theory-Experiment Comparison for Scotchply 1002, Glass-Epoxy Laminates

Nuismer and Whitney (9) measured the apparent critical stress intensity factor, K_Q, as a function of crack length, 2a, for $(0°/\pm45°/90°)_{2S}$ and $(0°/90°)_{4S}$ Scotchply 1002 glass-epoxy laminates. Their experimental results are compared with the theoretical predictions, based on Eq. 21, in Tables I and II.

The theoretical predictions for the $(0°/\pm45°/90°)_{2S}$ laminate were

TABLE I.   FRACTURE TOUGHNESS OF $(0°/\pm45°/90°)_{2S}$ GLASS-EPOXY LAMINATE

a	K_Q (ksi√in.)		
(in.)	Experiment	Eq. 4	Eq. 8
0.057	$13.8\ ^{+0.8}_{-1.0}$	15.0	15.1
0.152	$18.5\ ^{+1.3}_{-1.6}$	18.9	19.6
0.290	$20.6\ ^{+0.0}_{-0.4}$	20.6	21.2
0.488	$20.8\ ^{+0.6}_{-0.0}$	21.6	22.0

TABLE II.   FRACTURE TOUGHNESS OF $(0°/90°)_{4S}$ GLASS-EPOXY LAMINATE

a	K_Q (ksi√in.)		
(in.)	Experiment	Eq. 4	Eq. 8
0.0515	$16.0\ ^{+1.0}_{-0.7}$	17.3	18.4
0.1425	$20.9\ ^{+1.2}_{-1.6}$	22.9	24.0
0.2925	$25.0\ ^{+1.2}_{-1.7}$	25.9	26.7
0.509	$23.2\ ^{+1.6}_{-3.1}$	27.5	28.1

arrived at as follows. First, the elastic moduli of the $(0°/\pm45°/90°)_{2S}$ laminate

$$E_L = E_T = 2.75\times10^6 \text{ psi}, \quad G_{LT} = 1.08\times10^6 \text{ psi}, \quad \nu_{LT} = 0.2696 \tag{25}$$

were calculated using laminated plate theory from the single ply elastic moduli ($E_L = 5.6\times10^6$ psi, $E_T = 1.2\times10^6$ psi, $G_{LT} = 0.6\times10^6$ psi, $\nu_{LT} = 0.26$), reported by Nuismer and Whitney (9). The laminate elastic moduli were then substituted into Eq. 21 to give

$$K_C = 0.707\ c_o\ \sigma_{o\infty} \tag{26}$$

Since Nuismer and Whitney did not report $\sigma_{o\infty}$, the longitudinal strength of the basic ply was estimated to be 132 ksi by back calculating from σ_∞ = 46.4 ksi. With this value of $\sigma_{o\infty}$, Eq. 26 gives

$$K_C = 23.32 \text{ ksi } \sqrt{\text{in.}} \qquad (27)$$

Using Eqs. 23 and 24, we get

$$a_o = 0.08 \text{ in. }, \qquad a^* = 0.04 \text{ in.} \qquad (28)$$

Following the same procedure and taking σ_∞ = 52.5 ksi, we get

$$K_C = 30.20 \text{ ksi } \sqrt{\text{in.}} , \qquad a_o = 2a^* = 0.105 \text{ in.} \qquad (29)$$

for the $(0°/90°)_{4S}$ laminate. Finally, the theoretical K_Q values were calculated by using Eqs. 4 and 8.

As can be seen from Table I, the theory-experiment comparison is excellent for the $(0°/\pm45°/90°)_{2S}$ laminate. For the $(0°/90°)_{4S}$ laminate, the comparison is not as good. A possible explanation for the poorer theory-experiment agreement for the $(0°/90°)_{4S}$ laminate is the observation that Nuismer and Whitney did not take into account material anisotropy in reducing their test data. Finally, the theoretical K_Q's, corresponding to the intense energy region and average stress fracture theories (Eq. 4), are in better agreement with the experimental data than those corresponding to the point stress fracture theory (Eq. 8).

3.3 Theory-Experiment Comparison for T300/5208 Graphite-Epoxy Laminates

The apparent critical stress intensity factors, K_Q, as a function of crack length, 2a, were measured for T300/5208 graphite-epoxy laminates by Nuismer and Whitney (9), Konish and Cruse (8), and the author. Nuismer and Whitney (9) measured K_Q for $(0°/90°)_{4S}$ and $(0°/\pm45°/90°)_{2S}$ laminates using a center notched specimen; Konish and Cruse (8) measured K_Q for $(0°/\pm45°)_{nS}$ laminates using center notched and edge notched beam specimens; while, the author measured K_Q for $(0°/90°/90°/0°)_S$, $(90°/0°/0°/90°)_S$, and $(0°/\pm45°/90°)_S$

TABLE III.   FRACTURE TOUGHNESS OF $(0°/\pm45°/90°)$ GRAPHITE-EPOXY LAMINATES

LAMINATE	a (in.)	K_Q (ksi√in.)		
		Experiment	Eq. 4	Eq. 8
$(0°/\pm45°/90°)_{2S}$	0.0545	$26.1{+2.4 \atop -1.5}$	26.9	28.5
	0.1535	$34.0{+2.4 \atop -3.4}$	37.4	39.5
	0.2995	$42.9{+0.7 \atop -1.4}$	43.1	44.9
	0.495	$46.5{+0.9 \atop -0.0}$	46.4	47.8
$(0°/\pm45°/90°)_{S}$	0.198	$41.2{+1.5 \atop -0.8}$	39.7	41.7
	0.300	$45.7{+1.6 \atop -2.9}$	43.1	44.9
$(90°/\pm45°/0°)_{S}$	0.199	$37.9{+6.5 \atop -5.3}$	39.7	41.7
	0.300	$43.3{+1.4 \atop -2.5}$	43.1	44.9

TABLE IV.   FRACTURE TOUGHNESS OF $(0°/90°)$ GRAPHITE-EPOXY LAMINATES

LAMINATE	a (in.)	K_Q (ksi√in.)		
		Experiment	Eq. 4	Eq. 8
$(0°/90°)_{4S}$	0.0545	$28.5{+2.7 \atop -1.9}$	37.2	39.3
	0.152	$34.2{+2.5 \atop -3.8}$	46.8	48.6
	0.2995	$48.2{+8.9 \atop -11.6}$	51.1	52.3
	0.4975	$62.4{+3.5 \atop -5.8}$	53.2	54.1
$(0°/90°/90°/0°)_{S}$	0.200	$60.2{+1.1 \atop -1.7}$	48.7	50.3
	0.300	$48.6{+2.8 \atop -3.5}$	51.1	52.3
$(90°/0°/0°/90°)_{S}$	0.1985	$58.3{+6.0 \atop -6.2}$	48.7	50.3
	0.300	$46.5{+10.0 \atop -10.9}$	51.1	52.3

TABLE V.   FRACTURE TOUGHNESS OF $(0°/\pm45°)_{nS}$ GRAPHITE-EPOXY LAMINATES

n	a	K_Q (ksi$\sqrt{\text{in.}}$)		
	(in.)	Experiment	Eq. 4	Eq. 8
2	0.10	$25.4\ {}^{+0.4}_{-0.7}$	33.5	35.3
	0.20	$33.9\ {}^{+1.5}_{-1.2}$	38.9	40.5
	0.50	$39.7\ {}^{+3.4}_{-1.7}$	43.9	44.8
11	0.40	37.8	42.9	44.0

laminates using a double-edge notched specimen. The experimental results are compared with theoretical predictions in Tables III through V.

The theoretical predictions for the $(0°/\pm45°/90°)$ laminates correspond to

$$K_C = 53.3 \text{ ksi } \sqrt{\text{in.}}, \quad a_o = 2a^* = 0.159 \text{ in.} \tag{30}$$

which were predicted by using Eq. 22 and the basic ply elastic properties $(E_L = 23\times10^6 \text{ psi}, E_T = 1.6\times10^6 \text{ psi}, G_{LT} = 0.77\times10^6 \text{ psi, and } \nu_{LT} = 0.3)$ measured by the author. If Eq. 21 were used with $\sigma_{o\infty} = 234$ ksi, we would have

$$K_C = 41.4 \text{ ksi } \sqrt{\text{in.}}, \quad a_o = 2a^* = 0.096 \text{ in.} \tag{31}$$

Based on Eq. 4, this would give $K_Q = 24.9$ ksi$\sqrt{\text{in.}}$ for a = 0.545 in., $K_Q = 32.5$ ksi$\sqrt{\text{in.}}$ for a = 0.1535 in., $K_Q = 36.0$ ksi$\sqrt{\text{in.}}$ for a = 0.3 in., and $K_Q = 37.9$ ksi$\sqrt{\text{in.}}$ for a = 0.495 in. Hence, the experimental data for small crack lengths seems to be in better agreement with the theoretical predictions based on Eq. 21 than those based on Eq. 22, while the reverse is true for large cracks. This suggests that matrix crazing plays a major role in determining the fracture toughness of small cracks, while not affecting large ones.

The theoretical predictions for the $(0°/90°)$ laminates are based on

$$K_C = 57.05 \text{ ksi}\sqrt{\text{in.}} \quad , \quad a_o = 2a* = 0.074 \text{ in.} \tag{32}$$

which were predicted by using Eq. 21 with $\sigma_{o\infty} = 234$ ksi. As can be seen from Table IV, the theory-experiment agreement is poor. It should be noted that the scatter in the experimental data is not sufficient to explain the lack of agreement between theory and experiment. A possible explanation is the extensive longitudinal splitting observed in this class of graphite-epoxy laminates. The splitting normal to the crack changes the local stress fields at the crack tips and, hence, a fracture mechanics treatment of these laminates is not appropriate.

The theoretical predictions for the $(0°/\underline{+}45°)_{nS}$ laminates are based on

$$K_C = 48.4 \text{ ksi} \sqrt{\text{in.}} \quad , \quad a_o = 2a* = 0.109 \text{ in.} \tag{33}$$

which were predicted by using Eq. 21, $\sigma_{o\infty} = 204$ ksi and the basic ply elastic moduli ($E_L = 20.5 \times 10^6$ psi, $E_T = 1.37 \times 10^6$ psi, $G_{LT} = 0.752 \times 10^6$ psi, and $\nu_{LT} = 0.31$) reported by Konish and Cruse. $\sigma_{o\infty} = 204$ ksi is consistent with $\sigma_{\infty} = 82.8$ ksi, which is strength of the unnotched laminate reported by Konish and Cruse. Moreover, the raw data* was reduced using the maximum gross stress instead of the values used by the authors (8). At present, no explanation for the discrepancy can be offered.

4. DISCUSSION AND CONCLUSION

As can be seen from the theory-experiment comparisons presented in Sections 3.2 and 3.3, the theoretically predicted fracture toughness values are excellent for the quasi-isotropic laminates. For the $(0°/90°)$ laminates, the theory-experiment comparison is not as good. The discrepancy in the graphite-epoxy laminate data can be explained on the basis of the experimentally observed failure mode (longitudinal splitting of the $0°$-plies

---

* Courtesy of B. E. Kaminski, General Dynamics Corp.

followed by a net section failure). Since the sequence of events leading to failure was not reported in reference (8) for the $(0°/\pm45°)_{nS}$ T300/5208 graphite-epoxy laminates, it is impossible to determine whether the theoretical predictions are in error or the failure mode was such as to make a fracture mechanics reduction of the experimental data inappropriate.

The strain energy fracture theory presented in Section 3.1 predicts the critical stress intensity factors for large cracks in composite laminates. It does not predict the apparent fracture toughness for small cracks. K_Q is determined by using one of the other theories (intense energy region, point stress, or average stress) in conjunction with K_C predicted by the present theory. Strictly speaking, the theoretical values of K_Q based on Eqs 4 and 8 are only valid for a crack in an unbounded laminate. Since the experimental values of K_Q were determined from test data for finite width specimens, theoretical equations for K_Q that take into account the specimen geometry should be used instead of Eq.s 4 and 8. This is especially true for the double-edge notched specimens, where small net sections could cause the intense energy regions to overlap.

Finally, the present theory requires the following data for full verification:

(1) Basic ply elastic moduli and longitudinal strength;

(2) Laminate elastic moduli and longitudinal strength; and

(3) Apparent fracture toughness data for a minimum of two crack lengths.
Moreover, all data must be statistically valid. This requirement means that at least 10 replicates are needed for each test condition. We expect that the necessary data will become available in the near future.

5. REFERENCES

(1) O. L. Bowie, and C. E. Freese, "Central Crack in Plane Orthotropic Rectangular Sheet", International Journal of Fracture Mechanics,

Vol. 8, (March, 1972), pp. 49-57.

(2) G. S. Gyekenyesi, Elastostatic Stress Analysis of Finite Anisotropic Plates with Centrally Located Traction-Free Cracks, Ph. D. Thesis, Department of Metallurgy, Mechanics and Materials Science, Michigan State University, (1972), 158 pp.

(3) G. S. Gyekenyesi, and A. Mendelson, Elastostatic Stress Analysis of Orthotropic Rectangular Center-Cracked Plates, NASA Lewis Research Center, NASA TN D-7119, (November, 1972), 32 pp.

(4) M. D. Snyder, and T. A. Cruse, Crack Tip Stress Intensity Factors in Finite Anisotropic Plates, Department of Civil Engineering, Carnegie-Mellon University, AFML TR-73-209, (August, 1973), 130 pp.

(5) J. F. Mandell, F. J. McGarry, S. S. Wang, and J. Im, "Stress Intensity Factors for Anisotropic Fracture Test Specimens of Several Geometries", Journal of Composite Materials, Vol. 8, (April, 1974), pp. 106-116.

(6) S. N. Atluri, A. S. Kobayashi, and M. Nakagaki, "A Finite Element Program for Fracture Mechanics Analysis of Composite Material", presented at the ASTM Symposium on Fracture Mechanics of Composite Materials, Gaithersberg, Maryland, (September 25, 1974).

(7) M. E. Waddoups, J. R. Eisenmann, and B. E. Kaminski, "Macroscopic Fracture Mechanics of Advanced Composite Materials", Journal of Composite Materials, Vol. 5, (October 1971), pp. 446-454.

(8) H. J. Konish, Jr., and T. A. Cruse, "The Determination of Fracture Strength in Orthotropic Graphite/Epoxy Laminates", presented at the ASTM Composite Reliability Conference, Las Vegas, Nevada, (April, 1974).

(9) R. J. Nuismer, and J. M. Whitney, "Uniaxial Failure of Composite Laminates Containing Stress Concentrations", presented at the ASTM Symposium on Fracture Mechanics of Composite Materials, Gaithersberg, Maryland, (September 25, 1974).

(10) C. Zweben, "Fracture of Kevlar 49, E-glass and Graphite Composites", presented at the ASTM Symposium on Fracture Mechanics of Composite Materials, Gaithersberg, Maryland, (September 25, 1974).

(11) J. M. Whitney, and R. J. Nuismer, "Stress Fracture Criteria for Laminated Composites Containing Stress Concentrations", Journal of Composite Materials, Vol. 8, (July, 1974), pp. 253-265.

(12) T. A. Cruse, "Tensile Strength of Notched Composites", Journal of Composite Materials, Vol. 7, (April, 1973), pp. 218-229.

(13) E. M. Wu, "Strength and Fracture of Composites", in L. J. Broutman, ed., Fracture and Fatigue, Vol. 5 in Composite Materials, edited by L. J. Broutman and R. H. Krock, Academic Press, New York, (1974), pp. 191-247.

(14) G. C. Sih, and E. P. Chen, "Fracture Analysis of Unidirectional

388

Composites", <u>Journal of Composite Materials</u>, Vol. 7, (April, 1973), pp. 230-244.

(15) G. C. Sih, E. P. Chen, and S. L. Huang, "Fracture Mechanics of Plastic-Fiber Composites", <u>Engineering Fracture Mechanics</u>, Vol. 6, (September, 1974), pp. 343-359.

(16) G. C. Sih, and H. Liebowitz, "Mathematical Theories of Brittle Fracture", in H. Liebowitz, ed., <u>Fracture</u>, Academic Press, New York, (1968), Vol. 2, pp. 67-190.

(17) F. H. Chang, J. C. Couchman, J. R. Eisenmann, and B. G. W. Yee, "A Nondestructive Test Technique for the Detection of Damage near Stress Concentrations in Advanced Composite Laminates", presented at the ASTM Composite Reliability Conference, Las Vegas, Nevada, (April, 1974).

(18) J. F. Mandell, F. J. McGarry, J. Im, and U. Meier, "Fiber Orientation, Crack Velocity, and Cyclic Loading Effects on the Mode of Crack Extension in Fiber Reinforced Plastics", presented at TMS/AIME Conference "Failure Modes in Composites - II", Pittsburg, Pennsylvania, (May, 1974).

Crack Propagation in Viscoelastic,
Diaspastically Non-simple Solids: A Progress Report*

W. G. Knauss**

1. INTRODUCTION

The fracture of rate insensitive solids has been studied for many years
from the viewpoint of crack propagation or crack instability. By and large
the main thrust of this effort has been directed towards describing the fracture process for engineering purposes in terms of material parameters
which avoid details of the failure processes at the crack tip. As examples
of these "non-detailed" descriptions we cite the Griffith-Irwin theory of crack
instability in terms of the (plasticity modified) stress intensity factor [2, 3],
or the Barenblatt cohesion modulus [4], the crack-opening-displacement
idea (COD) [5], and the more recent concept of the path independent J-
integral [6]. These concepts try to predict the <u>onset</u> of fracture on the
argument that some average stress or deformation-related quantity such as
an energy release, a properly defined COD, etc. can form the basis of fracture mechanical design criteria.

Experience with the fracture behavior of polymers is much more limited
than with metals. Concepts developed for the fracture of rate independent
solids have been applied to rate sensitive materials with varying degrees of
success and such investigations have, in turn, raised many questions. The
new questions have invariably resulted from our lack of understanding the
effects contributed by the time-dependent material properties and how they
interact with the time-dependent loads on the structure under the influence
of a non-regular geometry (crack tip region). One might argue that the best
way of resolving these and similar questions is through experiments. However, that approach is successful only if one has a fairly clear idea of how

---

*A condensed version of this paper has been published as Reference 1

**Associate Professor of Aeronautics, California Institute of Technology,
  Pasadena, California 91125

to untangle the various observed phenomena; this requirement in turn calls
for an analytical apparatus that is not yet available. In this presentation we
hope to contribute towards the development of analytical tools that allow a
better planning and interpretation of experiments.

We shall presently review the analysis of experimental results on
steady crack propagation in a (viscoelastic) elastomer. From that development we pose questions which address the problem of non-steady crack
growth in some polymers.

As a partial answer to these questions a crack propagation model is
developed with a view towards the problem of initial crack growth. As an
interim step toward a more complete description for the time-dependent
problem it is evaluated for rate insensitive material properties. Crack
stability is discussed briefly in terms of the uniqueness of the
crack tip displacement in conjunction with a non-linear integral equation
resulting from the evaluation of the model for a rate independent material.

2. CONCEPTS OF THE FRACTURE PROCESS AND DEFINITIONS

Consider a two-dimensional structure containing a crack, and under
loading such that crack extension occurs in line with the original crack.
Upon application of a load history $\sigma_\infty(t)$ as indicated in figure 1 the crack
tip element is subjected to an elevated stress which we may think of as
purely tensile for the purpose of a simplified presentation. In fact, we may
think of the crack tip element as a small tensile specimen undergoing straining in a tensile tester, which tester is, however, not rigid, but the compliance of which is determined by the constitutive behavior and geometry of
the surrounding and nearly undamaged bulk material.

The locally elevated stress tends to disintegrate the material element
such that it progressively loses its strength of cohesion. A plausible

process for this strength loss is the production and growth of microscopic voids which coalesce, producing a foam-like or fibrillar structure which then fails by separation of the ligaments. We may think of this mechanical disintegration process as lowering the average density of the crack tip material. Once such an element has lost its cohesive strength completely, the crack has effectively advanced a small distance and the process repeats itself on the element that now is at the new crack tip except that the loading or straining history on this new element is different from that of the preceding one because its strain history is affected by the unloading rate of the previous element. In principle, we must, of course, consider that the crack propagation process may be a smooth process rather than an incremental process as adopted here momentarily for explanatory purposes.

Let us call the mechanically disintegrating material the cohesive material. From a continuum mechanical viewpoint it may suffice to represent these molecular forces by average forces and assign a constitutive law to the cohesive material. In general, this constitutive behavior depends on the deformation history and is strongly non-linear.

It is clear, then, that there are at least two dominant and material related sources of time dependence in the fracture process. One is the time dependence of the essentially undamaged bulk solid by which the boundary forces $\sigma_\infty(t)$ are transmitted to the crack tip region, while the other is the time dependence of the disintegrating cohesive material. Although one might argue that the molecular rheology of the bulk and of the cohesive material should be similar, this argument need not necessarily hold since the microstructure is, by definition, different; note also that the cohesive material is usually strongly drawn and molecularly oriented which makes its rate sensitivity distinctly different from that of the bulk, especially if the latter is amorphous and thus isotropic. For example, in polymethylmethacrylate the amorphous bulk polymer is converted into crazed material at the front of the crack [7, 8]. This crazed material exhibits a fibrillar structure complemented by voids so that the average density of the crazed material is lower than the bulk. The fibrils are aligned in the direction of tension (normal to the crack) and contain a large amount of similarly oriented molecule chains. Once the crazed material has been established at the crack tip, crack propagation occurs by the successive failure of the craze ligaments.

The thickness of the craze zone extends only a distance on the order of microns on either side of the crack axis. In many materials the zone in which the mechanical material disintegration occurs at the crack tip is very small compared to the other dimensions of the solid. This fact may be exploited to construct a simplified

3. CRACK PROPAGATION MODEL

Figure 2a represents a two-dimensional view of a crack tip propagating to the right. As the crack tip moves material points trace flow lines past the crack tip. We may use one particular flow line located a distance δ from the crack axis (in the undeformed state) as a separatrix between the material that will eventually sustain severe mechanical damage as it moves past the crack tip on the one hand, and the essentially undamaged material on the other.

In some materials such a marked separation between disintegrating and undamaged material is not possible, because the damage imparted to the material at the crack tip is not confined to a distinct layer. Rather, in many materials there exists a damage gradient in the material as the crack tip is approached which is roughly commensurate with the stress gradient. In fact, the strongly non-linear (viscoelastic) behavior of the material in the vicinity of the crack tip is a direct consequence of such damage (opening of voids, generation of microcrazes). We speak in this case of diffuse rather than localized damage and the placement of a separatrix as shown in figure 2a becomes debatable.

Let us call this layer the crack boundary layer. We assign to this boundary layer the function of providing the transition from continuum bulk material to ruptured material.

In figure 2b we show the boundary layer detached from the remaining solid which is denoted by R. The latter may, in principle, possess arbitrary non-linear constitutive behavior, while the boundary layer material may be endowed with its own characteristic properties. In the case of polymethylmethacrylate initial attempts at characterizing such properties (of craze material) have been made by Kambour [9]. Because the boundary layer is thin and we consider only the symmetric problem, it transmits only normal forces across the crack axis. With this understanding in mind we now formulate a boundary value problem for the region R which is to represent an approximation to the problem of the moving crack.

Let X = x-a(t) be the coordinate along the crack axis attached to the moving crack tip (crack tip travelling in the positive x-direction) with a(t) as the instantaneous crack length and y the ordinate normal to x at the crack tip X = 0. For points somewhere to the right of the crack tip the displacement in the y-direction of the points on the separatrix y = δ is nearly constant and small because δ is small. Let us say that these displacements are equal to zero and that this be true for X ≥ a. In the range 0 ≤ X ≤ a the boundary layer provides the cohesive forces which possess some maximal value in that range and then drop to approximately zero at X = 0 if the crack is traction free. For X < 0 we assume that the boundary layer attached to R provides zero surface tractions (for a traction free crack) because the boundary layer is thin.

We thus state the proposed boundary value problem in terms of the displacement component $v = v_y$ and the normal and tangential traction components T_n and T_t. If we further set $y = 0$ <u>on</u> the separatrix we have (cf. figure 3) on $y = 0$

$$
\begin{array}{llll}
T_t = 0 & -\infty < x < \infty & v = 0 & X \geqslant a \\
T_n = 0 & X < 0 & T_n = \lambda(X) & 0 < X \leqslant a
\end{array}
\tag{1}
$$

where $\lambda(X)$ may signify a prescribed function of X or represent a functional of the "average strain" history in the boundary layer $v(X, 0)/\delta$.

On the remainder of the boundary of R we prescribe the "far field" tractions $\sigma_\infty(t)$ which tend to drive the crack forward. We may consider now different types of materials and correspondingly assign different material properties to the boundary layer on the one hand, and to the region R on the other. The simplest tractable problems arise if the properties of region R conform to linearly viscoelastic behavior with linearly elastic behavior as a special case.

We review first the case when the boundary layer is vanishingly thin and represents only maximal, elastic forces; we may think of these elastic forces as the short range forces acting along molecule chains and let the remainder of the solid be linearly viscoelastic.[*] Under this assumption all the time-dependence of the fracture process is derived from the viscoelasticity of the bulk solid. If we refer to the crack boundary layer as the "interior" domain with R as the "exterior" domain of the solid, then this special case may be called the "exterior viscoelastic crack propagation problem."

As a second case we begin to discuss the complementary problem which is more appropriate for materials of the type like polymethylmeth-

---

[*] If we want to include long range forces by choosing δ finite rather than infinitesimal, we are likely to include viscous dissipation; we prefer to include that in the region R.

acrylate, polycarbonate and polystyrene; we assume the exterior domain R is linearly elastic while the boundary layer (of non-vanishing thickness) provides rate sensitive cohesive forces. Correspondingly this problem may be called the "interior viscoelastic crack propagation problem."

A third problem arises in the "mixed viscoelastic crack propagation problem," i.e., when both regions exhibit different rate sensitive properties. This case is, in general, very much more difficult than the two previous ones. We shall not discuss it here in any detail.

4. THE EXTERIOR VISCOELASTIC PROBLEM FOR A STEADILY MOVING CRACK

We review first the problem of a steadily moving crack in terms of a solution that is exact within the precepts of the linearized theory of visco-elasticity [10, 11, 12]. Following that development we discuss the approximate extension to crack propagation under non-steady conditions, and particularly the limitations of these approximations.

In accordance with the remarks at the end of the last section we assume that the constitutive behavior of the evanescent boundary layer allows attainment of a bounded stress distribution in the cohesive zone which is independent of the past deformation history. In addition we assume that the energy expended in de-cohering the thin boundary layer is also independent of the past deformation history. In connection with the assumption on the stress distribution this latter assumption is tantamount to saying that the cohesive material can only sustain a finite constant strain.

We call a material diaspastically* simple if its cohesive material attains a stress distribution that is insensitive to the deformation history under expenditure of a history-insensitive amount of work.

True diaspastical simplicity may be a fiction; however we shall see that materials exist whose fracture behavior can be explained under this

_____

*From the Greek διασπασισ: breakage

396

restriction.

The assumption of diaspastical simplicity allows us to prescribe the cohesive stress in the range $0 \leqslant X < a$ independent of the rate of crack propagation. It turns out that the piecewise linear prescription [10]

$$T_n = \begin{cases} (\sigma_o/\beta)X & 0 < X \leqslant \beta \\ \\ \sigma_o & \beta < X \leqslant a \end{cases} \tag{2}$$

is sufficient for our present purposes; here σ_o is a constant (maximum) cohesive stress and β is a length parameter of range $0 \leqslant \beta \leqslant a$. The final crack propagation solution is very insensitive to its value and therefore we set it equal to $\beta = a/2$.

An exact solution is readily attainable for a constant velocity $\dot{a}$. For our discussion of crack propagation behavior it is sufficient to be concerned with only the immediate vicinity of the crack tip, i.e., with a domain R whose dimensions are small compared to the crack size a.

It has been shown in an earlier analysis [10] of this problem that the solution is given in terms of a modified Kolosoff-Muskhelishvily potential ϕ^{*}

$$u(X, y, \dot{a}) + iv(X, y, \dot{a}) = \frac{3}{2} D(t) \left\{ \kappa \phi(\mathring{z}) - \phi(\overline{\mathring{z}}) - (\mathring{z} - \overline{\mathring{z}})\overline{\phi'(\mathring{z})} \right\}$$
$$+ \frac{3}{2} \int_o^t D(t-\xi) \frac{\partial}{\partial \xi} \left\{ \kappa \phi(z) - \phi(\overline{z}) - (z-\overline{z})\phi'(z) \right\} d\xi \tag{3}$$

$\kappa = 3 - 4\nu$ (plane strain); $\mathring{z} = x + iy$; $z = x - \dot{a}\xi + iy$

$D(t) = $ tensile creep compliance of the exterior region R.

Let K be the stress intensity factor which embodies both the geometry and the far field load $\sigma_\infty(t)$. In the present case of constant crack propagation K is required to be constant, too. Then the function $\phi(z)$ is given by

---

*For simplicity reasons we assume Poisson's ratio to be constant and equal to 1/2. This assumption is not necessary, but also not very restrictive. It is made primarily because the bulk (volume) deformation characterization of polymers are hardly ever available.

$$\phi(z) = \frac{K}{\sqrt{2\pi}} \sqrt{z-a} + \phi_c(z) \tag{4}$$

with $\phi_c(z)$ being determined in terms of the tractions (2) from

$$\phi_c'(z) = \frac{1}{2\pi i \sqrt{z-a}} \int_0^a \frac{T_n(\tau)\sqrt{\tau-a}}{\tau-z} d\tau \tag{5}$$

With $C = (1-z/a)^{\frac{1}{2}}$ and $C_0 = (1-\beta/a)^{\frac{1}{2}} = \sqrt{1/2}$ one finds

$$\phi_c(z) = i\frac{\sigma_0 a}{\pi} \left\{ \frac{\beta-z}{2a} \ell n \left| \frac{C+C_0}{C-C_0} \right| - C_0 C - \frac{a}{2\beta} \left[(1-C_0)(1+\frac{z}{a})C - \frac{1}{3}(1-C_0^3)C \right. \right.$$

$$\left. \left. + \frac{1}{2}\frac{\beta^2-z^2}{a^2} \ell n \left| \frac{C+C_0}{C-C_0} \right| + \frac{1}{2}\frac{z^2}{a^2} \ell n \left| \frac{C+1}{C-1} \right| \right] \right\} \tag{6}$$

The stresses are bounded at $X = a$ if the finiteness condition

$$\sigma_0 = \frac{1}{2}\sqrt{\pi/2} \frac{K}{\sqrt{a}} \frac{\frac{3}{2}\beta/a}{1-C_0^3} \left(= \frac{1}{2}\sqrt{\pi/2} \frac{K}{\sqrt{a}} \frac{3/4}{1-(1/2)^{3/2}} \right) \tag{7}$$

connects the maximum stress σ_0 in the cohesive zone and the zone length a
to the stress intensity factor K. The finiteness condition merely reflects
the fact that equilibrium exists at the crack tip without recourse to (inverse
square root) singular stresses as the point $X = a$ is approached.

5. FRACTURE ANALYSIS

Having available the distribution of stresses[*] and deformations in the
(two-dimensional) solid undergoing fracture does not _per se_ yield a relation
that determines how fast a crack grows as a function of the applied loads
(represented in the case at hand by the stress intensity factor K). To establish that relation we need an additional requirement commonly referred
to as a fracture criterion. We shall see later that the need for such a
requirement in addition to the stress analysis is a consequence of the rather
non-specific prescription of the cohesive forces which lead to diaspastic

_____

[*]For traction prescription on the boundary, the stress distribution turns
out to be exactly the same as for a Hookean solid, provided the crack does
not decrease in size. Of course the deformations are time dependent in
accordance with the linearized viscoelastic behavior of the exterior region.

simplicity. In fact the (additional) fracture criterion is implicit in the second precept of diaspastic simplicity, namely that referring to the energy expenditure or the ultimate strain of the cohesive material.

We assume that fracture occurs in such a way that the work done by the velocity or time independent tractions (2) against the displacement of the boundary y = 0 is a constant. Let us denote this work per unit of new surface created by Γ. Then

$$\int_0^a T_n(X) \cdot \dot{v}(X, 0, \dot{a}) \, dX = \Gamma \dot{a} \tag{8}$$

expresses the energy conservation locally on a per-unit-time basis. For the constantly moving crack the time derivative is transformed into a spacial one and integration by parts, under consideration of the boundary conditions (1) yield, in place of equation (8)

$$\int_0^\beta v(X, 0, \dot{a}) \, dx = \frac{a\,\Gamma}{2\sigma_o} \quad ; \qquad (\beta = \tfrac{a}{2}) \tag{9}$$

In order to display the further evaluation of this fracture criterion which relates the rate of crack growth to the applied loads (via K) we define the following terms, with E_∞ as the value of the relaxation modulus at infinite time, with D_o the creep compliance at zero time and $\Delta D(t) = D(t) - D_o$,

$$\vartheta\left(\tfrac{a}{\dot{a}}, \rho\right) = E_\infty \left\{ D_o F(\rho) - \int_\rho^1 \Delta D\left[\tfrac{a}{\dot{a}} (r-\rho)\right] F'(r) dr \right\} \tag{10}$$

$$\oplus\left(\tfrac{a}{\dot{a}}\right) = \frac{A(R)}{R} \int_0^R \vartheta\left(\tfrac{a}{\dot{a}}, \rho\right) d\rho \quad ; \quad R = \beta/a \ (= \tfrac{1}{2} \ \text{here}) \tag{11}$$

$$A(R) = \frac{3}{2} \frac{R}{1-C_o^3} \ ; \ C_o = (1-R)^{\tfrac{1}{2}} \ ; \ C(r) = (1-r)^{\tfrac{1}{2}} \tag{12}$$

$$F(r) = C + \frac{A(R)}{2} \left\{ \frac{R-r}{2} \ln \left| \frac{C+C_o}{C-C_o} \right| - C_o C - \left[(1-C_o)(1+r)C - \tfrac{1}{3}(1-C_o^3)C \right] \frac{1}{2R} \right.$$

$$\left. + \frac{1}{2R} \left[\frac{R^2-r^2}{2} \ln \left| \frac{C-C_o}{C+C_o} \right| - \frac{r^2}{2} \ln \left| \frac{C+1}{C-1} \right| \right] \right\} \tag{13}$$

It is important to recognize here that the functions ϑ and $\oplus$ embody the viscoelastic response of the (exterior) domain surrounding the crack tip.

The combination of (3), (4), (6) and (11) renders, after considerable algebra, $(\nu = \frac{1}{2})$

$$\frac{3}{4} K^2 \ominus (\frac{\dot{a}}{a}) = E_\infty \Gamma \tag{14}$$

which, after incorporation of the finiteness condition (7) becomes

$$\frac{3}{4} K^2 \ominus \left(\frac{\pi K^2 A^2(R)}{8\sigma_o^2 \dot{a}}\right) = E_\infty \Gamma \tag{15}$$

This relation determines the (constant) speed of crack propagation as a function of the (constant) stress intensity factor as the structural load parameter. It can be checked against experiments and we reproduce here only a limited portion of that work [10].

A constant stress intensity factor can be achieved for a rubbery solid formed into the strip geometry shown in figure 4; straining the strip by moving the long edges apart, parallel and normal to each other, and then holding them fixed, produces a gross strain ε_∞. If the crack propagates so slowly (all crack speeds observed were below 10 in/sec) that the rubber ahead of the crack has time to relax to its long time equilibrium value, then the crack moves into a constant stress field for which the stress intensity factor remains constant as long as the crack tip moves in the central portion of the strip.

The creep compliance of a polyurethane elastomer (Solithane 113) was measured to evaluate the function $\ominus$, equation (11), and crack propagation rates were measured. These data allowed comparison of equation (15) with the experimental results as shown in figure 5 (T is the absolute temperature and ϕ_T is the time-temperature shift factor of the thermorheologically simple materials.)

The solid line in figure 5 represents equation (15) for a particular choice of the fracture energy density Γ and the cohesive stress σ_o. These choices are

$$\Gamma = 0.014 \text{ lb/in.} \qquad (2.5 \times 10^3 \text{ erg/cm}^2)$$
$$\sigma_o = 2.1 \times 10^4 \text{ psi} \pm 20\%$$

Thus the assumption of a rate independent cohesive stress σ_o and fracture energy Γ achieves good agreement with experimental data. It is worth pointing out, however, that an assumption of a constant limiting strain for the cohesive material may be equally reasonable. Since we allowed the thickness of the boundary layer to vanish, the equivalent of a (constant) failure strain criterion would be a criterion whereby the displacement v at the point X = 0 achieves a constant value. Let the magnitude of this displacement be u_o. Then an evaluation of this criterion yields ($\nu = \frac{1}{2}$)

$$\frac{3}{4} K^2 \mathcal{S} \left(\frac{\pi K^2 A^2(R)}{8\sigma_o^2 \dot{a}} , 0 \right) = E_\infty \frac{\sigma_o u_o}{A(R)} \tag{16}$$

which differs from the energy criterion primarily by the material related function $\mathcal{S}$ instead of its integral $\circledR$. In principle, the two criteria yield different results; but it turns out that because D(t) is a relatively slowly varying function of (log) time equations (15) and (16) are virtually indistinguishable for the polyurethane elastomer represented in figure 5.

6. <u>LIMITATIONS OF DIASPASTIC SIMPLICITY</u>

Before turning from this brief review to a discussion of the limitations underlying diaspastic simplicity it would be appropriate to consider whether the agreement between the model as represented by equation (15) (and (16)) with the crack propagation data is reasonable. But in the interest of time and brevity of presentation we refer to an earlier presentation on that subject [10]. But it should be pointed out here that the choice of a rate dependent cohesive stress σ_o or fracture energy Γ would have led to a less satisfactory agreement. In fact, if one assumes for example that the size of the cohesive zone a were a constant it would follow that the cohesive stress σ_o would vary with the stress intensity factor and thus with the

velocity of crack propagation. The corresponding crack propagation prediction would disagree markedly with the experimental data.

Another important point regards the necessity of the size parameter
characterizing the extent of the zone in which the material disintegrates
mechanically. This size parameter a appears in the argument of the mater-
ial-time-dependent function $\oplus$ in equation (14) and in equations (15) and (16),
but represented in the latter cases via the finiteness condition 7 in terms
of σ_0 and the stress-intensity factor for the constantly strained strip
$(\sim \varepsilon_\infty)$. It seems clear that even on dimensional grounds this length parameter should occur in the argument of that function, which argument has the
dimension of time $(t \sim a/\dot{a})$. But we note that in contrast to the present
viscoelastic problem the fracture in rate-insensitive materials can be expressed without explicit reference to a. For, note that if $\dot{a} \to 0$ the function
$\oplus$ (and also δ) approaches a constant and the criterion (15) becomes independent of a. In fact, then the criterion becomes identical to one for rate
insensitive solids.* We conclude that while it may be possible to generate
a fracture criterion for rate independent solids without explicit reference
to a microstructural size parameter which characterizes the local fracture
process such is not true when viscoelastic solids are considered (12).

Having dealt with the problem of constant speed of crack propagation
we ask next how the problem is modified if the loading is such that the
stress intensity factor is a function of time, i.e., such that the crack
propagates with a variable speed. While the evaluation of the displacement
integral (3) is, of course, possible**, one can no longer cast it into a relatively simple form of the type associated with the definitions 10 and 11.
However one can show that if

---

*The same holds if $\dot{a} \to \infty$.

**We must then let $z = x - a(t) + iy$ in equation (3).

$$2a/\dot{a} \ll K/(dK/dt) \tag{17}$$

then the crack propagation equation (15) or (16) for constant propagation rate in a diaspastically simple solid is instantaneously valid. In that case equation (15) or (16) becomes a non-linear differential equation connecting the time-varying stress intensity factor K with the time-dependent velocity $\dot{a}(t)$. The inequality (17) expresses the limitation on the stress intensity factor history in the sense that the latter may not change much during the time that the crack propagates a distance a.

A further implication of the inequality and diaspastic simplicity is the following. A relation between the stress intensity factor K and the constant crack speed (figure 5) may be only known experimentally. Then the inequality (17) states the restriction under which such an experimental relation applies instantaneously for time-varying stress intensity factor.

For the viscoelastic "Griffith problem" of a crack in an infinite sheet under uniform and time-constant tension at infinity normal to the crack direction it can be shown that the inequality (17) is satisfied as long as the current crack size "a" is very large compared to a; alternately, this implies that the tractions at infinity must be small compared to the cohesive strength σ_o.

There is another thought connected with the inequality (17) which is implicitly satisfied by diaspastically simple solids: Changes in the load parameter (stress intensity factor) K should occur so slowly that the cohesive forces have enough time to adjust their distribution to the one experienced in steady crack propagation. Diaspastic simplicity assumes that the cohesive forces establish themselves independent of the local strain history, and adjust themselves, therefore, instantaneously, as if the cohesive material were endowed with elastic properties.

There are, of course, many materials for which the cohesive material cannot be assumed to possess such simple properties. But just as a deformation plasticity may be indistinguishable from non-linear elastic response under monotonically increasing loading, so diaspastically simple and non-simple behavior may be indistinguishable in certain load histories. The distinction becomes most pronounced when unloading, even partial unloading, of the cohesive material occurs. When unloading of the externally applied, far field traction occurs, a non-simple redistribution of forces within the cohesive zone takes place. This redistribution may result in producing new cohesive material and thus enlarge the cohesive zone upon external unloading. To see this we need only consider, for example, some cohesive material which loses all force transmission capability as soon as the strain is reduced by an arbitrarily small amount ("infinite modulus upon unloading"); the forces that have been eliminated by a small amount of external unloading in part of the cohesive zone may have to be made up by producing new cohesive material by cohesive zone enlargement in order to satisfy crack tip equilibrium without recourse to unbounded stresses.

7. THE INTERIOR VISCOELASTIC PROBLEM FOR CRACK INITIATION

It is clear from the preceding remarks that the problem in diaspastically non-simple solids becomes complex even when the exterior region (region R in figure 2b) exhibits linear constitutive behavior such as, e.g., linearly viscoelastic behavior. It is equally clear, however, that the problem of fracture in diaspastically non-simple solids must be examined in order to understand the fracture behavior of a large class of polymers under non-special load histories. This statement is particularly true in connection with the important problem of polymer fatigue*, since here the loading-unloading sequence is of fundamental impact.

----

*It should be recognized that the problem of crack-tip heating in fatigue is also a very fundamental and unresolved problem; but we do not address ourselves to that problem in this presentation.

But even the more transparent problem of fracture initiation is by no means trivial. In this connection we consider a solid containing a crack and subjected to increasing load of otherwise arbitrary history. The question arises then: At what time does the crack begin to propagate?

Upon load application[*] the material element at the tip of the crack is strained increasingly with time and a zone of highly strained cohesive material develops. During the loading process the forces at the tip of the crack, i.e., at the "open" end of the cohesive zone corresponding to the point $X = 0$ in figure 2a increase and then decrease as the maximal force transmission capability of the now developed cohesive material is exceeded (see figure 6 for a qualitative uniaxial stress-strain behavior of cohesive material). Certainly, when the strain at the crack tip increases to the point where the cohesive forces drop to zero (first at $X = 0$) any further increase in the external loads cannot be balanced by the cohesive forces and the crack must begin to move. This idea leads naturally to the concept of an ultimate strain or crack tip opening displacement criterion for fracture[**]. In general the strain rates at the crack tip depend interactively on the forces transmitted to the tip cohesive material via the exterior rate sensitive region and on the rate sensitive response characteristics of the cohesive material. It would be prudent to begin consideration of this complex problem by allowing the exterior region to be linearly elastic and to attribute rate sensitive properties to the cohesive material alone. This assumption should be representative for fracture initiation processes in polymethyl-methacrylate, polycarbonate and possibly polystyrene. We now consider the thickness of the boundary layer $\delta \neq 0$ though still very small, i.e., on the order of several microns.

---

[*]This may be in the form of a time step or in a continuous time history.
[**]We shall subsequently be led naturally to an alternate (in)stability criterion.

Let $P(x, t)$ represent the force distribution in $0 \leqslant x \leqslant a(t)$, where the time dependence is actually introduced via the rate of straining the cohesive material; this strain rate varies from point to point in the cohesive zone; furthermore, let $\xi = \frac{x}{a}$ (initially stationary crack tip). Then the normal displacement in the y-direction is

$$v(\xi, t) = \frac{\kappa+1}{2\mu} \left\{ \frac{K\sqrt{a}}{\sqrt{2\pi}} \sqrt{1-\xi} + \frac{a}{\pi} \left[\mathscr{K}(\xi, 0) P(0, t) + \int_0^1 \mathscr{K}(\xi, \zeta) \frac{\partial P(\zeta, t)}{\partial \zeta} d\zeta \right] \right\} \tag{18}$$

where μ = shear modulus of region $\mathcal{R}$ and

$$\mathscr{K}(\xi, \zeta) = \frac{\zeta - \xi}{2} \ell n \left| \frac{\sqrt{1-\xi} + \sqrt{1-\zeta}}{\sqrt{1-\xi} - \sqrt{1-\zeta}} \right| - \sqrt{(1-\xi)(1-\zeta)} \tag{19}$$

and K is again the stress intensity factor as the load parameter. The finiteness condition, corresponding to (7) for the diaspastically simple solid, becomes

$$K = 2\sqrt{2} \sqrt{\frac{a}{\pi}} \left\{ P(0, t) + \int_0^1 \sqrt{1-\zeta} \frac{\partial P(\zeta, t)}{\partial \zeta} d\zeta \right\} \tag{20}$$

If one takes $v(\xi)/\delta$ as the one-dimensional strain across the cohesive zone, then $P(\xi, t) = P[v(\xi, t)]$. Define the nondimensional quantities

$$N = \frac{\kappa+1}{2\mu} \frac{K}{\sqrt{2}} \quad \text{and} \quad \Lambda(\xi, t) = \frac{\kappa+1}{2\mu} P(\xi, t)$$

With the definition

$$M(\xi, \zeta) = \mathscr{K}(\xi, \zeta) + 2\sqrt{(1-\xi)(1-\zeta)} \tag{21}$$

the combination of (18) and (20) then yields

$$v(\xi, t) = \frac{a}{\pi} \left\{ M(\xi, 0) \Lambda[v(0)] + \int_0^1 M(\xi, \zeta) \frac{\partial \Lambda[v(\zeta, t)]}{\partial \zeta} d\zeta \right\} \tag{22}$$

while the finiteness condition (7) becomes, in nondimensional form,

$$N = 2\sqrt{\frac{a}{\pi}} \left\{ \Lambda[v(0)] + \int_0^1 \sqrt{1-\zeta} \frac{\partial \Lambda[v(\zeta, t)]}{\partial \zeta} d\zeta \right\} \tag{23}$$

Equations (22) and (23) may be written alternately, upon noting that $M(\xi, 1) = C$, and after integration by parts as

$$v(\xi, t) = - \frac{a}{\pi} \int_0^1 \frac{\partial M(\xi, \zeta)}{\partial \zeta} \Lambda \left[v(\zeta, t) \right] d\zeta \tag{22a}$$

$$N = \sqrt{\frac{a}{\pi}} \int_0^1 \frac{\Lambda \left[v(\zeta, t) \right] d\zeta}{\sqrt{1-\zeta}} \tag{23a}$$

Equation (22a) is a nonlinear integral equation similar to a Hammerstein equation and differs from the latter by the fact that the kernel $\partial M(\xi, \zeta)/\partial \zeta$ is not symmetric. These equations must be satisfied for any instant of time in accordance with the past history of the displacements $v(\xi, t)$. For the remainder of this presentation let us be concerned only with the restricted problem of a rate independent cohesive force*. This intermediate step seems desirable in order to gain confidence in a (numerical) solution scheme for the nonlinear integral equation. Having established the accuracy of such a scheme one would then proceed further with a time-dependent solution by applying it in some time-stepping procedure.

A solution of equation (22) may be obtained by Newton's method. For illustrative purposes we represent the cohesive forces by the curve in figure 6. Let $v_i(\xi)$ be an (initial) guess, and let $\Delta v(\xi)$ be a correction to that guess. Substitution into (22a) yields

$$v_i(\xi) + \Delta v(\xi) = - \frac{a}{\pi} \int_0^1 \frac{\partial M(\xi, \zeta)}{\partial \zeta} \Lambda \left[v_i(\zeta) + \Delta v(\zeta) \right] d\zeta \tag{24}$$

Linearization of the function Λ as $\Lambda = \Lambda \left[v_i(\zeta) \right] + \Lambda' \left[v_i(\zeta) \right] \Delta v(\zeta)$ allows casting (24) into, with $\Lambda' = \partial \Lambda(r)/\partial r$

$$\Delta v(\xi) + \frac{a}{\pi} \int_0^1 \frac{\partial M(\xi, \zeta)}{\partial \zeta} \Lambda' \left[v_i(\zeta) \right] \cdot \Delta v(\zeta) d\zeta = - \frac{a}{\pi} \int_0^1 \frac{\partial M(\xi, \zeta)}{\partial \zeta} \Lambda \left[v_i(\zeta) \right] d\zeta - v_i(\xi) \tag{25}$$

---

*This initial simplification makes the problem similar to those considered by Goodier and Kanninen [13] in a numerical scheme modelling non-linear atomic force interaction along the crack plane, and by Andersson and Bergkvist [14] who used a special cohesive property description (linear loading and linear unloading) with a different but otherwise nearly equally approximate formulation of the crack model.

A discretization on ζ leads to a matrix equation equivalent of (24) which can be solved for successively improved guesses of $v_i(\xi)$ as long as a/π is not an eigenvalue in conjunction with the finiteness constraint (23). So far only a limited study of convergence of the scheme has been conducted for n = 10, 16 and 22 intervals on $0 \leqslant \zeta \leqslant 1$; no marked difference was observed in the results as a function of n.*

Bearing in mind that the present work is a prelude to the case involving rate sensitive cohesive material properties, we record the results of a sample calculation for the properties represented in figure 6, which are indicative of the type of results expected for rate sensitive cohesive properties. However, where in the present case we record quantities as a function of the stress intensity factor N, the time-dependent case would call for time as the independent variable in order to display the time history of the fracture process when N is a prescribed function of time.

Figure 7 shows the distribution of the displacement $v(\xi)$ in the cohesive zone. Note that for increasing loading as represented by increasing N the displacement near $\xi = 0$ increases faster than near $\xi = 1$; this fact is the result of the material offering reduced resistance as the displacement (strain) at the crack tip increases.

Figure 8 records the stress and strain at the crack tip $\xi = 0$ as well as the size of the cohesive zone as a function of the stress intensity factor. As the value of N = 1.33 is approached the crack tip cohesive stress drops rapidly to zero while the "strain" tends to its limiting value of 1/2. Simultaneously the size of the cohesive zone, a, rises rapidly. Crack extension would thus occur near a value of N = 1.33 as indicated by the dashed boundary. Also shown in that figure as a dashed line for comparison purposes is the value of the crack tip displacement ($= a/\pi$) for the case that the non-

*An improved computational scheme is now being considered which takes account of the difficulties encountered as an eigenvalue of (25) is approached.

dimensional cohesive forces are equal to unity in the entire cohesive zone.

8. AN ALTERNATE CRACK STABILITY CRITERION

For the value of N beyond 1.33 the computational scheme did not converge. In fact at values of N on the order of 1.1 and 1.2 the convergence was slow and not thoroughly conclusive within the limited number of iteration allowed. Figures 7 and 8 must therefore be considered with some caution.

Indeed it is likely that an equilibrium state with the cohesive force at the tip vanishing cannot be achieved through a continuously increasing value of the stress intensity factor. Suppose we possess the equilibrium solutions for a set of increasing values of N or of a. Then the right hand side of (25) is zero and the then homogeneous integral equation (25) has only the zero solution for the variation $\Delta v(\zeta)$ except for an eigenvalue of $a = a*$ or $N = N*$. For these values the solution for $v(\xi)$ is no longer unique which can amount to a loss of stability. It is possible to visualize instability physically in the following way. Consider a variation $\Delta v(\xi)$ of the displacement about $v(\xi)$ corresponding to an equilibrium state. If such a variation is performed about an equilibrium state, say N_o, then a possible increase in $v(\xi)$ may cause unloading of region R while the boundary layer is strained at the expense of this released energy; this energy exchange process could occur without further work by the far-field tractions. Whether the latent (energy-based) instability at $N = N_o$ coincides with the loss of uniqueness at $N = N*$ needs to be further investigated.

It is clear from equation (25) and equation (23a) that not only the cohesive forces play a role in this "energy criterion" of crack stability, but also the gradient of these forces with respect to the boundary layer strain, and thus the displacement gradient in the cohesive zone. In view of this observation it would be of practical interest to establish for which kind of

cohesive "stress-strain" law (exemplified in figure 6) the energy criterion
of crack stability coincides with an ultimate strain criterion.   This will not
be done here.

9.   CONCLUDING REMARKS

Although the example presented for a diaspastically non-simple solid
has been carried out only to the stage of a time independent case (a special,
diaspastically simple solid), we emphasize that the intent is to present only
the direction in which work is being pursued.   Work in the fracture of metals
is similarly concerned with the contribution of high (rate independent) plastic
deformations in the crack tip region.   The approach taken in connection
with that problem is largely based on finite element numerical analysis,
which conveniently allows for diffuse plasticity rather than being confined
to a boundary layer model with a linearly elastic outer region.   To follow a
similarly numerically oriented method allowing for diffuse visco-plastic
material behavior around the crack tip does not seem very promising at
this time.

With respect to the effect of introducing time dependence into the interior viscoelastic and mixed-viscoelastic problem we can offer only some
educated guesses at this time.

Assume for demonstration purposes that a load is suddenly applied to
a cracked structure and then held constant.   The cohesive material will de-
grade and continuously lose strength until the crack tip is unloaded.   During this process the cohesive forces vary from point to point in the cohesive
zone in accordance with the prescribed (average) "stress-strain" behavior
of the cohesive material.   As long as the outer domain is linearly elastic
the problem for the rate sensitive cohesive material is tractable if not
computationally trivial.   But if different viscoelastic properties are prescribed for the boundary layer on the one hand and for the exterior region
on the other, the treatment of the resulting interaction problem becomes

difficult so that there may not be any advantage in formulating the crack growth problem in terms of a boundary layer model.

Once the crack tip has just begun to unload and the crack begins to move, new material must take on "cohesive properties" and feed into the leading tip of the cohesive zone while the material in the cohesive zone experiences a strain rate history that is due to both the rate of crack tip motion and rate of change of the far-field stresses. It is clear that within the framework of the boundary layer model both the fracture initiation as well as the crack propagation process are understood from a unified point of view without recourse to an additional failure criterion beyond the prescription of the cohesive properties of the material. The distinction between the initiation and propagation phases of a crack vanishes; this observation has been made previously in terms of a simpler model which permitted crack growth by discrete jumps [15].

It can be shown through rough estimates that if the time dependence of the cohesive material is simulated by behavior appropriate to linearly viscoelastic material, crack propagation follows a pattern that is similar to that experienced for the exterior problem. In fact, it may then be difficult to distinguish (experimentally) between the two cases in many crack growth histories.

However, for many materials a diaspastically simple representation will not be reasonable. Indeed, we have had various experiences with material for which the cohesive material had the following characteristics: At low rates of deformation the material offers low cohesive stresses and low ultimate strains. Both cohesive stresses and ultimate extensibility increased with deformation, except that after some particular rate has been achieved the ultimate strain capability decreases with increasing strain rate. This behavior is likely to lead to unstable crack propagation behavior

in the sense that crack propagation rates may experience (jump) discontinuities.

In conclusion it is appropriate to point out that the foregoing considerations and computations apply to the fracture of bonded joints as long as the
bonding layer can be considered as degrading uniformly across its thickness.

10. REFERENCES

(1)   Knauss, W. G.:"Crack Propagation in Diaspastically Non-simple
       Solids: A Progress Report." Proceedings of the IUTAM Sympos-
       ium on the Mechanics of Viscoelastic Media and Bodies. Sept.
       1974, Gothenburg, Sweden, J. Hult, ed. Springer, Berlin.

(2)   Griffith, A. A.: "The Phenomena of Rupture and Flow in Solids."
       Phil. Trans. Roy. Soc., London Series A. October, 1920,
       pp. 163-198.

(3)   Irwin, G. R.: "Analysis of Stresses and Strains Near the End of
       a Crack Traversing a Plate." J. Appl. Mech., vol. 24, 1957,
       pp. 361-364.

(4)   Barenblatt, G. I.: "The Mathematical Theory of Equilibrium
       Cracks in Brittle Fracture." Advances in Applied Mechanics,
       vol. 7, Academic Press, 1962, pp. 55-129.

(5)   Wells, A. A.: "Application of Fracture Mechanics at and Beyond
       General Yielding." British Welding Journal 10, 1963, pp. 563-
       570.

(6)   Rice, J. R.: "A Path Independent Integral and the Approximate
       Analysis of Strain Concentration by Notches and Cracks." J. of
       Appl. Mech., 35, 1968, pp. 379-386.

(7)   Kambour, R. P.: "Mechanism of Fracture in Glassy Polymers,
       II. Survey of Crazing Response during Crack Propagation in
       Several Polymers." Journal of Polymer Science, Part A-2, 4,
       1966, pp. 17-24.

(8)   Kambour, R. P.: "Mechanism of Fracture in Glassy Polymers,
       III. Direct Observation of the Craze Ahead of the Propagating
       Crack in Poly(methyl methacrylate) and Polystyrene. J. Pol.
       Sci., Part A-2, 4, 1966.

(9)   Kambour, R. P.; Kopp, R. W.: "Cyclic Stress-Strain Behavior
       of the Dry Polycarbonate Craze. J. Pol. Sci., Part A-2, 7,
       1969, pp. 183-200.

(10)   Knauss, W. G.: "On the Steady Propagation of a Crack in a
       Viscoelastic Sheet: Experiments and Analysis; in: Deformation
       and Fracture of High Polymers, Kausch, Hassel, Jaffee, eds;

412

Plenum Press, 1973, pp. 501-541.

(11)   Knauss, W. G.: "Mechanics of Fracture in Polymeric Solids."
         Proc. 3rd Int'l Conference on Fracture, Munich, Germany, 1,
         1973.

(12)   Knauss, W. G., Muller, H. K.:   Discussion of the Paper "On the
         Governing Equation for Quasi-static Crack Growth in Linearly
         Viscoelastic Materials", by R. J. Nuismer, J. Appl. Mech.,
         Sept. 1974. To appear in J. Appl. Mech.

(13)   Goodier, J. N., Kanninen, M.: See e.g. Chapter XIII of J. N.
         Goodier's Article "Mathematical Theory of Equilibrium Cracks"
         in: Fracture, 2, H. Liebowitz, editor, Academic Press, 1968.

(14)   Andersson, H., Bergkvist, H.: "Analysis of a Non-linear Crack
         Model," J. Mech. Phys. Solids, 18, 1970, pp. 1-28.

(15)   Williams, M. L.: "The Fracture of Viscoelastic Material, in:
         Fracture of Solids," Drucker and Gilman, etc., Interscience
         Publishers, New York, 1963, pp. 157-188.

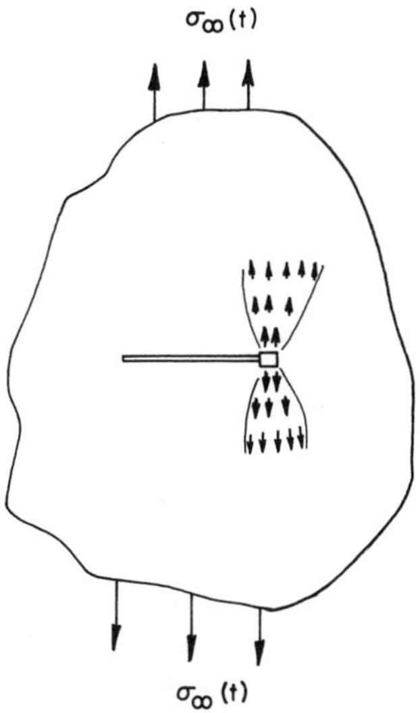

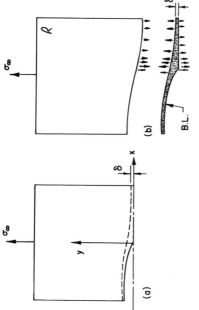

FIG. 2 DECOMPOSITION OF CRACK TIP DOMAIN INTO UNDAMAGED
REGION (R) AND BOUNDARY LAYER (B.L.)

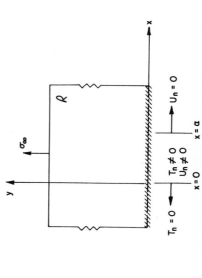

FIG. 3 EQUIVALENT BOUNDARY VALUE PROBLEM ON REGION R

FIG. I MATERIAL ELEMENT UNDER INTENSE STRAIN
AT THE TIP OF A CRACK

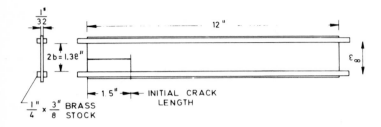

FIG. 4 TEST SPECIMEN GEOMETRY FOR CONSTANT CRACK SPEED EXPERIMENTS

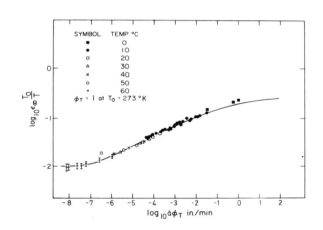

FIG. 5 COMPARISON OF EXTERIOR VISCOELASTIC SOLUTION WITH EXPERIMENTAL DATA

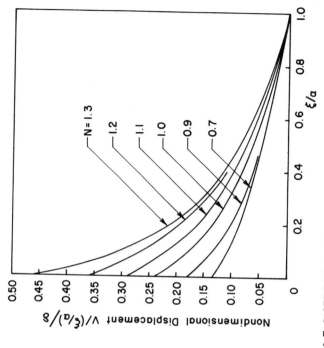

FIG. 7 DISTRIBUTION OF DISPLACEMENT IN THE COHESIVE ZONE
AS A FUNCTION OF NONDIMENSIONAL STRESS INTENSITY
FACTOR N

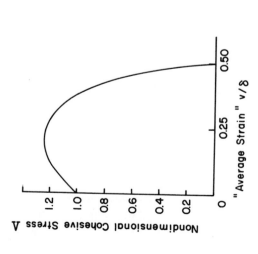

FIG. 6 RELATION BETWEEN COHESIVE STRESS
AND "STRAIN" IN THE COHESIVE ZONE

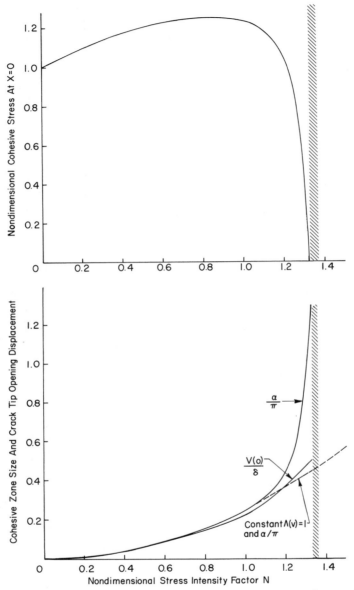

FIG. 8 FORCE AT CRACK TIP (ABOVE), DISPLACEMENT ("STRAIN")
AT CRACK TIP AND COHESIVE ZONE SIZE (BELOW) AS A
FUNCTION OF STRESS INTENSITY FACTOR; – – – REPRESENTS
COD AND α/π FOR CONSTANT COHESIVE STRESS

Experimental Study on a Slowly Advancing
Crack in a Viscoelastic Strip Plate

M. Takashi*, T. Kunio** and M. L. Williams***

1.   INTRODUCTION

Fracture mechanics based essentially upon the Griffith energy balance
concept has attained considerable success in fracture analysis, including
problems with viscoelastic materials. For example, Williams (1) performed
calculations predicting an exponential crack growth in a viscoelastic strip
using a maximum strain criterion and a Voigt model material representation.
He also extended the Griffith energy balance concept to the instability
analysis of a spherical flaw in a linear viscoelastic material subjected to
hydrostatic tension (2).

Knauss (3) investigated the unstable crack growth in a viscoelastic
material using the energy criterion and a critical transition crack rate.
Afterwards, he (4) (5) also supplemented his earlier work by the calculation
of the crack tip stress field and the derivation of the law for the growth of
a macroscopic crack in viscoelastic substances. Bennett, Anderson and
Williams (6), by an application of the thermodynamic power balance equation,
have studied the time-temperature dependent cohesive fracture energy in a
viscoelastic sheet. In the calculations, they included correction factors
for a finite aspect ratio plate specimen using previous analyses of
Westergaard (9) and Knauss (4), and showed that the value of specific
cohesive fracture energy, γ_c is not only time dependent but also time-

* Department of Mechanical Engineering, Aoyama University,
  6-16-1, Chitosedai, Setagaya, Tokyo, Japan
  (Presently, University of Pittsburgh, Pittsburgh, Pennsylvania)

** Department of Mechanical Engineering, Keio University,
   832 Hiyoshi-cho, Kohoku-ku, Yokohama, Japan

*** School of Engineering, University of Pittsburgh, Pittsburgh,
    Pennsylvania 15260 USA

temperature dependent in the WLF - shift (7) sense. From the viewpoint of
linear fracture mechanics, for example the stress field at the crack tip and
the stress intensity factor, Mueller (8) has also made an analysis for
certain boundary conditions.

From an experimental viewpoint, it is usually difficult but important to
detect the critical point of the threshold of crack propagation, particularly
in a viscoelastic strip which has a strong time-dependent mechanical property.
In this study, periodical photographs were carefully taken to monitor the
crack initiation time, to follow the crack extension, and to deduce the
compliance of cracked specimens having wide strips of 40 / 130 aspect ratio
and various lengths of the central crack. Then, some basic investigations
into the time dependent fracture energy corresponding to the variation of
crack size, and variation of boundary condition and an ad hoc prediction of
crack extension rate were performed.

Finally, the variation of nominal stress of a wide strip which contains
a slowly advancing crack was graphically explained using a compliance ratio
which changed with the increase of crack length.

2.   EXPERIMENTS

The material used in these experiments was made by mixing Epoxy resin
and triethylene-tetramine in the ratio of 10 to 1, and cured about 48 hours
at 80°C. All tests were carried out at 273°K under a constant crosshead
displacement speed. The glass transition temperature of the material is
about 240°K. The threshold of crack growth and subsequent crack growth were
periodically photographed in 2 second intervals. A fine grid pattern was
printed on the surface of the specimen to assist in the deformation and
crack length measurement.

3.   RESULTS AND DISCUSSION

The well known Griffith critical stress σ_{cr} for an infinite elastic sheet with a central crack (length 2C) subjected to uniform tension, is

$$\sigma_{cr} = \sqrt{\frac{2}{\pi} \frac{E \gamma_c}{C}} \tag{1}$$

where γ_c is the specific cohesive fracture energy, and E is the elastic modulus. Following the procedure of Bennett (6) with respect to certain corrections to this infinite strip case, e.g. (i) imcompressibility of the material, (ii) time dependent loading and modulus, (iii) symmetric growth of flaw, (iv) time dependence of fracture energy γ_c, and (v) geometrical correction factors for finite dimensions of the specimen, $F(\frac{C_o}{w}, \frac{C_o}{b})$ (4) (6) (9), a modified Griffith criterion for a viscoelastic plate strip (6) under constant displacement rate loading was hypothesized as

$$\sigma_{cr}^2(t_i) = [\frac{\pi}{2(1-\nu^2)^2} F(\frac{C_o}{w}, \frac{C_o}{b})]^{-1} \quad [2t^{-2} E_r^{(2)}(t_i)] \frac{\gamma_c}{C_o} \tag{2}$$

where $2C_o$ = initial crack length, t_i = crack initiation time and $E_r^{(2)}(t_i)$ is the second integral of the relaxation modulus up to t_i, i.e.,

$$E_r^{(2)}(t_i) = \int_0^{t_i}\int_0^\tau E_r(\xi) \, d\xi \, d\tau \tag{3}$$

Fig. 1 shows the experimental data for the crack extension curves and the corresponding overall average stress-strain curves. Marks designated as ●, □, Δ, O show the critical point of the crack threshold which were obtained by periodic photographs. The crack initiation time decreases with increase of initial crack length. It is interesting to note that increasing the initial crack size produces a more gradual crack extension curve. More precise measurement will, however, probably be required as mentioned afterwards.

Fig. 2 shows the relaxation modulus $E_r(t)$ and the second integral of the relaxation modulus up to the time t_i, $2t_i^{-2} E_r^{(2)}(t_i)$, as functions of the reduced time t/a_T. The WLF time-temperature shift factor is also shown in this figure. The quantitative difference between $E_r(t)$ and $2t^{-2}E_r^{(2)}(t)$ at the time scale tested, i.e., t_i, was less than 30%.

The correction factor for the effect of the finite width and finite height of the specimen $F(\frac{C_o}{w}, \frac{C_o}{b})$ was calculated separately as a product of two factors by $F_1(\frac{C_o}{w})$ using Westergaard's factor (9) for width, and by $F_2(\frac{C_o}{b})$ using Knauss' factor (4) for height. This product separation is valid over the small ranges of non-interacting height and width geometries, i.e.

$$F(\frac{C_o}{w}, \frac{C_o}{b}) \simeq F_1(\frac{C_o}{w}) \cdot F_2(\frac{C_o}{b}) \tag{4}$$

Figure 3 shows the separate correction factors $F_1(\frac{C_o}{w})$, $F_2(\frac{C_o}{b})$ and the multiplied correction factor $F_1(\frac{C_o}{w}) \cdot F_2(\frac{C_o}{b})$ which depends on the crack length of the specimen used in this experiments. For the geometry of our test specimen and crack size, $F_2(\frac{C_o}{b})$ factor is obviously dominant.

Using the data from Figure 1, 2 and 3, we can calculate using Eq. (2) the specific cohesive fracture energy, γ_c, as a function of the crack initiation time t_i. The relation between γ_c and t_i is shown in Figure 4. γ_c is seen to increase with increasing t_i, namely with decreasing initial crack size. This behavior of γ_c on the crack initiation time is just opposite that previously reported by Bennett, et al. (6), for a different material. No further comment upon this point can be made at this time, except to point out some probable pertinent factors such as the time dependence of the stress intensity factor, large deformation, blunting of crack tip and second order effect of the geometrical correction factor and so on. The order of values of γ_c, however, appear consistent with those found earlier as far as their order of magnitude.

Having noted the obvious difficulties in obtaining an exact formula for the fracture energy dependence, i.e. Eq. (2), it is fairly obvious that, relatively speaking, Bennett, et al. probably encountered similar difficulties. Hence, one should examine basic differences between the earlier data and our

results here. There are two possibilites: (i) differences in the measurement of crack initiation time, and (ii) the shorter range of initiation times
for which this present data was collected.

In spite of careful observations by periodic photographs of the crack
length, there could still exist a basic uncertainty in detecting the precise
threshold for crack propagation, particularly for a viscoelastic strip.
Hence, it would be desirable to develop a more accurate detection of the
initiation. Because the threshold of crack growth and, particularly, the
period between the threshold and the critical time for rapid growth of crack
advance, in which the crack grows very slowly and gradually, depend strongly
upon the initial crack length, fracture energy would be very sensitive to this
measurement. (Judging from the procedure in this experiment, we have to
recognize that the accuracy in detecting the crack threshold might be almost
0.2 ~ 0.5 mm increase in crack length).

As to the second time, measurements involving very cold specimens for
short t_i and hot specimens for long t_i could lead to a greater range of
fracture times and associated values of $\gamma_c(t_i)$. Such data could show that
the present monotonicity was an aberration due to measurement inaccuracies
over too short a time range, or establish that a fundamental difference with
the earlier Bennett (6) data actually exists. Further tests will be made to
verify the "hot" and "cold" end points of the time t_i data.

4. CRACK VELOCITY CHARACTERISTICS

In order to obtain velocity data for engineering purposes and evaluate
the magnitude of the crack growth rate, the distance that the crack travelled,
$\overline{\Delta C} = (C - Co)/C_o$ was plotted against the time $(t - t_i)$, as shown in Fig. 5.

For small increments of crack growth, say less than 10% of inital
length, the crack extension data were fit in an ad hoc manner to the equation

$$\overline{\Delta C} = \frac{C}{C_o} - 1 = B(t - t_i)^2 \qquad \text{1st stage} \quad (5)$$

where B is a load dependent constant. Also, when the crack grows more than 10% of initial length, fitting curves were taken as

$$\ln (\overline{\Delta C}) = \ln (\frac{C}{C_o} - 1) = A(C_o,\sigma)(t - t_i) \qquad \text{2nd stage} \quad (6)$$

$A(C_o,\sigma)$ is a function of initial crack length C_o, but is independent of time. The form of these latter two equations was suggested by Williams (1), in which he pridicted the crack growth in a viscoelastic material by using the maximum strain criterion and a succession of breaking Voigt model elements.

Now, we can see two diffent types of crack growth process in a viscoelastic strip of finite dimensions. Also, the transition of the type of crack growth from equation (5) to equation (6) occurs, in most cases, at about the time at which the maximum stress (load) was reached. Other expressions for crack propagation could also be considered, if necessary.

The crack growth rate $\Delta \dot{C}$ would be derived from equations (5) and (6), by simply differentiating respect to time,

$$\Delta \dot{C}/C_o = 2B (t - t_i) \qquad \text{1st stage} \quad (7)$$
$$\Delta \dot{C}/C_o = A \exp A (t - t_i) \qquad \text{2nd stage} \quad (8)$$

From equation (8), taking $\Delta \dot{C}/C_o \big|_{t = t_i}$ leads us to $A = \Delta \dot{C}(t_i)/C_o$, then we can obtain for 2nd stage crack growth,

$$\Delta C(t) = C_o \exp \{(\Delta \dot{C}(t_i)/C_o)(t - t_i)\} \qquad (9)$$

Also, it is interesting to check the growth rate and acceleration of crack extension at the time t_i, particularly for the 1st stage. From the equation (7), we get

$$\Delta \dot{C}/C_o \big|_{t = t_i} = 0 \qquad (10)$$
$$\Delta \ddot{C}/C_o \big|_{t = t_i} = 2B \qquad (11)$$

It is easy to conjecture that the crack will start to extend at the critical time t_i, in a very slow and gradual manner. On the other hand, if the crack growth rate $\Delta \dot{C}$ shows a finite value at the time t_i, then an infinite acceleration of crack growth would be expected and/or the crack might inevitably extend at $t_i = 0$, for any levels of stress. Recalling the former section about fracture energy measurement, and assuming that the 1st stage crack growth Eq. (5) hold for the threshold, i.e., $t - t_i$, the difficulties in detection of the intrinsic threshold of crack growth should be pointed out.

Now, turning back to Figure 1, let us consider the stress-strain curve of a wide strip which contains a slowly moving crack. Usually, the load response of a viscoelastic strip with a crack under a displacement input will be a complicated nonlinear function of time, geometry, crack length, and aspect ratio. But adopting the comparison of compliances of the strips which contain different lengths of crack to an uncracked strip, the compliance reduction ratio, $\lambda(\frac{C_o}{w}, \frac{C_o}{b})$ can be obtained experimentally. The tensile stress-strain curve of wide strip with a crack can be written as

$$\sigma_w(t) = \lambda(\frac{C_o}{w}, \frac{C_o}{b}) \int_o^t E_r(t - \tau) R \, d\tau \tag{12}$$

where R is a constant strain rate. Figure (6) shows the compliance reduction ratio $\lambda(\frac{C_o}{w}, \frac{C_o}{b})$ as a function of initial crack length. The thin curves (in Fig. 1) numbered $C_o = 20, \ldots \ldots 100$ were drawn using Eq. (12). As shown by dotted lines in the figure, nominal stress in a strip with a slowly moving crack can be estimated by the stress calculated from (12), using a stationary initial crack length.

424

5. REFERENCES

(1) Williams, M.L., "The Fracture of Viscoelastic Material," Fracture of Solids, Interscience Publishers (1963).

(2) Williams, M. L., "Initiation and Growth of Viscoelastic Fracture," International Journal of Fracture Mechanics, 1:292 (1965).

(3) Knauss, W. G., "The Time-Dependent Fracture of Viscoelastic Materials," Proc. of the 1st International Conf. on Fracture, Sendai, Japan (1965).

(4) Knauss, W. G., "Stresses in an Infinite Strip Containing a Semi-Infinite Crack," Journal of Applied Mechanics, 34:356 (1966).

(5) Knauss, W. G. "Delayed Failure - The Griffith Problem for Linearly Viscoelastic Materials," International Journal of Fracture Mechanics.

(6) Bennett, S. J., (with G. P. Anderson and M. L. Williams), "The Time Dependence of Surface Energy in Cohesive Fracture," Journal of Applied Polymer Science, 14:735 (1970).

(7) Williams, M. L., (with R. F. Landel and J. D. Ferry), "The Temperature Dependence of Relaxation Mechanisms in Amorphous Polymers and Other Glass-forming Liquids," Journal of American Chemical Society, 77:3701 (1955).

(8) Mueller, H. K., "Stress-Intensity Factor and Crack Opening for a Linearly Viscoelastic Strip with a Slowly Propagating Central Crack," International Journal of Fracture Mechanics, 7:129 (1971).

(9) Westergaard, H. M., "Bearing Pressures and Cracks," Journal of Applied Mechanics, A49 (1939).

(10) Rice, J. C., Discussion on Reference (4), Journal of Applied Mechanics, 35:248 (1967)

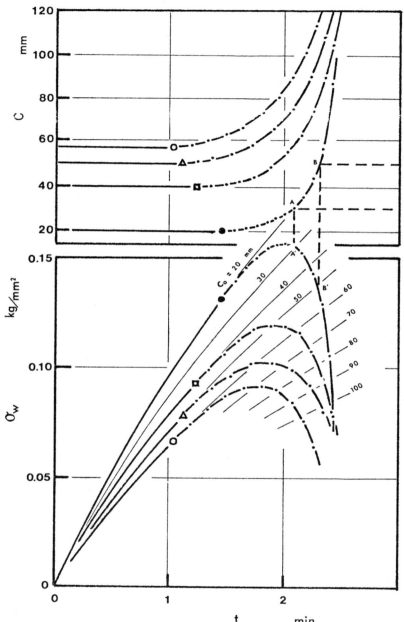

Fig. 1 Time and initial crack length dependent stress-strain relations and crack extension curves.

426

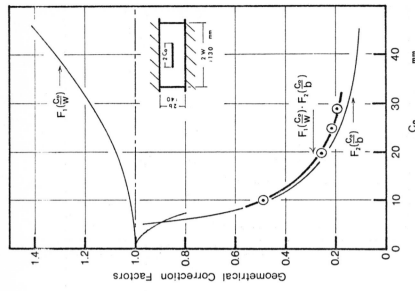

FIG. 3 GEOMETRICAL CORRECTION FACTORS FOR FINITE DIMENSION; $F_1\left(\frac{C_o}{W}\right)$: BY WESTERGAARD[9] AND $F_2\left(\frac{C_o}{b}\right)$: BY KNAUSS[4], (10).

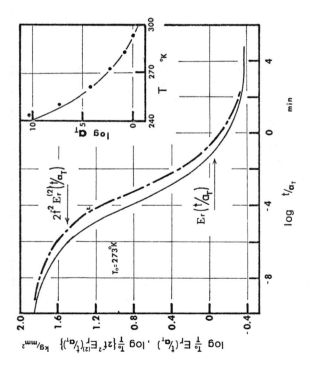

FIG. 2 RELAXATION MODULUS, SECOND INTEGRAL OF RELAXATION MODULUS AND WLF TIME-TEMPERATURE SHIFT FACTOR.

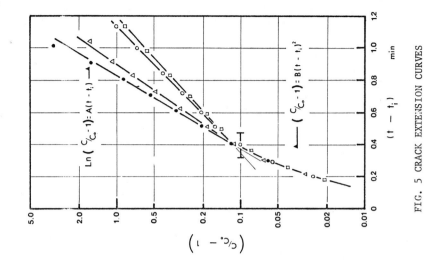

FIG. 5 CRACK EXTENSION CURVES

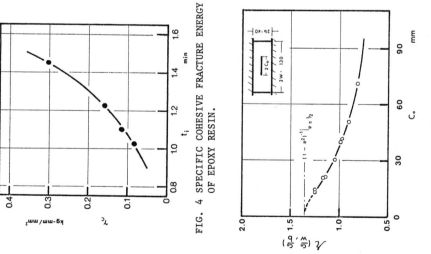

FIG. 4 SPECIFIC COHESIVE FRACTURE ENERGY
OF EPOXY RESIN.

FIG. 6 COMPLIANCE REDUCTION RATIO

Fatigue of Rigid Plastics and

Plastics Composite

by   Shoji Shimamura [*]

Hiroo Miyairi [**]

1. Introduction

Interests in the endurance property, particularly as regards fatigue behaviour, of rigid plastics and plastics composites have been growing strong as their use for structural components has remarkably diversified of late. To improve the fatigue strength of plastics composites, it is essential to clarify the various effects of reinforcement on the fatigue strength.

Although the basic studies such as of extended application of the rule of mixtures, formulation of various types of composite structures, respectively, have been energetically conducted by many, the designing of these materials against fatigue has not yet been studied widely.

This paper will discuss the result of studies on the fatigue strength in Japan, centering on the following subject matters.

(1) A general description of the study of fatigue of rigid plastics and plastics composites, and its recent trend in Japan

(2) Guidelines for designing these materials against fatigue, prepared by Shimamura

(3) Study of the hybrid structure of CF/GF (Carbon Fiber/Glass Fiber)

 * Dr., Chief of Material Properties Section, Mechanical Engineering Laboratory, AIST, MITI, Tokyo

** Dr., Associate Professor, Institute for Medical and Dental Engineering, Tokyo Medical and Dental University, Tokyo

2. General Trend of the Study of Fatigue in Japan

The study of the fatigue of rigid plastics and plastics composites in Japan began in the middle of the 1950s, approximately 15 years later than the United States, where the same type of study started in the early 1940s by Lazan and Yorgiadis[*]. In Japan, the first report on the results of the study was made by Enomoto et al[1] on the softening phenomena of polyester and phenolic laminates with the rise of temperature during the progress of fatigue. Later as Shimamura[2]-[4], [6] and [8] commenced his study under the following subjects, other researchers who were interested began actively participating in the study.

(1) Existence of the transition point (a knee) on the S-N curve

(2) Recognition of the absence of endurance limit before 10^7 cycles of repeated stress

(3) Dependence of temperature increase on the stress amplitude and rate of stress applied

(4) Decrease of rigidity during fatigue testing

(5) Effects of heat treatment and moulding conditions

(6) Effects of stress concentrations

(7) Changes in the degree of polymerization

Selected articles on these subjects, out of approximately one hundred which have been published since then, will be shown in the References at the end of this paper.

Some of these papers treating the subject phenomenally reported effects of moulding pressure on the fatigue strength of phenolic resin laminate

---

[*] B.J. Lazan & A. Yorgiadis, "The Behaviour of Plastics under Repeated Stress", ASTM, STP 24 (1944), 66-94.
B.J. Lazan & A. Yorgiadis, "The Behaviour of Plastics Materials under Repeated Stress", Mod. Plast., 1944-8, 119-128.

under rotary bending and effects of the type of matrix of FRP (Fiber Glass Reinforced Plastics) and the glass content on the notch effect under repeated tensile and bending loads, respectively. Some demonstrated that there is a relationship between damping by internal friction and fatigue strength of unsaturated polyester and this resin is virtually unaffected by moisture[12].

Study of the mechanism of fatigue failure of rigid plastics and plastics composites has also been made by observing their fatigue behaviour and setting up models to simulate their mechanical properties. In a paper, a combination of a model of fiber and matrix arranged in series with another model of the same materials in parallel arrangement was successfully employed to simulate directional characteristics of fatigue strength[13]. In other papers, a more complicated model was used incorporating spring, dashpot, slider to investigate static behaviour and fatigue strength of the materials in question[15],[28].

Yokobori[25] conducted a comparative study of the fatigue failure of both high-polymers and metals, and made clear how the differences in the microstructures of both materials, despite many common characteristics ascertained, would be reflected on the formulas of failure mechanism. This study can be expected to contribute to the understanding of fatigue behaviour by bridging the gap between the studies of microstructure and macrostructure of the materials. Some reports have clarified the relationship between fatigue strength and moulding method particularly in relation to surface finish and fatigue strength of dental plastics[29] and also the fatigue of bonded joints of FRP[45].

Recent trend in the study of the fatigue of rigid plastics in Japan is as in the following.

(1) An increasing number of researches apply the method of fracture

mechanics for metal to plastics.

(2) With a wide use of scanning electron microscopy, fractographic studies have also been increasing, this becoming a useful tool for the study of the mechanism of fatigue failure of plastics.

(3) Recently, more studies have been made on the fatigue of FRTP (Fiber Reinforced Thermo-Plastics), which can be attributed to the fact that FRTP can be treated as a viscoelastic material because the matrix is made of thermo-plastics.

(4) Studies of the fatigue of FRP and FRP of hybrid structure with new reinforcements and matrix have increased.

On the other hand, however, studies for a wider scope of designs using plastics have been few, which may be due to the inevitable tendency of any research to shift towards particular subjects from a general subject.

3. Designing of Rigid Plastics and Plastics Composites Against Fatigue

One of the authors, Shimamura[48] has recently presented a method of designing plastics materials against fatigue in his thesis containing the results of experiments and studies conducted and organized by him for years. However, because of the limited space in this paper, only a flow-sheet as shown in Fig. 1, is given. It presents a designing method against fatigue considering moulding conditions, service conditions, stress concentration, etc. of rigid plastics and plastics composites.

4. Fatigue of Hybrid Composites

Although CFRP (Carbon Fiber Reinforced Plastics) are known to have excellent properties as structural composites, they show brittle fracture despite of their high strength and high elasticity, which, combined with their high price, accounts for its limited use today. As a solution of this, a FRP of hybrid structure with CF and GF, i.e., GFRP and CFRP, can be expected. Following types of hybrid structure of glass fiber and carbon fiber are tested now.

(1) Mixed hybrid

(2) Laminated hybrid

The mixed hybrid is composed of carbon fiber and glass fiber as reinforcements, uniformly mixed by some relevant method and moulded, while the laminated hybrid is composed of layers of carbon fiber and glass fiber laminated alternately.

This paper will report the fatigue properties of mixed hybrid composites varying in the mixture ratios in 8 steps between carbon fiber and glass fiber. The details, however, will be omitted here as they are shown in the Bulletin of Mechanical Engineering Laboratory (No. 12)[49], which will be distributed separately in the Seminar. Tests are being conducted on the laminated hybrid, and a report will be made on these tests at a later date on some other occasion.

434

References

(1) N.Enomoto and S.Sugiyama, "Repeated Bending of Polyester and Phenol Resin Laminates", J. of Railway Engineering Research, 12-13 (1955), 328.

(2)[*] S.Shimamura, "On Fatigue Properties of Rigid Engineering Plastics with Emphasis on the Effect of Stress Raisers", JSME 1957 Semi-Int'l Symp. Ex. Mech. II.

(3) S.Shimamura, "Fatigue Testing on a Polyester / Glass-Mat Laminate at Temperatures Ranging +30 ~ -30 deg C", Zairyō-Shiken 8 (1959), 868.

(4)[*] S.Shimamura & H.Maki, "On Fatigue Properties of Some Plastics Materials", Proc. JCTM, 4 (1960), 120.

(5) Y.Hori, "An Interesting Point of Fatigue Diagram of High-polymer Materials", Kōbunshi 10 (1961), 606.

(6) S.Shimamura & H.Maki, "On Fatigue Properties of Some Thermoplastic Materials (I)", Proc. JCTM, (1962), 136.

(7) T.Kuroda, T.Ishibashi, "Rotating Bending Fatigue Test of Phenolic Resin Laminates" Zairyō-Shiken 10 (1961), 485.

(8)[*] S.Shimamura & H.Maki, "On Fatigue Properties of Some Thermoplastic Materials (II)", Proc. JCTM, 5 (1963), 136.

(9)[*] T.Fujii & S.Otsuki, "The Effect of Glass Fiber Contents on the Fatigue Strength of Fiber Reinforced Plastics", Proc. JCTM, 6 (1963), 131.

(10)[*] T.Fujii, S.Otsuki, K.Mizukawa & Y.Masuda, "Flexural Fatigue Strength of Fiber Reinforced Plastics under Constant Deflection and Load", Proc. JCTM, 7 (1964), 133.

(11) T.Kuroda, K.Komaki, "Fatigue Testing of Thermoplastics", Zairyō-Shiken 14-138 (1965), 172.

(12) K.Endo, S.Ishihara, "Some Studies on Fatigue Strength of Unsaturated Polyester", 15-155 (1966), 561.

* printed in english

(13)[*] K.Endo et al., "Studies of Fatigue Strength of Fiber-glass Reinforced Plastics", Proc. JCTM, 9 (1966), 104.

(14)[*] S.Okuda & S.Nishina, "On the Chemical Fatigue of Rigid-PVC under low Frequency Repeated Stress", Proc. JCTM, 9 (1966), 109.

(15) T.Fujii, "Fracture Phenomena in Glass Fibre Reinforced Plastics(Strength of Glass Fiber Reinforced Plastics) Zairyō 16-163 (1967), 257.

(16) S.Amijima, M.Sugimura, "The Fatigue Properties of Satin Woven Glass Fabric Reinforced Plastics", Zairyō-Shiken 17-172 (1968) 52.

(17)[*] K.Endo "Fatigue Properties of Fiberglass Reinforced Plastics, Effects of Stress Direction and Mean Stress", Proc. JCTM, 11 (1968), 131.

(18) F.Ohishi, "Classification of Fatigue Behavior of Plastics", Zairyō-Shiken 17-182 (1968), 1016.

(19) T.Tateishi, H.Miyamoto, "Fatigue Fracture Surface of Polycarbonate", Zairyō Kagaku 5-2 (1968), 124.

(20)[*] T.Fujii & K.Mizukawa, "The Effect of the Combination of Roving Glass Cloth and Mat upon the Fatigue Strength of Reinforced Polyester Laminates", Proc. Tech. Conf. SPI, RP/CM Div., 24 (1969), 14-D.

(21)[*] K.Endo & M.Watanabe, "Fatigue Properties of Fiber Glass Reinforced Plastics (Effect of Stress Direction)". Proc. JCMR, 14 (1970), 120.

(22) Y.Kawada, H.Kobayashi, "On the Low Cycle Fatigue Strength of Glass Fiber Reinforced Plastics", 19-206 (1970), 961.

(23) K.Endo, "Fatigue Strength of FRP of Combined Laminate" Zairyō-Shiken 19-206 (1970), 956.

(24) M.Higuchi, Y.Imai, "Rheological Interpretation of Heat Generation Associated with Fatigue of Polycarbonate" Zairyō-Shiken 19-199(1970), 362.

(25) T.Yokobori, "The Similarities and Dissimilarities of Fatigue Fracture between Polymers and Metallic Materials "Zairyō-Shiken 20-211(1970),453.

(26) E. Jinen, **M.Suzuki**, "Effect of Glass Fiber Contents on the Temperature
      Dependence of Dynamic Viscoelasticity of Glass Fiber Reinforced Nylon",
      Kōbunshi Kagaku, 28-314 (Oct. 1971), 815.

(27) M.Suzuki, E.Jienen, H.Ueda, "Plane Bending Fatigue of Glass Fiber
      Reinforced Polycarbonate", Kōbunshi-Kagaku, 28-320 (Dec. 1971), 958.

(28)[*] M.Zako, T.Fujii, K.Mizukawa & T.Fukuda, "Nondestructive Testing Method
      and Fatigue Model for F.R.P.", Proc. Tech. Man. Conf. SPI, PR/CM Div.,
      26 (1971), 10-B.

(29) H.Miyairi, A.Muramatsu, "Studies on the Fatigue Properties of Dental
      Plastics Materials", Zairyō-Shiken 20-211 (1971), 531.

(30) M.Suzuki, M.Shimizu, E.Jinen, M.Maeda, and Y.Sasaki, "The Effect of
      Glass Fiber Reinforcement on How Cycle Fatigue of Nylon Resin", Zairyō
      20-21 (1971), 1050.

(31)[*] T.Fujii, K.Mizukawa, M.Zako, "On the Cumulative Fatigue Damage of Glass
      Fiber Reinforced Plastics Subjected Tensile Impact Load", Proc. JCME,
      15 (1972), 154.

(32)[*] T.Kurobe & H.Wakashima, "On the Fatigue Crack Propagation in Polymeric
      Materials", Proc. JCMR, 15 (1972), 137.

(33) M.Suzuki, T.Yamashita, K.Nishimura, "A study on the Transition of
      Fracture Mode during Fatigue Crack Propagation in Polycarbonate Plate",
      Zairyō 23-244 (1974-1), 52.

(34) C.Asahara, Y.Ando, H.Syozi, "Plane Bending Fatigue of Thermoplastic
      Materials", Zairyō, 23-245 (1974), 122.

(35) Taichi Fujii, F.Yamada, "A Study on the Flexural Fatigue and the Change
      of Rigidity for C-GFRP Material", 17th JCMR, 17 (1974-3), 159.

(36) Y.Masuda, "Effects of Watery Environment on Fatigue Strength of FRP",
      JCMR, 17 (1974-3), 172.

(37) M.Suzuki, M.Yada, N.Mabuchi, "Comparison of the Mechanism of Fatigue
      Crack Propagation in Polyvinyl Chloride with those of Polypropylene
      and ABS Blend Polymer", Kōbunshi-Kagaku, 29-327 (1972-7), 490.

(38) A.Zinen, M.Suzuki, "Crack Propagation by Bending Fatigue of Glass Fiber Reinforced Nylon 6 Plastics (The case of notched specimen)", Kōbunshi-Kagaku, 30-344 (1973-12), 727.

(39) K.Yamada, M.Suzuki, "Effect of Mean Stress on the Fatigue Crack Propagation in Low Density Polyethylene", Kōbunshi-Kagaku, 30-336 (1973-4), 206.

(40) K.Yamada, M.Suzuki, "Observation of Fatigue Crack in Polyethylene", Kōbunshi-Kagaku, 30-333 (1973-1), 28.

(41) M.Suzuki, A.Zinen, "Application of the Nucleation Theory of the S-N Relation on Plane Bending Fatigue of Nylon 6 Plastics", Kōbunshi-Kagaku, 30-339 (1973-7), 395.

(42) T.Fujii, K.Mizukawa and Z.Maekawa, "Influence of Cycle Ratio on the Elastic Modulus of Glass-fiber Reinforced Plastics Subjected to Repeated Tensile Load", JCMR, 16th (1973), 248.

(43) S.Fujiwara, T.Hirai, S.Ōki, and M.Yamamoto, "Tensile Fatigue Properties of FRP", Trans. J.S.M.E. 39-319, (1973-3), 805.

(44) M.Kitagawa, "Fatigue Crack Growth in High Polymers (Polycarbonate)", Trans. J.S.M.E., 39-324 (1973-8), 2299.

(45) H.Miyairi, A.Muramastu, H.Fukuda, "Low Cycle Fatigue Strength of the FRP Adhesive-Bonded"(Joints-Endurance of Adhesive-Bonded Joints and Hysteresis Diagram), Zairyō, 22-239, (1973-9), 769.

(46) H.Hyakutake, "Low Cycle Fatigue Tests of Polycarbonate under Push-Pull Loading", Zairyō, 22-240 (1973-9), 858.

(47)[*] S.Shimamura, H.Furue, "Effect of Stress Raisers on Fatigue Strength of Rigid Plastics and Plastics Composites", Fukugo Zairyo, 2-3(1973-9),21.

(48)[*] S.Shimamura, "Researches on Designing Rigid Engineering Plastics against Dynamical Loading", Technical Report of Mechanical Engineering Laboratory, No. 80 (1974).

(49)[*] S.Shimamura, H.Furue, M.Nuka, "On Flexural Fatigue Properties on Hybrid Composites(Carbonfiber/Glassfiber Reinforced Epoxy Laminates)", Bulletin of Mechanical Engineering Laboratory, No. 12 (1974).

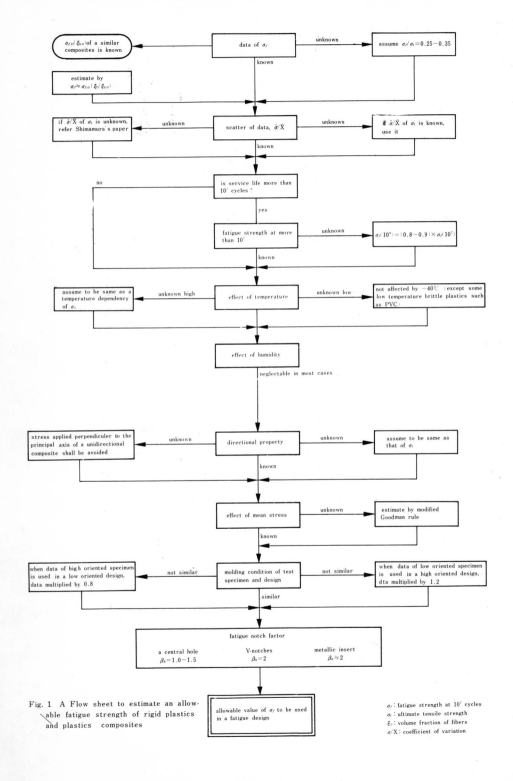

Fig. 1 A Flow sheet to estimate an allowable fatigue strength of rigid plastics and plastics composites